MEETING UNDER THE INTEGRAL SIGN? ∮

The Oslo Congress of Mathematicians on the Eve of the Second World War

HISTORY OF MATHEMATICS VOL 44

MEETING UNDER THE INTEGRAL SIGN?

The Oslo Congress of Mathematicians on the Eve of the Second World War

Christopher D. Hollings
Reinhard Siegmund-Schultze

in collaboration with Henrik Kragh Sørensen

AMERICAN MATHEMATICAL SOCIETY
Providence, Rhode Island

Cover image: National Library, Oslo, unprocessed 115 Carl Størmer Archives.
Back cover image: *Dagbladet*, 14 July, 1936, p. 3.

2010 *Mathematics Subject Classification*. Primary 01A60.

For additional information and updates on this book, visit
www.ams.org/bookpages/hmath-44

Library of Congress Cataloging-in-Publication Data

Names: Hollings, Christopher, 1982– author. | Siegmund-Schultze, R. (Reinhard), author. | Sørensen, Henrik Kragh, author.
Title: Meeting under the integral sign? : the Oslo Congress of Mathematicians on the eve of the Second World War / Christopher D. Hollings, Reinhard Siegmund-Schultze ; in collaboration with Henrik Kragh Sørensen.
Description: Providence, Rhode Island : American Mathematical Society, [2019] | Series: History of mathematics, 0899-2428 ; volume 44 | Includes bibliographical references and index.
Identifiers: LCCN 2019039704 | ISBN 9781470443535 (hardcover) | ISBN 9781470455156 (ebook)
Subjects: LCSH: International Congress of Mathematicians (1936 : Oslo, Norway) | International Congress of Mathematicians–History. | Mathematics–Congresses–History–20th century.
Classification: LCC QA1 .H65 2019 | DDC 510.7–dc23
LC record available at https://lccn.loc.gov/2019039704

♾ The paper used in this book is acid-free and falls within the guidelines
established to ensure permanence and durability.
Visit the AMS home page at `https://www.ams.org/`

10 9 8 7 6 5 4 3 2 1   25 24 23 22 21 20

Contents

List of Figures

List of Tables

Preface

When the International Congress of Mathematicians convened in Oslo between 14 and 18 July 1936, Europe was in uproar. The Nazis had been in power in Germany since 1933 and their infamous law of 7 April that same year had led to the expulsion of thousands of scholars. The anti-Semitic "Nuremberg Laws" of September 1935, i.e., the "Law for the Protection of German Blood and German Honour" and the "Reich Citizenship Law", exacerbated the situation, particularly for those German scholars who were of Jewish origin. Many dismissed mathematicians were looking for jobs abroad or tried at least to maintain contacts with the international community. Some of them used the congress in Oslo for this purpose, as difficult and expensive as it was for them to visit the venue. On 9 March 1935, Hermann Göring had announced the existence of a German Air Force and one week later Hitler proclaimed the General Draft. These measures and the remilitarisation of the Rhineland in March 1936 went against the military stipulations of the Treaty of Versailles, but were tolerated by the Western Allies. The Nazis enhanced their "international success" (in the eyes also of German participants in Oslo such as Erhardt Schmidt and Helmut Hasse) by organising the Olympic Games in Berlin in the first half of August 1936. This was shortly after the Oslo congress, and was meant to present Germany as a peace-loving nation. Already in November 1935, the League of Nations, which included among its members such countries as Norway that had remained neutral during the First World War, had proclaimed sanctions against Italy for its aggression against and subsequent occupation of Ethiopia (then Abyssinia). Although the sanctions were officially lifted on 15 July 1936 — during the Oslo congress — this came too late for the Italian mathematicians who had been barred by Mussolini from participating. While the representation of algebraic geometry at the congress suffered in particular due to the absence of the Italians, topology was affected by the non-appearance of the Russians, who had been prevented from coming to Oslo by Stalin and his Politburo. Exactly at the time of the congress in the Norwegian capital, the leading Soviet communist newspaper *Pravda* ran a campaign against Russian mathematicians, such as N. N. Luzin, whose behaviour was deemed to be politically unacceptable. The fact that Stalin's main opponent Leon Trotskiy had been granted asylum in Norway in 1935 by the newly established Social Democratic government in all likelihood did not increase the dictator's willingness to let the USSR's leading scientists visit that country. As to German-Norwegian relations, they had been burdened by the affair surrounding the Nobel Peace Prize for 1935. It was finally awarded in November 1936 — one year late and *in absentia* — to the German journalist Carl von Ossietzky who had been taken to a concentration camp by the Nazis and died from the after-effects in 1938. The fearless pacifist had warned the world about German remilitarisation. In the Norwegian Nobel Committee, internal disagreement prevailed, and two of its members stepped down for fear of German reactions: one of these members was the foreign minister Halvdan Koht, who had addressed the Oslo ICM in July that year. The press debate was fierce, in Norwegian, German, and international newspapers. In 1937 Hitler issued a decree forbidding Germans from accepting Nobel Prizes in any category "for all future times". Under these circumstances, it is no wonder that the German delegation to the Oslo congress was politically selected and supervised. Its leader, Walther Lietzmann, had to contact the Nazi authorities in Oslo during his stay and had to send a political and mathematical report to the government afterwards.

The politics of appeasement towards Nazi Germany and Fascist Italy by the Western Powers, exemplified by the inconsequential sanctions against Italy, continued in the years to come. Political events, such as the beginning of the Civil War in Spain a few weeks after the congress, the occupation of Austria by the Nazis, the "Reichskristallnacht" and the Munich Conference (all 1938), the occupation of Czechoslovakia in March 1939, and finally the invasion of Poland in September that year, reveal the Oslo congress as having been "on the eve of the Second World War". Indeed the next ICM, agreed in Oslo for 1940, would not convene until 1950. We are aware that the historical actors involved in the events of 1936, the mathematicians in Oslo, could not possibly anticipate the war or the delay of the next congress. But history has to be written with a view also to consequences, and it is in this sense that we understand the subtitle of our book.

We will go in detail into this political background, but we are aware that beside disruption and discontinuity there was also continuity, in particular developments both within mathematics and in the social infrastructure of science, which cannot be reduced to contemporary political events.

The history of the various ICMs from their inauguration in the nineteenth century, but in particular in the changed international environment following the First World War, has to be taken into account as well. The final decades of the nineteenth century saw an increase in internationalist sentiments in much of Europe and North America. In the academic sphere, this manifested itself in the establishment of regular discipline-specific international conferences. Although they appeared quite late in the period, the International Congresses of Mathematicians provide us with a prime example of such a movement. Starting from the initial meeting in Zurich in 1897, the attendant mathematicians have usually claimed the separation of mathematics from world politics. As one might expect, however, the situation was rather more complicated from the start. Even in the choice of neutral Zurich as the venue for the first ICM, less than thirty years after the Franco-Prussian War, we see a shading of the congress organisation by wider world events. Further subtle effects appeared in the staging of the following early congresses, such as the relative lack of French representation at the Heidelberg congress of 1904, but for much more dramatic consequences of political events on the ICMs, we must look to the First World War, which led to the cancellation of the planned meeting of 1916 (Stockholm), and to the congresses of the early 1920s (Strasbourg and Toronto) with their exclusion of mathematicians from the defeated nations. Particular political importance for mathematics may be found in the ICM that was held in Bologna in 1928, this being the first international congress after the war at which Germany and its former wartime allies were allowed to take part with equal rights. At the same time, Bologna 1928 was the first ICM to take place under dictatorial conditions

— in Mussolini's Italy. Thus, despite the readmission of German mathematicians, the tensions of the decade may have contributed to the decision there to return to politically neutral Switzerland and Zurich in 1932.

Despite belonging to a nation that had been neutral during the First World War, the Norwegian organisers were not neutral in their desire to promote the mathematical interests of their nation, and of Scandinavian mathematicians more generally. The result was that some of the new mathematical developments of the 1920s and 1930s were clearly represented at the congress, while others that we might expect also to see were not, partly influenced by political factors, as described above.

Thus, the Oslo congress appears as a mathematical and political melting pot, standing between the academic fall-out of the First World War and the further disruptions of the Second. In the present volume, we describe the ways in which the political situation shaped the staging of the congress, and the consequences that this had for the development of mathematics in the 1930s. The congress itself provides us with a snapshot of the mathematics of that decade, made the more interesting by the fact that this was the first ICM at which the now well-established Fields Medals were awarded.

The book is organised in four parts. Part 1 contains Chapter 1 (Introduction). It concerns the prehistory of the 1936 ICM, with only brief references to the time before the First World War. This terrible event a hundred years ago, referred to by some as the original catastrophe of the twentieth century, shaped much of the ensuing general political and scientific environment. Looking at the Oslo congress in particular, but having broader issues of mathematical communication in mind, we identify certain "strategies" that were either consciously adopted by the various countries and their representatives, or are retrospectively observed by us as historians. Finally in this opening chapter we give an overview of the sources that have been used when writing this book.

Part 2 of our book is devoted to The Politics of the Congress. Its first chapter (Chapter 2) concerns the preparation of the congress by the Norwegian organisers. It is also in this chapter that we include some remarks about the general events of the congress itself, including social gatherings. However, we try to restrict the narrative to information which cannot be easily drawn from the published proceedings of the congress. The background to the selection of the German delegation, as well as their subsequent interaction with other mathematicians at the congress, is the subject of Chapter 3. This is followed by two chapters (Chapters 4 and 5) that deal, respectively, with the absence of Soviet and Italian mathematicians from the congress, while Chapter 6 examines the ways in which the congress was reflected in contemporary Norwegian newspapers. The many interviews with refugees in these newspaper articles, which are partly reproduced in the appendices of this book, provide new information about the conditions for emigration of mathematicians from Nazi Germany in the 1930s, arguably the most important social and political factor influencing the global development of mathematics in that decade. At the same time, the newspaper reports give insight into Norwegian attitudes towards the refugees and Nazi Germany. In Chapter 7, we conclude our discussion of political issues in a more narrow sense by looking briefly at the aftermath of the congress and at international mathematics shortly before, during, and after the Second World War. We discuss the plans among American mathematicians for the ICM in 1940, although this fell victim to the war. The failed German attempt to "restructure" European mathematics during the war is discussed, and efforts to reconnect European and American mathematics, as revealed in a conference in Princeton in 1946, are indicated. The imminent Cold War between the West and its former allies in Russia, the re-establishment of the IMU, and the delayed ICM in Cambridge, Massachusetts, in 1950 can only be hinted at.

Part 3 of the book is about The Mathematics of the Congress. While the first two parts of the book, as well as the fourth (see below), are hopefully accessible and interesting to a general readership including historians and natural scientists, the third part is aimed primarily at mathematicians, although without going too much into the details of the various disciplines. Chapters 8 and 9 set the congress into its mathematical context, and assess its impact upon twentieth-century mathematics via a detailed discussion of the content of the plenary lectures. Remarks on the ICMI session at the congress and thus on the international state of mathematics education can be found in the very brief Chapter 10, and Chapter 11 contains conclusions regarding the mathematics of the congress.

The final part of the book consists of a series of appendices. It contains transcriptions and English translations of a range of relevant archival materials and newspaper articles, often shedding light on the situation of mathematical refugees. Naturally, these appendices relate to various previous chapters in the book, in particular to Chapters 6 and 7. The list of mathematicians to be invited to the failed 1940 ICM (Appendix H), as well as discussions within the American organising committee about recent tendencies in several modern subdisciplines, stand out. We also present the rather obscure, but politically influential figure of Eberhard Kern, the German Nazi representative in Oslo during the time of the congress in 1936, through an anonymous satirical article written about him in a Norwegian daily in 1937. This and other appendices therefore deepen the specific Norwegian dimension of the book and thus, we hope, contribute to the further unveiling of the still under-explored history of Norwegian and Scandinavian science and mathematics.

In recent years and decades political developments have taken place which nobody would have anticipated. Nationalism in various regions of the world is once again on the rise and the danger of global conflict is far from being banished. It seems also from this perspective that another look at the troubled political situation around the Oslo congress of 1936 is warranted and the potential and the need of mathematics to uphold rationality and to foster international understanding and "to meet under the integral sign" should be always on our minds.

Acknowledgments

The authors wish to express their special thanks to Henrik Kragh Sørensen, not only for granting permission to use his text for Chapter 2, but also for providing the majority of the newspaper clippings and several transcriptions that we include in the appendices, as well as several of the pictures (particularly Figures 1.2 and 2.6), and for his useful comments and help with references. David E. Rowe (Mainz) and Norbert Schappacher (Strasbourg) read the entire manuscript and gave a great deal of helpful advice.

Thanks must also go to Annalisa Capristo (Rome), to Liliane Beaulieu (Paris), and to Alexei Kojevnikov (Vancouver). The latter kindly alerted us to the important source (**Esakov, 2000**).

Abbreviations

AMS	=	American Mathematical Society
AUMA	=	Ausstellungs- und Messe-Ausschuss der Deutschen Wirtschaft
CNR	=	Consiglio Nazionale delle Ricerche (Italy)
CPSU	=	Communist Party of the Soviet Union
DKZ	=	Deutsche Kongreßzentrale
DMV	=	Deutsche Mathematiker-Vereinigung
ICAM	=	International Congress for Applied Mechanics
ICM	=	International Congress of Mathematicians
ICMI	=	International Commission on Mathematical Instruction
ICTAM	=	International Congress for Theoretical and Applied Mechanics
INdAM	=	Istituto Nazionale di Alta Matematica (Italy)
IRC	=	International Research Council
IEB	=	International Education Board
IMU	=	International Mathematical Union
IMUK	=	Internationale Mathematische Unterrichtskommission (=ICMI)
IRC	=	International Research Council
KUD	=	Kirke- og undervisningsdepartementet (Norway)
LMS	=	London Mathematical Society
MAA	=	Mathematical Association of America
REM	=	Reichserziehungsministerium (Germany)
SCM	=	Scandinavian Congress of Mathematicians
UMI	=	Unione Matematica Italiana

Conventions for Transliteration from Cyrillic Characters

Аа	=	Aa	Кк	=	Kk	Хх	=	Kh kh
Бб	=	Bb	Лл	=	Ll	Цц	=	Ts ts
Вв	=	Vv	Мм	=	Mm	Чч	=	Ch ch
Гг	=	Gg	Нн	=	Nn	Шш	=	Sh sh
Дд	=	Dd	Оо	=	Oo	Щщ	=	Shch shch
Ее	(see	below)	Пп	=	Pp	ъ	=	”
Ёё	(see	below)	Рр	=	Rr	ы	=	y
Жж	=	Zh zh	Сс	=	Ss	ь	=	’
Зз	=	Zz	Тт	=	Tt	Ээ	=	Ee
Ии	=	Ii	Уу	=	Uu	Юю	=	Yu yu
Йй	=	Yy	Фф	=	Ff	Яя	=	Ya ya

Ее = Ye ye (initially, after a vowel, or after й, ъ, or ь); Ee (otherwise)
Ёё = Yë yë (initially, after a vowel, or after й, ъ, or ь); Ëë (otherwise)

Part 1

Introduction

CHAPTER 1

Introduction: Four Strategies, the Prehistory of the Oslo Congress, and Sources

1.1. Four strategies for international mathematical communication

Oral communication, and congresses in particular, remain a crucial element within mathematical communication — even in the current age of electronic mail. This can be seen both on the national level (at meetings of the LMS, DMV, AMS, etc.) and internationally, with huge mathematical congresses ever growing in size. Conferences and other meetings serve many more purposes than simply communicating information about recent mathematical research, and this has always been the case, with each particular historical period setting different priorities.

The title of our book alludes to an article reporting on the 1936 ICM with a pun in its title: "Oslo under the integral sign" (**Dunnington, 1936a**). The article stresses the unifying and harmonising aspects of international mathematical communication and sees them exemplified in the Oslo congress of 1936; the congress had chosen the sign for a contour integral as its logo — this appears also on the title page of the congress proceedings (Figure 1.1).[1] We know of course that international communication at that time was very much influenced by the rise of the Nazis in Germany and dictatorial regimes in other countries as well. However, the report's author, G. Waldo Dunnington (1906–1974), who was known for his boundless admiration of German mathematics, and who wrote an influential biography of Gauss (**Dunnington, 1955**), is probably not a very good source for a critical judgement of the darker aspects of that congress and of the particular historical conditions under which it was held.[2]

[1]Dunnington also wrote in his report that "the delegates to the Congress wore a badge in the form of an integral sign, which entitled them to ride free on the street cars and busses in Oslo and vicinity [*sic*]" (**Dunnington, 1936a**, p. 93).

[2]Dunnington lavished praise upon the report that Walther Lietzmann from Göttingen gave at the congress to the International Commission on Mathematical Instruction (ICMI) (**Lietzmann, 1937**). He described this report, which strongly reflected the implications of political changes in Germany for school mathematics, as "exceedingly interesting and enlightening as to mathematics in the schools of present-day Germany" ((**Dunnington, 1936a**, p. 88); see also the remarks in Chapter 10 of the present book). Dunnington also accepted an invitation to attend the bicentenary celebrations of Göttingen University in June 1937, even planning to bring with him a "group of American students in order that this youth acquires a better understanding of and good friendship with Germany" (quoted from a German letter from Dunnington to the rector of Göttingen University, before April 1937, kept as a copy in University Archives Göttingen: Sekr., IB2). Participation in these celebrations was controversial among American mathematicians, as had been that of George David Birkhoff (1884–1944) in the 550th anniversary celebrations of Heidelberg University in 1936, prior to the Oslo congress, which Dunnington mentioned in his report without judgement (**Dunnington, 1936a**, p. 89). The AMS, as an organisation, finally declined politely to participate in the Göttingen event after

COMPTES RENDUS
DU
CONGRÈS INTERNATIONAL
DES MATHÉMATICIENS
OSLO 1936

Tome I

Procès-Verbaux
et
Conférences Générales

A. W. BRØGGERS BOKTRYKKERI A/S

OSLO 1937

FIGURE 1.1. Title page of volume I of the proceedings of the Oslo congress (Source: `https://www.mathunion.org/icm/`)

The congress was the only one on this level to be held during the period of the Nazi regime in Germany (1933–1945) and after the wave of emigrations from it. The Nazis and the war they unleashed led to the postponement by a decade of the next ICM, initially planned for 1940 in Cambridge, Massachusetts. Thus we choose to adapt the title of Dunnington's report for our purposes, adding, however, a question mark. Indeed, along with the integrating interests of the mathematicians, one has also to take into account the differences and differentials (to stay in the mathematical picture) between the goals of the various participants, which are clearly stressed in the report on the congress that was subsequently produced by

internal discussions which had pointed to the fact that mathematics in Göttingen had been all but destroyed by the Nazis (AAMS, box 26).

the German mathematician Walther Lietzmann (see below). Lietzmann drew a line between "us" and "them", the latter referring in particular to those mathematicians who had been expelled from Nazi Germany.[3]

We find the following four major strategies for international scientific communication manifesting themselves in the organisation and the staging of the congress in Oslo. These are exemplified by the actions of several nations, and by various members of the international mathematical community as a whole; we denote them for short by the words in square brackets:

- the newcomer strategy of the ambitious and developing nation [newcomer];
- the expansionist and propagandistic strategy of the (mathematically) established nation [expansionist];
- the isolationist and often fearful strategy of the (politically) ousted nation [isolationist];
- the scientific, for the most part internationalist, strategy of the research mathematician [scientific].

It goes without saying that most of the strategies that were actually pursued had to be mixtures of these four "ideal-type" strategies, even if this made for some contradictions in adopting them. Mathematicians, of course, would usually aspire for unhampered international communication, but at the same time they often shared the political goals of their nation. There were also mixed approaches in national strategies. To be sure, the interests of an ambitious and rising nation (such as Norway, or of Scandinavia as a whole) were served best by preventing monopolies in mathematical research. However, national strategies could be isolationist and expansionist at the same time (particularly in Nazi Germany), and even isolationist strategies, exemplified by the Soviet Union and Italy, were sometimes mellowed by concerns, such as for instance among the Russians, over losing international political impact.

In addition to the four strategies mentioned above, other strategies came into play at least in a rudimentary form, such as private or public economic interests supporting the pursuit of mathematics. However, the economic usefulness of pure mathematical research beyond mathematical teaching, for instance in industry, was less visible during the 1930s, which resulted in a lack of state funding even in nations such as the United States. Also, the predominance of pure over applied mathematical disciplines on the agenda of the Oslo congress (there had been parallel international congresses for applied mechanics since the 1920s) made the "economic strategy" a rather marginal one, at least as far as the congress in Oslo was concerned. One is certainly tempted to identify the desire for international peace and mutual understanding as the significant strategy lying behind the movement for the ICMs, the latter having had their first major event in Zurich in 1897.[4] However it seems to us that "love of peace" is too fundamental, general, and agreed-upon a category to be of much use in the analysis of historical motives, not least due to

[3]We could not of course go so far as to add a swastika to the signs under which the congress was being held because the Norwegians certainly did not adopt it, although the swastika figured visibly on the wreath laid down by the German delegation at the Abel monument in Oslo (see Figure 3.4 on p. 80).

[4]On scientists and internationalism in the late-nineteenth century more generally, see (**Fox, 2016**).

FIGURE 1.2. Group photograph at the Oslo congress, taken in the square in front of the historical aula of the university, beneath the statue of the famous Norwegian historian Peter Andreas Munch (1810–1863); Carl Størmer appears in the middle of the front row. (Source: National Library, Oslo, unprocessed 115 Carl Størmer Archives)

its propagandistic misuse even by dictators such as Hitler, or its potential to lead to appeasement and misguided collaboration with oppressive regimes.

Looking in detail at the Oslo ICM (see Figure 1.2) as a special event first requires the realisation that it was the result of a longer trend in the development of mathematical communication.[5]

1.2. The prehistory of the Oslo congress

In the following section we give a short overview of the history of the International Congresses of Mathematicians from their inauguration in 1897.

There exist several more global[6] and more specialised[7] accounts of the history of the various ICMs, but few in-depth analyses of any particular congress.[8] In addition, there has so far been little analysis of the *mathematics* of the ICMs, the one paper by Bernard Bru (**Bru, 2003**) on the stochastic work at the Bologna congress

[5]See, for example, (**Lehto, 1998a**). Note also the parallels drawn between Oslo 1936 and Bologna 1928 in our Section 1.2.3.

[6](**Albers *et al.*, 1987**), (**Curbera, 2009**), and (**Lehto, 1998a**).

[7](**Parshall and Rowe, 1994**), (**Barrow-Green, 1994**), and (**Riehm and Hoffman, 2011**).

[8]In fact, we are aware of only one such (published) in-depth analysis: (**Guerraggio and Nastasi, 2008**) on the 1908 Rome ICM. The 1950 ICM at Harvard features heavily in (**Barany, 2016**).

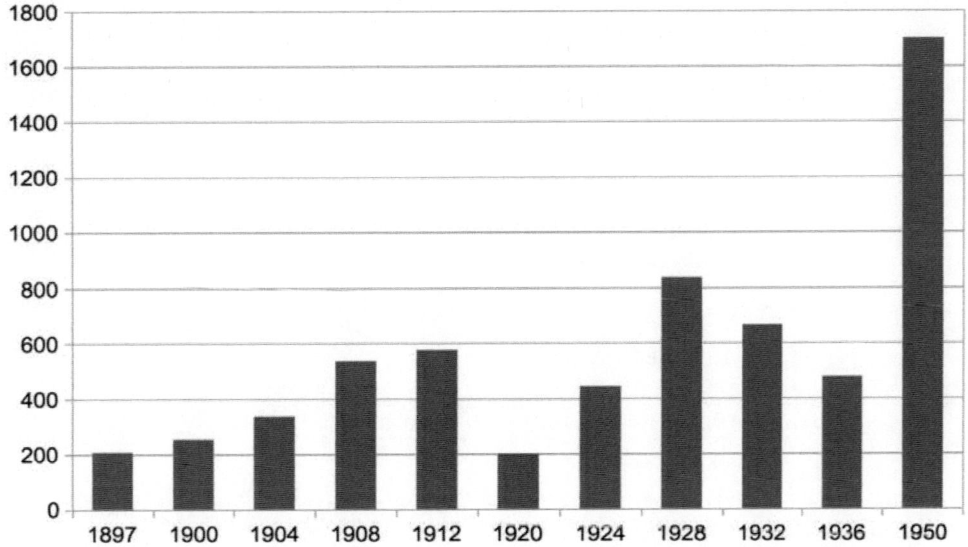

FIGURE 1.3. Indication of the growth in attendance of the ICMs, 1897–1950

of 1928 being an exception. We note, however, that the proceedings of all the ICMs are freely available online via the website of the International Mathematical Union (IMU).[9]

In the following three subsections, we will focus on those events that are most important for the understanding of the circumstances and conditions surrounding the Oslo congress of 1936, and to which later passages in our book will therefore have to refer repeatedly. The three subsections deal in particular with

- the history of the ICMs before the First World War, in particular Rome 1908;
- the so-called boycott of German science, organised in mathematics by the IMU, which was founded in Strasbourg in 1920;
- the ICM in Bologna in 1928, the only such congress before the Second World War that was organised under a dictatorial political regime, and which foreshadowed several communications problems that arose in connection with Oslo 1936.

For overall perspective, we include Figure 1.3, which gives a rough indication of the numbers of mathematicians attending the ICMs from their inception until the 1950 congress. For reference, we also list the dates and venues of the first eleven ICMs in Table 1.1.

1.2.1. The five ICMs before the First World War, in particular Rome 1908. We discuss these early ICMs only very briefly because, as we will see, the war changed the conditions for communication considerably.

[9]At `https://www.mathunion.org/icm/proceedings` (accessed 22 January 2019).

TABLE 1.1. Dates and locations of the first eleven International Congresses of Mathematicians

9–11 August 1897	Zurich
6–12 August 1900	Paris
8–13 August 1904	Heidelberg
6–11 April 1908	Rome
22–28 August 1912	Cambridge
22–30 September 1920	Strasbourg
11–16 August 1924	Toronto
3–10 September 1928	Bologna
4–12 September 1932	Zurich
13–18 July 1936	Oslo
30 August–6 September 1950	Cambridge, MA

A preliminary meeting was organised in Chicago in connection with the World's Fair in 1893.[10] Probably the most-quoted indirect outcomes of this Chicago meeting were the beautiful lectures delivered in English by Felix Klein (1849–1925) at Northwestern University in Evanston, Illinois, a few weeks after the meeting (**Klein, 1894**). The Göttingen mathematician had attended the Chicago meeting together with his former student Oskar Bolza (1857–1942), and had used it strategically to stimulate exchange with the then still young mathematical culture in the United States.

The first of the series of regular ICMs was held in Zurich in 1897. After that, a famous one (because Hilbert's problems were presented there) was organised in Paris in 1900. The following ICMs, before the outbreak of the First World War, took place in Heidelberg (1904), Rome (1908), and in Cambridge, England (1912). There are three traditions from the pre-First World War congresses that are worth discussing in some detail, and which are connected in particular with the ICM in Rome in 1908. These are:

- the foundation of ICMI,
- the first ideas concerning the foundation of an international mathematical organisation, and
- the first award of an international prize in mathematics.

The meeting's section IV (Philosophical, historical and didactical issues) led to the formation of the International Commission on Mathematical Instruction (ICMI), which would have a strong presence at future congresses (in the case of the Oslo ICM, see Chapter 10):

> The Congress, having recognised the importance of a thorough examination of the programmes and methods of the teaching of mathematics in the secondary schools of the various nations, entrusts to Professors KLEIN, GREENHILL and FEHR the task of constituting an international committee that will study the issue and report on it to the next Congress.[11]

[10]On this preliminary meeting, see above all (**Parshall and Rowe, 1994**).

[11]Translated from the Italian in (**ICM, 1908**, vol. I, p. 33). These discussions had a precursor in a 1905 article by Henri Fehr and David Eugene Smith's response thereto: see (**Lehto, 1998a**, p. 13).

The second point of note is that the assembled mathematicians approved a resolution put forward by Professor Alberto Conti of Bologna:

> The Congress votes that the constitution of an International Association of Mathematicians be placed upon the agenda for the next Congress.[12]

However, whereas ICMI did indeed report to the next congress in Cambridge in 1912,[13] the idea of forming an international mathematical organisation was not taken up.[14] It may be that only a political event like a war could lead to such an organisation of mathematics, but then with problematic traits (see below).

The third and final remark to make with regard to internationalism at the Rome ICM is that this was the first congress at which an international mathematical prize was awarded. Four years earlier, in Heidelberg, it had been announced that Giovanni Battista Guccia (1855–1914), the founder of the Circolo matematico di Palermo, had made 3,000 francs available for the award at the next congress of a prize, to be named the Guccia medal, for a work making substantial progress in the theory of algebraic curves.[15] A judging panel consisting of Max Noether (1844–1921), Henri Poincaré (1854–1912), and Corrado Segre (1863–1924) was established, and the decision was to be made in the manner of the prizes traditionally awarded by the Paris Academy: through the assessment of submitted memoirs. However, since none of the memoirs received were deemed of sufficient merit to warrant the prize, it was instead awarded to Francesco Severi (1879–1961) for his overall contributions to geometry.[16] This was the only time that the prize was given, though its single award may later have provided some impetus to the establishment of the Fields Medal.

All four internationally leading nations, plus neutral Switzerland, had thus been organisers of an ICM before the war. The ICM in Cambridge in 1912 figured officially and in the proceedings as the fifth international congress. It would be the last ICM to carry a number explicitly, and the reason lay in the political problems that followed the war.[17]

1.2.2. The IMU and the boycott of the 1920s. In 1918 and 1919, starting just before the end of the First World War, representatives of the scientific academies in the Allied (*Entente*) powers began discussing the future of scientific cooperation after the war. The result was the creation of the International Research Council (IRC), and the decision to exclude Germany and its allies from

[12]Translated from the Italian in (**ICM, 1908**, vol. I, p. 33).

[13]See (**ICM, 1912**, vol. I, pp. 54–56); see also the comments in (**Lehto, 1998a**, p. 14).

[14]Indeed, remarks made by the president of the Cambridge congress, Sir George Darwin, suggest that he was resistant to the idea, deeming the existing arrangements for the organisation of the ICMs to be sufficient (**ICM, 1912**, vol. I, p. 40).

[15]See (**ICM, 1904**, p. 53), and also the remarks in (**Guerraggio and Nastasi, 2008**, pp. 75–76).

[16]See (**ICM, 1908**, vol. I, pp. 209–216); see also the comments in (**Brigaglia, 2002**, p. 195), as well as those in (**Guerraggio and Nastasi, 2008**, pp. 115–119).

[17]On a more anecdotal level, this led to a lasting controversy over the numbering of the congresses; in 1932, Hermann Weyl (1885–1955) — then the president of the Deutsche Mathematiker-Vereinigung (DMV) — would, with a combination of sternness and idiosyncratic humour, remark that the 1932 Zurich congress was the nth International Congress of Mathematicians for some $7 \leq n \leq 9$, and that axiomatic mathematics was insufficient to decide exactly which number was the right one (**ICM, 1932**, vol. I, p. 71).

international science. Under the auspices of the IRC, whose early president was the French mathematician Émile Picard (1856–1941), an International Mathematical Union (IMU) was set up in 1920 in Strasbourg. There, in the newly recaptured region of Alsace on the border between France and Germany, an ICM with restricted internationality took place at that time.[18] The first president of the IMU was the Belgian mathematician Charles de la Vallée-Poussin (1866–1962). The official goal of the Union was to facilitate international cooperation of mathematicians and to organise international congresses. However, conceived of in the immediate post-war atmosphere, the IRC and the IMU were heavily influenced by nationalistic arguments among politicians and scientists alike. On a practical level, the IMU served only as a union for Allied mathematicians.

Neutral countries — including the Scandinavian ones — were invited to join the IRC and the IMU and did so in 1919 and 1920. Soon thereafter, some neutral countries, such as the Netherlands and Denmark, began lobbying for an opening towards the Axis powers, but strong Belgian and French opposition counteracted British and American willingness to normalise affairs.[19] Thus the exclusion of Germany and its wartime allies lasted until the mid-1920s and also prevented the participation of their mathematicians in the ICM in Toronto in 1924. But the fact that this congress had to take place in Canada[20] and not in the United States (which had been chosen in Strasbourg as the next venue) was the result of existing American resistance to the boycott. In the proceedings of the Toronto ICM, many American and British mathematicians, as well as some from neutral countries, voiced their concerns about the boycott. That the mathematicians gathered in Toronto did not automatically confirm Brussels as the next venue for an ICM was one of the results of these concerns. Indeed, some mathematicians came to regret their previous participation in the Strasbourg congress in 1920.[21]

The brothers Niels and Harald Bohr (1885–1962 and 1887–1951), the physicist and the mathematician, respectively, showed their impatience with the situation in an interview in Copenhagen in September of 1925 with Augustus Trowbridge, a functionary of the Rockefeller Philanthropies (he was from the International Education Board). They announced that the Scandinavian countries would most likely

---

[18]On the IMU, see (**Lehto, 1998a**); on the role of M. Fréchet in the organisation of the Strasbourg congress, see (**Siegmund-Schultze, 2005**). The number "six" appears nowhere in connection with the Strasbourg congress — not even in the speeches.

[19]We will see in Chapter 2 that this did not mean that all Scandinavian mathematicians were united in this endeavour: the president of the Oslo congress, Carl Størmer, leaned more to the French side.

[20]The Canadian John Charles Fields (1863–1932) stepped in as organiser (**Lehto, 1998a**, pp. 33–34). We owe the creation of the Fields Medal to this twist of events, and thus also to political reasons: see below.

[21]For example, Norbert Wiener (1894–1964), who later wrote: "In many ways this was an unfortunate decision. It was one which later led me to regret my little share in sanctioning the meeting by my presence. The Germans were excluded as a sort of punitive measure. In my mature, considered opinion, punitive measures are out of place in international scientific relations" (**Wiener, 1956**, pp. 49–50). Attitudes some years later are recorded in the protocol of the IMU session in Zurich in 1932, in which Wiener and Oswald Veblen (1880–1960) attacked the IMU "vigorously" as "useless". The two Americans were supported by Harald Bohr; see (**Fehr, 1932**, p. 277).

withdraw from the IRC if the boycott continued. Trowbridge quoted them indirectly in the following way:

> Scientifically, the Germans are as important to us as any nation. We are not politically unanimously pro-German, but scientifically we are. ... [Trowbridge] was rather surprised at the heat with which the Scandinavian view was presented.[22]

Purely scientific reasons were not sufficient to convince French and Belgian mathematicians, who had suffered personally the most from the war, to resume international contacts — or at least these sufferings put a considerable delay on normalisation.

Brigitte Schroeder-Gudehus, in her analysis of the German reaction to the scientific boycott in the 1920s (**Schroeder-Gudehus, 1973**), goes one step further. She finds that many scientists, among them mathematicians, were not leading international reconciliation after the war but obstructing it. Even when the German government became sufficiently politically active to join the League of Nations in the mid-1920s, many of those scientists who had felt offended by the boycott, and who were not convinced of Germany's main responsibility for the outbreak of the war, still hesitated.

At an extraordinary meeting of the IRC in Brussels in June of 1926, the exclusion clause was repealed, and Germany was invited to adhere to the various international scientific unions (**Schroeder-Gudehus, 1973**, pp. 110–111). However, the German government failed then in its efforts to convince leading German scholars to join the unions. Moreover, the mathematicians on both sides were still divided about reconciliation when it came to the next congress, the ICM in Bologna in 1928.

1.2.3. The ICM in Bologna 1928 between various nationalist and internationalist motivations. In the aftermath of the war, Italy had retained its prominent position within the international mathematical community.[23] On the international stage, Italian scientists, as members of one of the victorious nations, were heavily involved in the foundation of the IRC (**Greenaway, 1996**, Chapter 2), and consequently also of the first IMU (**Lehto, 1998a**, §1.4). Thus, leading Italian mathematicians such as Vito Volterra (1860–1940) were complicit in the exclusion of Germany and her wartime allies from international scientific activities during the early 1920s.

In a more ambiguous position than Volterra stood Salvatore Pincherle (1853–1936), president of the IMU between 1924 and 1928. He was at the same time the organiser of the 1928 Bologna ICM, which was the first congress after the First World War at which the mathematicians of all nations were once again welcome. The legal and political sleight of hand employed by Pincherle in order to bring this about, and the consequent split between the IMU and the ICMs, is well described elsewhere, so we do not dwell on it here.[24] We must stress, however, the (political) parallels between Bologna 1928 and Oslo 1936: for instance, in both cases background tensions led to boycotts of each congress by certain parties (see below).

[22]Quoted in (**Siegmund-Schultze, 2001**, p. 59).

[23]For more about Italian mathematics after the First World War, see Chapter 5.

[24]See instead (**Lehto, 1998a**, §2.4), (**Guerraggio and Nastasi, 2008**, Chapter 6), (**Riehm and Hoffman, 2011**, Chapter 10), or (**Siegmund-Schultze, 2016a**, pp. 58–60).

Scientific internationalism was protested, but German and Italian nationalism came to the fore, both in the preparation of the congress and during the event itself.

Bologna saw the first return of a German mathematical delegation of considerable size (76 mathematicians), led by David Hilbert (1862–1943), who was known for his internationalist position. Hilbert planned to give a political talk in Bologna. Although, for unknown reasons, it appears that the presentation was not delivered, the manuscript has survived and contains an often-quoted passage:

> It is a complete misunderstanding [of our science] to construct differences or even incompatibilities according to peoples and races, and the reasons for which this has been done are very shabby ones. Mathematics knows no races.[25]

The participation of German mathematicians had been disputed in the run-up to the congress, when even the influential Dutch topologist and mathematical intuitionist L. E. J. Brouwer (1881–1966) had advised his German colleagues not to take part for the sake of their honour. In the end, a number of German mathematicians boycotted the congress in protest at their earlier exclusion (**Lehto, 1998a**, p. 46). After the congress, on 20 November 1928, the mathematical physicist from Göttingen, Max Born (1882–1970), who was in Bologna, wrote about this in a letter to his friend Albert Einstein (1879–1955):

> Hilbert is not politically very left-wing ... But when it comes to the question of the intercourse between scientists of different countries, he has a very sharp eye for detecting what is best for the whole. Hilbert considered, as we all did, that Brouwer's behaviour in this affair, where he was even more nationalistic than the Germans themselves, was utterly foolish. But the worst of it all was that the Berlin mathematicians were completely taken in by Brouwer's nonsense. ... I can understand this in Erhard Schmidt's case, for he always did lean to the right in politics, as a result of his basic emotions. For Mises and Bieberbach, however, it is a rather deplorable symptom. (**Born and Einstein, 1971**, p. 98)

It is worth mentioning that there were among the boycotters of Bologna such widely different characters as the future Nazi Ludwig Bieberbach (1886–1982), who did not dare show up in Oslo, and the other mathematicians from Berlin, Richard von Mises (1883–1953) and Erhardt Schmidt (1876–1959). The future Jewish refugee von Mises, in 1936 shied away from the effort involved in long-distance travel from his place of refuge (Istanbul), while the German nationalist Schmidt gave a rather naive propaganda speech for the Nazis in Oslo (see Chapter 2).

Indeed, the Bologna ICM was as politically eventful as that in Oslo eight years later. This makes the congress particularly interesting for the historian of mathematics, although it has perhaps not been studied as thoroughly as it might have

[25]See (**Siegmund-Schultze, 2016a**), in which the complete text of Hilbert's planned speech is published both in the German original and in English translation. It is argued that no known contemporary source (proceedings, printed reports, or private letters) alludes to Hilbert's talk; the biography of Hilbert (**Reid, 1970**) claims that it was actually delivered, but without giving a source.

been.[26] The Italian organisers were not just driven by internationalist motivations, and from the French both Émile Picard and Émile Borel were absent, probably for different political reasons.[27] As (**Curbera, 2009**, p. 88) has noted, the ICM was "a great international exhibition for the Fascist state of Benito Mussolini". The fascist regime certainly made itself felt at the congress, not least through the state patronage sought by Pincherle, who had joined the National Fascist Party at the end of 1926 (**Capristo, 2016**, pp. 292–295). If we browse the congress proceedings, we find frequent indications of state involvement, notional or otherwise: for example, Mussolini himself is named as president of the congress' "honorary committee" ("comitato d'onore") (**ICM, 1928**, vol. I, p. 23), while the proceedings record "the cordial hospitality" ("la cordiale ospitalità") (**ICM, 1928**, vol. I, p. 18) offered by Bologna's Casa del Fascio, as well as the greetings of the mayor of Bologna:

> Fascist Bologna is proud to welcome you and to be able to show you what it has become under the vivifying impulse of Fascism.[28]

A further, quite inconspicuous, indication of a fascist presence may also be seen on the title page of the proceedings (Figure 1.4), where the dates of the meeting are given as "3–10 settembre 1928 (VI)", the "VI" indicating that this was the sixth year of the regime.[29]

Despite the readmission of German mathematicians, the tensions of the decade were palpable in Bologna, which may have contributed to the decision there to return to the neutrality of Switzerland and Zurich for the 1932 ICM. International tensions rose once again in the years after Bologna, not least under the influence of the world-wide economic crisis of 1929.[30]

In Zurich, as in Bologna before, the political problems of international mathematical collaboration were discussed openly, for instance in the meeting of the IMU (see below). It was also in Zurich in 1932 that the president of the congress, Rudolf Fueter (1880–1950), informed the assembled mathematicians about the decision made by the recently deceased John Charles Fields, to donate a "gold medal for young mathematicians".[31] The Fields Medal would be awarded for the first time at the ICM in Oslo in 1936; see Section 9.22.

At the meeting of the IMU at the ICM in Zurich on 11 September 1932, the IMU was declared liquidated ("mise en liquidation") and its assets were for the

[26](**Capristo, 2016**, p. 289) observes that "[t]o date, the Congress has been studied almost exclusively as an episode in the history of the International Mathematics Union (IMU) and its internal conflicts", and calls for an analysis of the Bologna ICM similar to that advanced in the present book for the Oslo congress. Some recent steps in this direction have been taken by (**Cioli, 2016**). On the mathematical side (namely, stochastics), note also (**Bru, 2003**).

[27]Picard was quite explicit in boycotting the congress because German mathematicians would be present (**Capristo, 2016**, pp. 300–304), although Borel's position is more ambiguous; (**Capristo, 2016**, pp. 297–300), following (**Bru, 2003**), suggests that an opposition to fascism may explain Borel's absence. The plenary lecture that he was due to deliver at the ICM was given instead by Cartan (**ICM, 1928**, vol. I, p. 67).

[28]Translated from the Italian in (**ICM, 1928**, vol. I, p. 69).

[29]It is worth reiterating the observation made in (**Siegmund-Schultze, 2016a**, p. 59) based on information from Norbert Schappacher, namely that the "VI" does not indicate that the editors deemed this congress the sixth ICM (thereby discounting Strasbourg and Toronto), as has sometimes been suggested.

[30]In this connection, the German mathematician Hermann Weyl noted in an address to the Zurich congress: "Many an international congresses were cancelled last year; the congress of mathematicians takes place." (Translated from the German in (**ICM, 1932**, vol. I, p. 73).)

[31]Translated from the German in (**ICM, 1932**, vol. I, p. 43).

ATTI

DEL

CONGRESSO INTERNAZIONALE

DEI MATEMATICI

BOLOGNA 3-10 SETTEMBRE 1928 (VI)

TOMO I.

RENDICONTO DEL CONGRESSO

CONFERENZE

BOLOGNA

NICOLA ZANICHELLI

EDITORE

FIGURE 1.4. Title page of volume I of the proceedings of the Bologna congress (Source: `https://www.mathunion.org/icm/`)

time being ("provisoirement") deposited at the Banque de France. This decision followed lively discussions about the function and tasks of the Union (**Fehr, 1932**). The mood then reigning is probably best indicated by the fact that although the session was led by the French mathematician Georges Valiron (1884–1954), the decision was nevertheless taken against the French-Belgian foundation of 1920. Interestingly enough, it was the Swiss educator Henri Fehr (1870–1950), who, when publishing Valiron's report in the journal *L'enseignement mathématique*, described the dissolution of the IMU as a "gross ingratitude", pointing to the financial contribution of the IMU to the creation of the "two medals" to be awarded by the congress and acknowledging the financial support "by a great number of governments" (**Fehr, 1932**, p. 278). This might also indicate that the relative merits of the IMU were greatest in the context of mathematical education, where government support was crucial. *L'enseignement mathématique* had become the official organ of ICMI, the commission founded in 1908 (see above). The Americans present in Zurich, however, were more focused on research and the unhampered exchange of ideas than educators such as Fehr. They held that it

> would appear that a permanent international organization had no problems important enough to warrant its existence; [and] that if an international organization in mathematics were to continue, it should be divorced from connection with national governments. (**Richardson, 1932**, p. 773)

According to Valiron's protocol, the liquidation of the IMU was considered provisional, subject to the deliberations of a commission that was to be founded. Indeed, the Zurich congress decided the following on its concluding session on 12 September 1932:

> An international commission is formed in order to re-study the question of international collaboration in the sphere of mathematics and to make propositions with regard to its reorganisation at the next congress. (**ICM, 1932**, vol. I, p. 61)

The following mathematicians were elected members of that commission: Severi (president), P. S. Aleksandrov (1896–1982), H. Bohr, Lipót Fejér (1880–1959), Gaston Julia (1893–1978), Louis J. Mordell (1888–1972), Esteban Terrades i Illa (1883–1950), de la Vallée-Poussin, Oswald Veblen (1880–1960), Hermann Weyl (1885–1955), and Stanisław Zaremba (1863–1942).

On the same day, it was also decided that Oslo should be the venue for the 1936 congress (**ICM, 1932**, vol. I, p. 60). Although the proceedings are silent on the matter, one has to assume that the choice of Oslo was once again influenced by the desire to have the congress in a politically neutral venue.

However, before the Oslo congress could take place, other unexpected and catastrophic political events, as described in the preface, took place; these, above all the seizure of power in Germany by the Nazis in 1933, would shape the future congress considerably. Under the immediate impression of the refugee crisis, the secretary of the AMS, R. G. D. Richardson (1878–1949), wrote the following to Harald Bohr:

> Since the war, we have been constantly compelled to think of colleagues as nationals and not as citizens in the international domain.[32]

1.3. Sources

In compiling this book, we have used a mixture of printed and unpublished sources, the latter pertaining in particular to the policies of the Norwegians and the Germans in relation to the congress. Many of these sources are being exploited in this context for the first time.

[32]This is in a letter from Richardson to Bohr of 26 June 1934, which is quoted in (**Siegmund-Schultze, 2001**, p. 16).

The most important printed source is, of course, the *Comptes rendus de congrès international des mathématiciens, Oslo 1936* (i.e., the congress proceedings), which appeared in two volumes in Oslo in 1937, and was reprinted in one volume in 1967.[33]

Caution is necessary when using the congress proceedings for historical analysis, even though the proceedings appeared a year after the event and should therefore have been an accurate record. The list of "members", that is, the "Liste des Membres du Congrès" (**ICM, 1936**, vol. 1, pp. 29–38), gives all the Russian (eleven) and Italian (five) mathematicians who originally registered, but who, according to other contemporary sources, did not take part (as discussed in Chapters 4 and 5). The breakdown of the members of the congress by country (**ICM, 1936**, vol. 1, p. 39) is additionally marred by the vaguenesses stemming from the uprooting and the mass emigration of mathematicians at that time, and from the principal problem of whether to count the country of origin of a mathematician or that of their present residence. Also, the mathematical contributions printed in the proceedings do not necessarily reflect the physical attendance of the respective author at the congress; in particular, the proceedings contains a very short note by the Moscow topologist L. S. Pontryagin (1908–1988) (**ICM, 1936**, vol. 2, p. 140), despite the fact that he did not attend the congress.[34]

Finally, the reports of the various countries to the International Commission on Mathematical Instruction (ICMI) were not printed in the proceedings, although they were part of section VIII (Pédagogie) of the congress. They were published for the most part in the Swiss journal *L'enseignement mathématique*, volumes 36 (1937) and 37 (1938), with the exception of Lietzmann's report, which had already appeared in Germany as (**Lietzmann, 1937**).[35]

To the best of our knowledge, the most detailed published reports on the congress are (**Lietzmann, 1936**) and (**Dunnington, 1936a**).[36] Some political background for the German participation is given in personal recollections written by Lietzmann around 1942 and published posthumously in (**Lietzmann, 1960**, pp. 60–62). Written from the perspective of one participant, the recollections in (**Behnke, 1978**, pp. 247–249) provide a vivid picture. Contemporary Norwegian newspapers also provide some reports on aspects of the congress: we reproduce and discuss some of these in Chapter 6 and Appendix A.

The only printed source which gives an explanation for the non-participation of the Russians in the congress is (**Esakov, 2000**), which publishes the decision

---

[33]The proceedings are also available online: see footnote 9 above. We therefore omit the original text of the translated quotations from these proceedings that appear throughout the present book. Indeed, in the interests of saving space, we similarly omit the original text of any *published* source that is quoted here in translation.

[34]In fact, (**Fehr, 1936**, p. 382) mentions that the note was read by Solomon Lefschetz (1884–1972) from Princeton.

[35]In *L'enseignement mathématique*, volume 36 (1937), p. 236, there is a reference to (**Lietzmann, 1937**), but also a remark that this publication was now obsolete due to very recent educational reforms in Germany, and could therefore not figure as the official German report to ICMI for this journal. The ICMI reports that were presented in Oslo are discussed further in Chapter 10.

[36]In addition, there are two much shorter contemporary reports, (**Fehr, 1936**) and (**Morse, 1936**), although the latter is unreliable since it lists the planned talks by Gel'fond and Khinchin without mentioning that they were not actually given. As far as we have been able to find, there is no report in *Science* or *Nature*, nor in the *Jahresbericht der Deutschen Mathematiker-Vereinigung.*

by the Soviet Politburo not to send a delegation. Corresponding materials for the Italian case comprise the few archival sources that we use in Chapter 5.

Not unexpectedly, some important *unpublished* sources at our disposal concern the preparation for the congress by the Norwegian organisers and the German delegation. The Norwegian sources that see particularly heavy use here are the archives of the Norwegian Ministry of Church and Education (Kirke- og undervisningsdepartementet: KUD) and a range of letters to and from the major figures in the organisation of the congress, held in the Manuscript Collection of the National Library in Oslo.

With regard to unpublished German sources, one that is very useful for us is a 15-page type-written report for the German Ministry of Education (Reichserziehungsministerium: REM), which the head of the German delegation, Walther Lietzmann, sent to various German, mostly political, authorities in mid-August 1936, after participating in the congress.[37] This report (WLR 1936) consists of two parts, a political part (pp. 1–5) and a mathematical one (pp. 6–15). There is some overlap between this unpublished report and (**Lietzmann, 1936**), particularly in the mathematical part. Nevertheless, we include the unpublished report here as an appendix (Appendix D), in both transcription and English translation, not least because we quote from it occasionally, but also because even the mathematical part contains many political side-remarks. The political part also reports on the preparation of the German delegation, in particular, on political criteria for its composition, which can be traced in more detail in two other unpublished sources: the files of the Nazi-dominated REM (Bundesarchiv Berlin, REM 2905), and in the Walther Lietzmann Papers in the Göttingen Municipal Archives (Depositum 89). A third, somewhat less important, source for German participation in Oslo is the files of the "Deutsche Kongreßzentrale" (DKZ; see Chapter 3), held at the Hoover Institution at Stanford University. The files in Berlin and Stanford both contain copies of Lietzmann's unpublished report. Finally, the correspondence of the German-Jewish emigrant to America, Richard Courant (1888–1972), who participated in Oslo, with German and other mathematicians, including the Russian P. S. Aleksandrov, is another valuable source (Richard Courant Papers: RCP)

[37]From the standpoint of the historian of mathematics, it is fortunate to have an internal report such as the one by Lietzmann, although it was produced against the backdrop of a serious curtailment of the participation of Germans in Oslo and has to be read carefully given its political context.

Part 2

The Politics of the Congress

CHAPTER 2

The Norwegian Hosts: The Newcomer with Historical Traditions

Henrik Kragh Sørensen
in collaboration with
Christopher D. Hollings and **Reinhard Siegmund-Schultze**

In this chapter, we describe the organisation and implementation of the International Congress of Mathematicians in Oslo from the perspective of the Norwegian organisers. We detail many of their preparations in positioning Scandinavia, and Oslo in particular, on the map of international mathematics, so as to be able to attract the congress. We analyse the particular internationalist aims that the organisers sought to achieve for the congress in a period when internationalism in the sciences was extensively debated and contested. We provide examples of how such debates were visible in the planning and public rhetoric of the congress, and we discuss some instances in which national and international values contrasted.

2.1. Prehistory of the ICM in Scandinavia

Although the 1936 ICM was the first such event to take place in Scandinavia, mathematicians in the Northern countries (Denmark, Sweden, Norway, and to some extent Finland) had a history of attending to the developments in the central mathematical nations. Thus, in order to situate the organisation of the 1936 ICM within the Scandinavian context, it is first relevant to describe how Scandinavian mathematicians viewed themselves and their international relations in the first decades of the twentieth century, then to describe some of the details of the organisation of the congress, and finally to address a number of more specific themes.

2.1.1. The 1916 ICM that wasn't. Preparation for the Cambridge congress in 1912, coupled with the meeting of Scandinavian mathematicians that was held in Copenhagen in 1911, had bolstered a feeling of unity and a common identity among Scandinavian mathematicians. The Scandinavian Congress of Mathematicians (SCM) in Copenhagen was the second (after Stockholm 1909) in a series of usually biannual meetings that proved to be an important institution for Scandinavian mathematics (**Turner and Sørensen, 2013**). In 1912, the influential and charismatic Swedish mathematician Gösta Mittag-Leffler (1846–1927) invited the 1916 congress to Stockholm (**ICM, 1912**, vol. I, p. 42), where he was laying the foundations for the research institute that would be established in his name. The ICM was meant to coincide with a meeting of Scandinavian mathematicians, with Mittag-Leffler's 70th birthday, and with the presentation of the will of Mittag-Leffler and his wife in which they bestowed their home at Djursholm in the outskirts of Stockholm as a mathematical research institute. However, the outbreak of war in Europe interfered with Mittag-Leffler's plans, and in November 1915, the congress had to be postponed until the arrival of more peaceful times. Mittag-Leffler was insistent that when the time came to revive the ICMs, Stockholm should be given the possibility of hosting the congress, as agreed before the war.

Throughout his life, Mittag-Leffler deliberately cast himself as a truly international Scandinavian mathematician,[1] and he was certainly not alone in pursuing an identity for Scandinavian mathematics and its relation to continental mathematics. This became a pressing question after the First World War when Scandinavian identity and internationalist outlooks combined to provide opportunities for Scandinavian mathematicians. In mathematics, where the centres of innovation and research had long lain with the German universities, the breakdown of communication during and after the war was severely felt and criticised by groups of internationalist mathematicians such as G. H. Hardy (1877–1947) and Veblen in Anglophone countries, and Scandinavians such as Harald Bohr and Mittag-Leffler (**Dauben, 1980**).

In the aftermath of international conflict, two strands of development were particularly important to Scandinavian mathematics: one was the founding of the IMU, as described in Chapter 1, and with it the resumption of international meetings, and the other was the role that Scandinavian mathematicians could play, individually and collectively, in normalising relations between the international community and the German mathematicians who were, at the time, a very prominent and important contingent of colleagues. These two lines of argument fused together over attracting an ICM to Scandinavia after the First World War; these came to a first confrontation around the 1920 ICM in Strasbourg.

2.1.2. Scandinavians and the debate over the 1920 ICM. In Chapter 1, we reported on the scepticism not only among Scandinavian but also British and American mathematicians about the anti-German boycott. When mathematicians from the victorious parties eventually convened in Strasbourg in 1920, the absent Mittag-Leffler was still upset. Because Stockholm had been passed over for the 1920 ICM, he refused to recognise the Strasbourg congress as the sixth ICM — and it certainly did not match his ideas of what an international congress should be.[2]

The controversial nature of the IMU itself, with its exclusion of German mathematicians and the decision to reject Stockholm as the venue for the 1920 ICM, prompted a debate elsewhere in Scandinavia: Danish mathematicians, for example, were undecided as to whether or not to reject membership of the IMU when Denmark, as a neutral country, was invited to join. Two different factions emerged within Danish mathematics, arguing for different means of steering the international mathematical community towards a more complete normalisation of international cooperation. Niels Erik Nørlund (1885–1981), who had attended the ICM in Strasbourg, argued in favour of Danish membership so as to influence the organisation from within. On the other hand, Harald Bohr argued that nothing short

[1]See (**Stubhaug, 2010**) and (**Turner, 2011**).

[2]See (**Dauben, 1980**) and (**Turner, 2011**). In a letter to Edmund Landau, Mittag-Leffler had noted an unsuccessful attempt by his student Nørlund to change the title of the Strasbourg congress from "des mathematiciens" to "de mathematiciens", reflecting the fact that not all mathematicians were welcome: Mittag-Leffler to Landau, 23 March 1923, as quoted in (**Siegmund-Schultze, 2011**, pp. 150–151).

of a complete return to pre-war cooperation was satisfactory and that the Danish Mathematical Society (*Dansk Matematisk Forening*) should remain outside the IMU. In 1921, the Danish Mathematical Society effectively took Bohr's line towards the IMU after a heated debate showed no possibility of achieving a consensus about entering it.[3] The day after the Mathematical Society had abstained from taking any joint decision, the matter was taken up at the Royal Danish Academy of Sciences and Letters (*Kongelige Danske Videnskabernes Selskab*). It was announced there that an independent committee with Nørlund and other prominent Danish mathematicians would be formed and would join the IMU (**Ramskov, 1995**, pp. 208ff). Thus, during the 1920s, the Danish mathematical community was split over the means of fighting the exclusion of their Axis-power colleagues. And this struggle against the boycott was certainly not restricted to Scandinavia, or even to the neutral countries. In 1928, within discussions about the impending ICM in Bologna, Bohr wrote to Brouwer in the Netherlands:

> We [Hardy and Bohr] have also fought against membership in the Union with all means (in Denmark, I would surely have been successful in preventing this membership if Nørlund had not built a committee in favour of joining independent of our Academy and our Mathematical Society).[4]

We note that the wider purpose of Bohr's letter here had been to dissuade Brouwer from his proposed boycott of the Bologna congress (see p. 12).

It appears that Carl Størmer (1874–1957), one of the most influential of all Norwegian mathematicians at the time (see Figure 2.1), and pioneer in the physical and mathematical analysis of the Northern Lights, was much less critical of the French-led boycott than some Danish mathematicians such as Bohr. Størmer, who had started as a number theorist of some fame, had studied partly in Paris (1898–1900), with Picard, among others, and briefly also in Göttingen in 1902. He was then quickly promoted to full professor at the University of Kristiania (now Oslo) (**Chapman, 1958**). He would later play a leading role in the preparation of the Oslo congress (see Section 2.5), of which he ultimately became the president, and at which he delivered a plenary lecture (see Section 9.2). In his letters after the First World War to Mittag-Leffler, with whom he was in close contact as co-editor of the journal *Acta Mathematica*, Størmer showed no qualms about going to the Strasbourg congress in 1920 as an official representative of Norway. He gave a talk there on his new method for the numerical solution of second-degree ordinary differential equations (**ICM, 1920**, pp. 243–257).

After the Strasbourg congress, Størmer gave an interview for the Norwegian daily *Aftenposten* in which he tried to take a neutral position but showed that his sympathies lay primarily with the French:

> From the Central Powers nobody appeared, because nobody was invited. In the interest of science it would have been desirable if mathematicians from the Central Powers had not been excluded. However it is very plausible that the French and the Belgians who have suffered from the horrors of invasion, do not yet wish to

[3]See (**Crone, 1923**, p. 74); on the history of the Danish Mathematical Society, see (**Ramskov, 2000**).

[4]H. Bohr to Brouwer, 3 July 1928, as translated from the German by (**Ramskov, 1995**, p. 209); see also (**van Dalen and Brouwer, 2011**, pp. 332–333).

FIGURE 2.1. Carl Størmer (1874–1957) (Source: Nasjonalbiblioteket–Flickr: Portrett av Carl Størmer. https://commons.wikimedia.org/wiki/File:Portrett_av_Carl_St%C3%B8rmer.jpg. CC BY 2.0 https://creativecommons.org/licenses/by-sa/2.0/)

meet Germans personally, which cannot be avoided at a congress. Several of the most important French mathematicians have lost their sons in the war. Jordan, for instance, two sons and a son-in-law. Picard, a son and a daughter who was a nurse, and the great mathematician Hadamard lost a son who was a very promising mathematical genius.[5]

[5]Translated from the Norwegian in *Aftenposten*, 27 October 1920, p. 4. The final remark relates to Hadamard's second son Étienne (1897–1916).

Informed by Mittag-Leffler[6] about the latter's critical attitude towards the Strasbourg congress, Størmer, however, had something critical to say in return. He said that only one "Swede", the Danish-born Nørlund, had been in Strasbourg, which was probably due to the rejection of Stockholm as a venue. Størmer also mentioned the protest by the Deutsche Mathematiker-Vereinigung (DMV) against the congress, but he did not take a position. On 26 December 1923, Størmer wrote to Mittag-Leffler about the impending ICM in Toronto (see Chapter 1). He told Mittag-Leffler that he had recently given a talk in Paris[7] and that the French mathematician Gabriel Koenigs (1858–1931) (who had been among the most fervent supporters of the boycott and played a leading role in the Strasbourg congress) had asked him to send a list of all Scandinavian mathematicians to be invited to Toronto, which Koenigs apparently wanted to pass on to Fields. Størmer then added expressly in his letter to Mittag-Leffler:

> The invitations are personal and he asked me explicitly to ensure that no Germans end up at the list because, in fear of the consequences, Frenchmen do not wish to meet personally with the Germans.[8]

Although Størmer knew about Mittag-Leffler's opposition to the boycott, he seems to have been rather insensitive in this respect and passed the request on to the latter without commentary and thus implicitly approving.

2.1.3. Behind-the-scenes efforts to bring the ICM to Scandinavia. The ICM in Toronto in 1924 had left it open as to whether Stockholm or Bologna should be chosen as the venue for the next ICM (**Lehto, 1998a**, p. 44). The decision was left to the next IMU meeting in 1926. However, the fact that the Italian Pincherle had been elected IMU president in Toronto seemed to favour a decision for Bologna. Still, in 1926, efforts were once again made by Mittag-Leffler to assert Stockholm's claim to host an international congress. During the summer and autumn of 1926, he corresponded with the leading American mathematician George David Birkhoff (1884–1944), who was then on a trip to Europe; their correspondence was both scientific — concerning a paper by Birkhoff for *Acta mathematica* — and organisational, as Birkhoff was centrally positioned with the International Education Board (IEB) of the Rockefeller Philanthropies.[9] In early June that year, Birkhoff spent a few days in Stockholm.[10] On 30 June 1926, he informed Mittag-Leffler in a handwritten letter from Brussels about his participation in a meeting of the IRC the day before:

> You would have been much interested in the meeting of the International Research Council here yesterday. The meeting lasted just one hour, and in that time a vote had been unanimously

[6] For instance in a letter dated 19 August 1920 (IML).

[7] After which he had allegedly "received much praise by Emile Picard" ("fikk høre mange rosende lovord af Emile Picard").

[8] "Inbydelserne er personlige og han bad mig udtrykkelig sørge for at der ingen tyskere kom med på listerne da franskmandene for konsekvensernes skyld ikke ønsket at møtes med tyskerne." (IML: Størmer to Mittag-Leffler, 26 December 1923)

[9] For more on the role of the Rockefeller Philanthropies in the internationalisation of mathematics, see (**Siegmund-Schultze, 2001**).

[10] All the following information comes from Birkhoff's letters, kept in the Mittag-Leffler Institute in Stockholm.

> passed to strike out certain parts of statutes etc. which might be objectionable, and another vote had been unanimously passed to invite Germany, Austria, Bulgaria to join the International Research council. It seemed to me a constructive step of importance. The general attitude of everyone was very fine.

One month later, on 30 July 1926, Birkhoff responded to another request by Mittag-Leffler:

> Certainly it would be greatly to the advantage of mathematics and of international good will among mathematicians if an international mathematical congress, properly arranged, could be held soon in Sweden. I will certainly see that the merits of the situation are clearly brought before the Board, although I feel uncertain of the reaction.

Thus, machinations were taking place between Mittag-Leffler and his extended international network in order to promote and reinstate the claim for an ICM in Scandinavia. Such claims had been recurring ever since the planned 1916 Stockholm ICM had been abandoned. As Mittag-Leffler had stated at the meeting of Scandinavian mathematicians that replaced the planned ICM, American colleagues had enquired about the possibility of alleviating war-torn Europe of the burden of organising the next congress after the war: "The Scandinavian North and Stockholm first, then America" (**Turner and Sørensen, 2013**, p. 401). And in 1926, after the 1924 congress had been held in Toronto, the Americans were no longer competitors, but rather collaborators, in bringing the ICM "back" to Scandinavia if not to Stockholm. However, Mittag-Leffler himself seems to have felt unable to pursue the project because in another letter to him, Birkhoff wrote on 8 September 1926 that he had

> informed Professor Trowbridge [the man responsible at the IEB: see p. 10] of the change in the situation concerning the possibility of a mathematical congress at Stockholm.

Exactly what this change in situation consisted of is not clear from the letters sent by Birkhoff. There had been financial troubles lingering for Mittag-Leffler and his institute, at the latest from 1924, and the IEB was apparently not willing to step in with sufficient funds. This might also have overshadowed the prospects for an ICM in Stockholm (**Siegmund-Schultze, 2001**, pp. 178–180). At any rate, it is likely that the now elderly Mittag-Leffler no longer saw fit to invite the ICM to Sweden, but was suggesting other potential venues instead. In July 1927, Mittag-Leffler died after decades as the central father-figure of Scandinavian mathematics. But rather than split and dissolve as a community, Scandinavian mathematicians soon began looking for activities and initiatives to promote further and to exploit their standing in the international community in general, and with the ICM in particular.

2.2. 1928: Norwegians enter the international stage in new numbers

In Bologna, the Scandinavian contingent included about a dozen Danish and Norwegian mathematicians, many of whom would later play a central role in the organisation of the 1936 ICM in Oslo (for details of the Norwegians, see Table 2.1). Among the Norwegians, 62-year-old Alf Guldberg (1866–1936; see Figure 2.2) was the senior member of the delegation. Guldberg came from a family with strong

FIGURE 2.2. Alf Guldberg (1866–1936) (Source: (**Meidell, 1936**))

mathematical traditions in Norway and he himself was an expert on actuarial mathematics and a trusted advisor to the insurance industry.[11] Second in seniority within the Norwegian delegation was the Danish-born Poul Heegaard (1871–1948; see Figure 2.3), who had emigrated to Oslo to take up a professorship there in 1917 (**Munkholm and Munkholm, 1998**), (**Munkholm and Munkholm, 1999**). Between them, they represented the established generation of Norwegian mathematicians: for instance, at the time of the congress, Guldberg had served as the chairman of the Norwegian Mathematical Society (*Norsk Matematisk Forening*) since 1925, a position that Heegaard would then take over and hold until 1935 (**Thalberg, 1943**). We have no information as to why Størmer, probably the

[11]For biographies of Guldberg, see (**Lange-Nielsen, 1931**) and (**Meidell, 1936**). On his influence on the study of statistics in France, see (**Catellier and Mazliak, 2012**).

TABLE 2.1. The Norwegian delegation at the 1928 Bologna ICM (**ICM, 1928**, pp. 38–59). In addition to these mathematicians listed with affiliations in Norway, the Norwegian emigrant to the USA, Øystein Ore (1899–1968), was also present.

Brun, Viggo	(1885–1978)
Guldberg, Alf	(1866–1936)
Heegaard, Poul	(1871–1948)
Lange-Nielsen, Fredrik	(1891–1980)
Meidell, Birger	(1882–1958)
Nagell, Trygve	(1895–1988)
Stephansen, Elizabeth	(1872–1961)
Tambs Lyche, Ralph	(1890–1991)

most influential of all Norwegian mathematicians, did not attend the Bologna congress. Another important, though much younger, Norwegian mathematician who was missing was Thoralf Skolem (1887–1963).

In Bologna, a new idea was taking shape: during the congress, Harald Bohr, together with his Norwegian-American colleague Veblen, tried to persuade Heegaard to invite the next congress to Oslo.[12] Bohr was accompanied to the congress by his brother, the physicist Niels; together, the two saw themselves as "ambassador[s] from the international world" (**Reid, 1976**, pp. 122–123) in travelling to many German universities and establishing contacts with like-minded internationalists in other countries, and in attempting to renormalise international collaboration and abolish the IRC. Bringing the ICM to a neutral and international venue like Oslo would be part of such a plan for this group of mathematicians. However, at the time, Heegaard apparently was not sufficiently confident of the backing at home (see below). Instead, Swiss mathematicians invited the 1932 ICM to Zurich. The next congress would thus go to a neutral country anyway, something that had been a major motivation for the internationalist group around Bohr and Veblen in proposing Norway as a venue.

2.3. The 1932 Zurich congress and the Norwegian bid for 1936

When Heegaard returned to Norway from Bologna, he apparently found wide support for the idea of organising an ICM in Oslo. In 1930, he wrote to Veblen that "although he did not dare to issue the invitation at the time [1928] he found when he got home that he might well have done so".[13] This request spurred the letter quoted above, in which Veblen also asked Birkhoff whether any American interests would conflict with Heegaard's wish; and, indeed, Birkhoff confirmed that he was contemplating inviting either the 1936 or the 1940 ICM to the USA.[14] Thus, a new competition between Scandinavia and the USA seemed to be in the making, and it

[12] Veblen Papers, container 2, file GDP (George David Birkhoff) 1929–1947: Veblen to Birkhoff, 5 March 1930.

[13] *Ibid.*

[14] *Ibid.*: Birkhoff to Veblen, 17 March 1930.

FIGURE 2.3. Poul Heegaard (1871–1948) (Source: Oslo Museum, image no. OB.F05964b, via www.digitaltmuseum.no, https://commons.wikimedia.org/wiki/File:44154_Poul_Heegaard.jpg. CC BY-SA 4.0 international, https://commons.wikimedia.org/wiki/License/by-sa/4.0.deed.en)

would be important for the Norwegians to showcase their bid for the 1936 ICM as soon as possible.

In Zurich, the Norwegian contingent was the largest among the Scandinavian nations. In total, 8 mathematicians[15] and 5 accompanying persons constituted the Norwegian delegation (see Table 2.2), as compared with 6 + 2 Danish and 5 + 2 Swedish participants. In addition, the following Norwegian mathematicians were listed as delegates from foreign institutions: Trygve Nagell (Uppsala) and Øystein Ore (Yale). Thus, the Norwegians backed their bid for the next ICM by a clear presence in numbers under Guldberg's leadership.

[15]In fact, one delegate (Schieldrop) was more of a physicist than a mathematician (see below).

TABLE 2.2. Norwegian participation in the 1932 Zurich ICM (**ICM, 1932**, pp. 29–35). Official delegates are marked with an asterisk. Spelling is that of the proceedings.

Fjeldstad, J. E.	Meidell, Birger
*Guldberg, Alf	*Schieldrop, Edgar B.
Guldberg, Gustav	*Störmer, C.
Johansson, ?, Engineer	*Tambs Lyche, Ralph

However, when the mathematical world gathered in Zurich, the Americans had agreed to postpone their invitation until the 1940 ICM. Instead, a competing bid from Greece was developing. When it came to the final session, the two bids were announced and Guldberg elaborated on the Norwegian proposal. It was then accepted with great enthusiasm and ovation.[16] At the beginning of the congress, Guldberg had been elected as one of its vice presidents at the instigation of Fueter, the president of the congress. This was, of course, itself a declaration of status and importance, as the other vice presidents comprised such high-profile and centrally positioned mathematicians as Jacques Hadamard (1865–1963), Hilbert, N. M. Krylov (1879–1955), Pincherle, Teiji Takagi (1875–1960), de la Vallée-Poussin, Veblen, Wilhelm Wirtinger (1865–1945), W. H. Young (1863–1942), and Zaremba (**ICM, 1932**, vol. I, p. 42). Noticeably, no representatives from the Greek delegation were put forward for such honorary roles, something that suggests that the Norwegian delegation was more senior and better-connected than their Greek counterparts. Thus, it seems fair to say that the Norwegian bid was implicitly supported and furthered by the Swiss organisers and the American and British participants, as well as by the increasingly influential Scandinavian delegates. Whatever the precise machinations, it is also likely that by the choice of Oslo over Athens, the mathematicians voted again to have the next ICM in a country that had not been involved in the First World War.

In any case, for the 1932 ICM to accept Oslo as its next venue four years later was a remarkable decision and, at least in part, a tribute to the rising importance of Scandinavian (and Norwegian) mathematics. The overriding factor, however, was probably, as it had been for both Swiss congresses, the host nation's political neutrality.

2.4. The 1929 Abel celebrations in Oslo

Before Oslo 1936, the Norwegians had already been involved in staging one international mathematical event. At the end of the 1925 SCM in Copenhagen, Størmer, on behalf of the Norwegian delegation and the Norwegian community of mathematicians, invited the next SCM to Oslo in August 1929. Apparently in order to avoid three international congresses in three successive years 1927–1929, the Scandinavians deviated from the usual biennial pattern of the SCMs and focused on the year 1929, which would coincide with the centennial of the death of Norway's greatest mathematical hero, Niels Henrik Abel (1802–1829). In various ways, these

[16](**ICM, 1932**, vol. I, p. 60); on the renewed bids of the Americans and the Greeks in 1936, see also the letter from Heegaard to Engel in Appendix B.

celebrations played a role in preparing for the 1936 ICM and served as a first test in staging an international event.

Although initially planned to coincide with the official celebrations of the Abel centennial undertaken by the University of Oslo, the two events became separate when the university chose the precise date of Abel's death (6 April) as the time for its event, while the SCM opted for August 1929 in order to line up with international summer vacations. Thus, the two events unfolded independently, yet with a considerable overlap in interests and people. In particular, the 1929 university celebrations became an international event that celebrated Abel's legacy and drew connections back to 1902 when the centennial of Abel's birth had been celebrated in the midst of national political agendas, shortly before Norwegian independence from Sweden in 1905 (**Sørensen, 2006b**). However, the university celebrations in April 1929 were now smaller and more local in nature, as this was first and foremost a Norwegian event. Nevertheless, Norwegian mathematicians did not hesitate to utilise this celebration to extend and consolidate their professional networks. In 1902, honorary doctorates had been bestowed upon 29 prominent foreign mathematicians, and this tradition was developed further in 1929, when another 16 international mathematicians were honoured in this way.[17] Among them were again some of the world's leading mathematical lights and, in particular, some of the most powerful mathematical contacts of the Norwegians, including Fueter, Hardy, Landau, and Veblen (see Table 2.3). However, not all of these dignitaries were present in Oslo in April 1929. In fact, apart from the Scandinavians Christian Sophus Juel (1855–1935) and Lars Edvard Phragmén (1863–1937), only the Germans Edmund Landau (1877–1938) and Friedrich Engel (1861–1941) made the trip — all four were recorded as being present at the meeting of the Norwegian Academy of Science and Letters (*Det Norske Vitenskaps-Akademi*) that was also part of the celebrations.

Whereas the university and the Academy of Sciences organised the celebrations in April, the organisation of the 1929 SCM was put into the hands of a committee of 13 prominent mathematicians, including Størmer and Heegaard (**SCM, 1929**, pp. V–VI). From within this committee, a working group was established, comprising, among others, Richard Birkeland (1879–1928), Guldberg, Fredrik Lange Nielsen (1891–1980), and Thorvald Richardt, and given the responsibility of an executive committee. When Birkeland died in April 1928, Edgar B. Schieldrop (1891–1965) and Skolem were added to the working group. Guldberg was chosen as the president of the congress, and his wife was also involved in the women's social programme.

In his opening address on Monday 19 August 1929, Guldberg thanked the university, the Academy of Science, the Municipality of Oslo, and the National Theatre for their hospitality, and also the Norwegian life insurance companies for their financial generosity, which made the congress possible.[18] This constellation of partners would again prove to be a useful network for the ICM in 1936. Guldberg was himself involved with one of these companies (*Norske Liv*), and two other members of the working group, namely Lange-Nielsen and Richardt, also had strong ties to

[17]See, for example, (**Anon, 1929a**, pp. 583–584) and (**Anon, 1929b**, p. 741); note that the former records the date of the bestowals erroneously as 6 March.

[18]Guldberg's opening lecture is reproduced in (**SCM, 1929**, pp. X–XVI); the reference here is to p. X.

TABLE 2.3. Honorary doctorates conferred at the Oslo University celebrations in 1929

Brouwer, L. E. J.	(1881–1966)
Engel, Friedrich	(1861–1941)
Fueter, Rudolf	(1880–1950)
Hadamard, Jacques	(1865–1963)
Hardy, G. H.	(1877–1947)
Hensel, Kurt	(1861–1941)
Juel, Christian Sophus	(1855–1935)
Landau, Edmund	(1877–1938)
Lindelöf, E. L.	(1870–1946)
Painlevé, Paul	(1863–1933)
Phragmén, Lars Edvard	(1863–1937)
Pincherle, Salvatore	(1853–1936)
Takagi, Teiji	(1875–1960)
de la Vallée Poussin, Charles Jean	(1866–1962)
Veblen, Oswald	(1880–1960)
Weyl, Hermann	(1885–1955)

the industry, being, respectively, director of the statistical office of the Norwegian life insurance companies, and director of the supervisory agency The Insurance Council (*Forsikringsrådet*). At the banquet, speeches were delivered by many of the participants, as well as by the president of the Academy of Science and Letters, Halvdan Koht (1873–1965), a member of the Stortinget and future foreign minister who later became involved with the ICM in 1936 (see Section 2.9).

Thus, the 1929 SCM can be seen as a precursor — even as a testing ground — for the bigger international event, although the ICM had not yet been assigned to the Norwegians at that time. The central actors — the mathematicians Størmer and Guldberg (professors at the University of Oslo from 1903 and 1913, respectively), the university officials, the insurance companies, and the politicians — were all mobilised for this Scandinavian gathering, building a network, and a working group, that would be called upon to organise the 1936 ICM.

2.5. Setting up the organisation of the 1936 ICM

As we have seen, securing the 1936 ICM for Oslo had been the achievement of Guldberg in Zurich in 1932. Guldberg was, due to his age (he was born in 1866), the Nestor of Norwegian mathematicians.[19] He worked mainly in insurance mathematics and thus, as we have noted, had strong connections to the Norwegian insurance companies that sponsored much of Norwegian mathematical life, including the foundation of the Norwegian Mathematical Society in 1918.[20] Guldberg is today almost forgotten among mathematicians; he is not included in the MacTutor History of Mathematics archive[21] and is only briefly mentioned in (**Birkeland,**

[19]For detailed information written during his lifetime, see (**Lange-Nielsen, 1931**).

[20]See (**Thalberg, 1943**) and (**Birkeland, 1993b**).

[21]`http://mathshistory.st-andrews.ac.uk/` (accessed 9 December 2018).

1993a), which focuses mainly on pure mathematics. Nevertheless, due to his experience of organising international meetings, and his good connections with some of the major funding agencies, Guldberg was the natural pivotal figure and driving force of the organising committee for the ICM. Around him, a large number of other individuals became involved with the organisation, forming several subcommittees tasked with planning specific aspects of the congress. In particular, a large part of the practical organisation came to rest on Schieldrop, who was appointed general secretary of the congress. Schieldrop was an applied mathematician at the University of Oslo who, in the 1910s, had studied at the Technical University in Trondheim before going abroad in the 1920s to work with such figures as Élie Cartan (1869–1951) in France, and Einstein and Max Planck (1858–1947) in Germany. In 1936, he became the president of the Norwegian Polytechnical Society.[22]

The congress was soon scheduled for mid-July 1936 — a choice that both reflected the customary season for congresses and also allowed for some outdoor activities. In October 1935 the preliminary invitations were sent out to institutions and individual mathematicians all over the world: some 6000 copies were printed and distributed.[23] On 15 February 1936, Guldberg died at the age of 70 after a short illness. His death came as an abrupt disturbance to the Norwegian planning of the ICM and the organising committee had to adjust quickly after the loss of its leading figure. It brought in even more of the younger Norwegian mathematicians at various institutions in Oslo, in Trondheim, and in Bergen, and the overall responsibility for organising the ICM fell to Størmer. Together with the Dane H. G. Zeuthen (1839–1920), Størmer had been one of Mittag-Leffler's initial supporters in creating the SCMs as a Scandinavian forum for mathematics.[24] Even before this, Størmer had experience of organising international scientific gatherings. In 1902, at about the time that he was appointed professor of mathematics, he had been one of the major practical organisers of the Abel centennial celebrations in Oslo. With the exception of Bologna in 1928, Størmer had been a regular participant in the ICMs.

The other senior figure in Norwegian mathematics at the time, Heegaard, assumed a slightly less central role in the organisation of the congress, and would in fact sometimes look upon the organisation with criticism. For example, four days before the start of the congress, he wrote the following to Engel:

> Unfortunately, the good people [presumably the organisers][25] lack the *bon sens* required to direct such a large assembly. I fear that there will be a lot of fuss. I long to have this behind me. It wears me down to see stupidities being committed.[26]

[22]On Schieldrop, see `https://nbl.snl.no/Edgar_B_Schieldrop` (accessed 9 December 2018). He is not mentioned in (**Birkeland, 1993a**).

[23]A little later, on 2 March 1936, the conference organisers requested that the Ministry of Education send out invitations to *official* delegations. Initially, invitations were to be sent to the list of countries with diplomatic representatives in Oslo, as well as Canada, the Irish Free State, Yugoslavia, Latvia, and Palestine (KUD). Later, it was recognised that they had forgotten to invite Egypt and wished to invite the British Dominions and Southern Rhodesia.

[24]See the remarks above and in (**Turner and Sørensen, 2013**), as well as (**Sørensen, 2006a**, p. 4).

[25]In both German and Danish, the term "good people" has the ironic sense of "naive people".

[26]"Dessverre mangler de gode mennesker den bon sens som skal til for å lede en så stor forsamling. Jeg frykter det kommer til å knirke på mange steder. Jeg lenges efter å ha det hele overstått. Det sliter sådan på mig å se dummheter bli begått." (EN: NE150304, Heegaard to

However, Heegaard was a member of the executive committee, serving together with Størmer, Schieldrop, and three members of an eminent generation in Norwegian mathematics, namely Viggo Brun (1885–1978), Lange-Nielsen, and Skolem. And during the congress itself, Heegaard would also take on a prominent role, acting on behalf of Sem Sæland (1874–1940), the rector of the university.

Brun had studied mathematics in Oslo before he went to Göttingen in 1910, specialising in number theory.[27] Three years later, he returned to Norway as professor at the newly founded Technical University in Trondheim. Lange-Nielsen was an actuarial mathematician who would go on to occupy a number of central positions in the Norwegian insurance industry, becoming the CEO of the company *Statens Pensjonskasse* from 1938 to 1945 before moving on to a similar position with *Norske Liv*. Skolem had also studied mathematics in Oslo, where he concentrated on logic and received his doctorate in 1913 before going to Göttingen. He returned to Norway to assume a newly created position as docent at the University of Oslo before relocating to the independent research institution the Chr. Michelsen Institute in Bergen during 1930–1938. Feeling professionally isolated in Bergen, Skolem would return to the capital in 1938, but throughout his time away, he was very engaged with the Norwegian Mathematical Society and its activities. Thus, all three younger members of the executive committee had had comparable career paths up to this point: all three had studied both in Oslo and abroad, in particular in Göttingen, before returning to Norway to pursue academic careers in new institutions or to take up new positions that had been created, in part, to maintain what had been the high international standard of Norwegian mathematics during the 1910s. One mathematician who could not be saved for Norway was the slightly younger algebraist Ore,[28] who, after graduating from Oslo under Skolem in 1922, studied at Göttingen for a while and held temporary positions in Oslo (**Anon, 1970**). However, in 1927 he was invited to Yale University in the USA, where he quickly made a career for himself as an eminent mathematician. Although he was not formally involved with the organisation of the ICM, the executive committee consulted him on a number of issues, and Ore's strong position in the US mathematical community was a recognised asset of the Norwegian organisers.

2.6. Efforts to secure funding and advertise the event

The organisers of the Oslo ICM recognised that the biggest item of expenditure for such a congress was the publication of its proceedings. Traditionally, the proceedings had included full-length versions of all the papers presented. However, in 1932, the Swiss organisers had decided to limit the inclusion of session papers in the proceedings to brief abstracts and to reproduce only the plenary sessions in full papers. This had drastically reduced the cost of printing, so the Norwegians decided to do the same in 1936.

Engel, 10 July 1936. It seems the letter is somewhat apologetic because Engel was not invited to deliver a major talk.)

[27]On Brun and Skolem, see (**Birkeland, 1993a**), and also the online *Store Norske Leksikon* `https://snl.no` (accessed 9 December 2018). On Lange-Nielsen, see the latter, as well as the online *Norsk Biografisk Leksikon* `https://nbl.snl.no` (accessed 9 December 2018).

[28]Ore is not included in (**Birkeland, 1993a**), apparently because the authors counted him as American. Another promising emigrant was the number theorist Trygve Nagell, who is almost forgotten today; for a biography of Nagell, see `https://nbl.snl.no/Trygve_Nagell` (accessed 9 December 2018).

Nevertheless, the cost of printing the proceedings would be substantial. Based on a breakdown of the financial report of the 1932 ICM in Zurich, the Norwegian organisers estimated the cost of the 1936 congress in Oslo at approximately 70,000 Norwegian Crowns (*kroner*).[29] This was a very considerable figure, but, for an ICM, it was planned as a low-cost congress, as is evident from the budget (see Table 2.4).[30]

Even before presenting the invitation in 1932, Guldberg had obtained a guarantee from the association of insurance companies that they would provide 25,000 Crowns for the congress.[31] As we have noted (p. 32), this organisation had a history of supporting mathematics in Norway; the insurance companies had also subsidised the journal *Norsk Matematisk Tidsskrift*, issued by the Norwegian Mathematical Society since 1919 (**Thalberg, 1943**). And, of course, Guldberg was well connected in the industry through his own involvement with actuarial mathematics (see p. 27). As a more general remark, we note here that, at that time, insurance mathematics was in many countries a material basis for research: in Austria, Germany, and Poland, many mathematicians were forced to take jobs as actuaries, for lack of teaching jobs at universities.[32]

The number of participants was of course difficult to estimate with any precision in advance, but in applying for support, the organisers projected about as many participants in Oslo as four years previously in Zurich, i.e., some 650 full members of the congress and 200 accompanying persons, as listed in the official proceedings. As it turned out, the number of participants in Oslo would be considerably lower. Nevertheless, at a price of 40 Crowns for full membership and 20 Crowns for an accompanying person, even a slightly conservative guess of 500–600 participants would bring in something like the 20,000 Crowns that figured in Guldberg's budget (for more on this, see below).

But even with fees from the participants and the guarantee from the insurance companies, a considerable margin was still left to be covered, and the committee sought to secure funding from the state and from private organisations. In order to obtain support from Norwegian banks and the state, the organisers prepared a budget which is summarised in Table 2.4, together with the numbers from the 1932 ICM, which were also provided to prospective sponsors for reference.

In 1933, Guldberg approached major Oslo banks with a request for support for the congress. As part of his argument, he reiterated the fact that Norwegians had repeatedly been guests at congresses abroad without inviting the international mathematical community to Norway, and that Scandinavia had recently hosted scientific congresses in fields closely related to mathematics.[33] His request was for 15,000 Crowns, and in order to interest the banks, he promised that a section

[29]The website `http://www.historicalstatistics.org/Currencyconverter.html` (accessed 28 February 2019) provides historical currency conversion rates for a range of currencies up to 2015. According to this source, 1 Norwegian Crown at the 1936 rate is equivalent to approximately 3,17 € or US$3.52 at 2015 rates. In connection with Table 2.4, we note also that 1 Swiss Franc at 1932 rates is equivalent to approximately 3,50 € or US$3.88 at 2015 rates.

[30]The information in Table 2.4 is drawn from a typewritten letter from the organising committee (Guldberg, Størmer, Schieldrop, Heegaard, Lange-Nielsen) to the Norwegian government, dated Oslo, 22 December 1934 (KUD).

[31]As mentioned in the typewritten letter cited in note 30 above.

[32]For the Austrian case, see (**Sigmund, 2004**).

[33]For instance, an economics conference held in Stockholm in June 1931 (**Lotsch, 1932**, p. 619).

TABLE 2.4. Budget of the 1936 ICM

	1936 ICM budget (predicted) (1936 Norwegian Crowns/ 2015 US$)	1932 ICM budget (realised) (1932 Swiss Francs/ 2015 US$)
Total projected costs	70,000/19,886	80,000/20,619
National government support	10,000/2,841	10,000/2,577
Regional government support		10,000/2,577
Municipal support	10,000/2,841	10,000/2,577
Support from banks	15,000/4,261	20,000/5,155
Insurance companies and organisations	25,000/7,102	18,000/4,639
Participants	20,000/5,682	
Other	5,000/1,420	
Total income	85,000/24,148	68,000/17,526

at the congress would be devoted to the application of mathematics to statistical and economical problems.[34] Eventually, a smaller amount of 10,000 Crowns was granted, and director Nicolai T. N. Rygg of *Norges Bank* and president Erling Sandberg of the Norwegian Association of Banks (*Den norske Bankforening*) were added as honorary members of the congress.[35]

In December 1934, the organisers formally approached the government with a request for support.[36] The petition was signed by the original organising committee, consisting of Guldberg, Størmer, Schieldrop, Heegaard, and Lange-Nielsen, and thus included all the nation's most central mathematical figures. In their application, they repeated the above argument for an international congress in Oslo, and noted that their suggestion in 1932 to bring the 1936 event there had been very positively received. They noted that they had only been able to extend that invitation because they had had an advance guarantee of 25,000 Crowns from the association of life insurance companies.

For the application to the government, the organisers drew heavily on the rhetoric of international collaboration, arguing that the scientific outcomes that followed directly from such congresses were considerable, and noting their congeniality, "a fact which weighs heavily, not least in our days".[37] And, they argued, the smaller and neutral nations had a particular role to play here. Yet the outcome would not be exclusively for mathematics or for world peace: there would also be a stimulating effect on Norwegian tourism, as an accompanying letter from the tourist board explained (see Section 2.7).

The funding application was sent ultimately to the Department of Finance (*Finansdepartementet*), together with input from the Ministry of Church and Education (KUD), which stressed the potential outcome in terms of international collaboration, as well as the national pride of welcoming foreign scientists to Norway. The requested 10,000 Crowns were subsequently granted, the decision being ratified by the King on 8 February 1935.[38] Thus, the funding of the congress was secured with 25,000 Crowns from the insurance companies, 10,000 from the banks, and 10,000 from the state (**ICM, 1936**, vol. I, p. 9). With 487 full members and

[34]KUD; Norges Bank.

[35](**ICM, 1936**, vol. I, p. 11). Rygg had an academic background as professor of economics and statistics at the University of Oslo; Sandberg on the other hand was educated in banking and at the time was also the director of the *Christiania Bank og Kreditkasse*, the oldest bank in Oslo.

[36]Letter dated 22 December 1934 (KUD).

[37]"... dertil kommer at kongressene bidrar til å fremme samfølelsen mellem videnskapens dyrkere verden over, et moment som ikke mindst i våre dager må tillegges betydelig vekt." (*Ibid.*)

[38]KUD: St. prp. nr. 1, Tillegg nr. 1, kap. 226B.

FIGURE 2.4. Contemporary postcard featuring a view of Oslo with the sailors' school in the foreground (Source: Universitätsarchiv Gießen, Nachlass Friedrich Engel)

182 accompanying persons, another 23,120 Crowns would be generated in registration fees, bringing the combined incomes to 68,120 Crowns — very close to the projected 70,000. The precise number of actual participants in Oslo is difficult to estimate for reasons to be discussed later (Section 2.10); a good guess would be around 425. Even though it was still a considerable number and enough to balance the budget, it was still rather less than the 667 participants in Zurich or the 836 in Bologna.

2.7. Promoting tourism in Norway

When the organisers approached the government for financial support, they advanced two different arguments in favour of state subsidy of the congress. The first concerned Norwegian national pride and the furthering of international scientific contacts, while the second emphasised the possibility of attracting foreign interest in Norway beyond science and mathematics, including showcasing Norwegian culture and nature to potential tourists. The first line of this argument was explicitly picked up by the government officials in their response to the application:

> In order to support international collaboration in this field, it is viewed as very important that the State offers a financial contribution now that the congress is for the first time held in Norway; similarly it must be seen as a matter of national pride for the country that the many scientists from different countries find their stays here most satisfactory.[39]

In support of the latter argument, the organisers had previously approached the tourist authorities in Norway. The director of the National Association for

[39] "Av hensyn til det internasjonale samarbeide på dette område anser man det av stor betydning at statsbidrag blir gitt når kongressen nu for første gang holdes i Norge, likesom det må anses som en æressak for landet at opholdet her for de mange videnskapsmenn fra de forskjellige land blir mest mulig tilfredsstillende." (KUD: St. prp. nr. 1, Tillegg nr. 1, kap. 226B)

Tourism (*Landslaget for Reiselivet i Norge*), G. B. Lampe, responded positively, thereby strengthening the application for support from the national government and the Oslo municipality:

> We believe that the prospect alone of gathering 5–600 women and men from different countries in Oslo for such a scientific congress will not be completely without significance for the reputation of our country abroad. Those messages and reports that will be spread around the world will direct attention to our country and thereby be an advertisement for it. To that is added that the personal acquaintance with our country, which the participants of this congress will acquire, must be considered considerable weight as advertisement for future tourists.[40]

In 1936, Norwegian tourism was still a relatively minor industry, although it was advertising internationally, promising relaxing, healthy, and beautiful holidays (see Figure 2.4). For instance, an extensive advertisement in the Danish newspaper *Politiken* during the summer of 1936 advertised Norway as "[a]n approachable and affordable country from where you return in good health" and named Norway the "promised land of tourists" which had recently invested substantially in infrastructure and tourism.[41] Norwegian tourism, the advertisement argued, provided a sanctuary for busy modern people who would benefit in "health and nerves" from two weeks in the Norwegian mountains. And Oslo was also a place for sightseeing, where Gustav Vigeland's monument to Abel in the palace park was highlighted as a tourist attraction (Figure 2.5).[42]

And, indeed, an alliance was formed with the travel and tourist industry. Getting to Oslo meant sailing either from Britain or from the Continent, and special reductions in the fares for getting to Oslo were negotiated with the two major companies: Fred Olsen & Co., sailing from Newcastle, Rotterdam, and Antwerp, and *Søndenfjelds-Norske Dampskibsselskab*, sailing from Kiel and Hamburg (**ICM, 1936**, vol. I, p. 9). Many of the foreign guests at the ICM also made time for an additional stay in Norway, recovering and enjoying the magnificent Norwegian nature. They were helped in this by a favourable deal struck by the congress organisers with one of the major British tourist agencies working on the Norwegian market, Bennett's Travel Bureau, offering four different sightseeing packages to other parts of Norway (KUD). And in some reports by mathematicians from the congress, sightseeing figured prominently — for example, the US *National Mathematical Magazine* carried a short notice on the Abel monument by Dunnington, in the issue following the one in which his congress report appeared (**Dunnington, 1936b**). A few other visitors used the occasion to stay with friends and colleagues in and around Oslo, working on joint projects. Thus, for example, Max Dehn (1878–1952), who had a

[40] "Vi er av den opfatning at alene den ting at 5–600 kvinner og menn fra de forskjellige land, samles i Oslo til en slik videnskapelig kongress, ikke er helt uten betydning for vårt lands anseelse i utlandet. De meddelelser og referater som blir spredt rundt omkring i landene, vil henlede opmerksomheten på og derigjennem være til reklame for vårt land. Dertil kommer at det personlige kjendskap til vårt land som deltagerne i denne kongress vil få, må tillegges adskillig vekt som fremtidig turistreklame." (KUD: Lampe to Schieldrop, 17 December 1934)

[41] *Politiken*, 26 July 1936.

[42] This was at a time when the Vigeland Park in Oslo, which now has many more sculptures, was not yet completed: it was finished in 1939 in a location separate from the palace park.

FIGURE 2.5. The Abel monument in Oslo (Source: http://www.abelprize.no/)

longstanding collaboration with Heegaard in topology, remained until mid-August to work with him and intended to return in time for Christmas.[43]

2.8. Setting up the programme of plenary speakers

Organising the scientific programme of the congress was a complicated but obviously very important task that was dealt with in the first instance by a smaller committee consisting initially of Størmer and Schieldrop; Lange-Nielsen was later

[43]Størmer Archive: Heegaard to Størmer, 8 August 1936. The letter does not reveal the nature of the collaboration, but given that Heegaard had given a historical talk at the ICM and that Dehn was very much interested in history too, their shared topic could have been in this field. David E. Rowe (Mainz) alerts us to the fact that there were no signs of a revitalisation of the old contacts in topology between the two mathematicians which dated back to the turn of the century. Eventually, in 1940, Dehn would flee from Nazi persecution via Norway before arriving in the USA after an eventful journey: see (**Dawson, 2002**).

added to this group (**ICM, 1936**, vol. I, p. 12). They were responsible for organising the so-called "general lectures" which were held in plenum; each of the eight sections of contributed papers would also be organised in turn by a special group of one or two mathematicians. However, as the plenary lecturers[44] were invited from among the most prominent contemporary mathematicians, planning was of great importance. And it involved balancing not only disciplinary scales, but also national interests, and also providing a mixture of well-known scholars and some of the newest results in mathematics. The resulting process was a complicated one with at least one early meeting to brainstorm names and draft lists of mathematicians to approach, a second gathering in which a preliminary programme was prepared for circulation with the call for registration, and a final stage in which late changes occurred before the programme was implemented at the congress.[45] Some of the names on the first drafts were annotated with affiliations and even possible topics for the talks; these pieces of information were then included in the preliminary programme. Even though the planning was centralised in the two- or three-man committee, the latter would sometimes consult with colleagues, in particular with Brun and Skolem, as well as with Ore in the USA.[46]

2.8.1. From preliminary ideas to the final programme. During the preliminary brainstorming of Størmer and Schieldrop, at least two draft lists of potential speakers were prepared, discussed in the organising committee, and subsequently revised and elaborated upon. In the first draft, different typographies clearly distinguish two stages of notes, presumably the initial ones by Størmer and Schieldrop, with additions then made during a meeting of the entire organising committee. Between the first and second drafts, candidates were approached to enquire whether they would be willing to deliver a plenary lecture at the ICM, and the second draft was updated accordingly, with new suggestions sometimes added.

The first drafts were organised according to seven major themes and suggested a prioritised list of speakers and reserves. In the second draft, the categories were:

(I) Algebra and number theory;
(II) Analysis;
(III) Geometry and topology;
(IV) Probability theory, mathematical statistics and actuarial mathematics;
(V) Astronomy;
(VI) Mechanics and mathematical physics;

---

[44]We note that it is not certain that the "main" lectures in Oslo were always delivered to the full congress: in some places in the proceedings, several lectures are indicated for the same time slot. Thus, one can at least conclude that there were "plenary lectures" of different ranks, probably longer and shorter ones. In the interests of simplicity, we retain the term "plenary lecture" throughout to refer to the main lectures of the congress. We note however that the proceedings of the following congress (**ICM, 1950**, vol. 1, p. 88) are more specific in this respect, indicating different rooms for parallel one-hour "addresses invited by the organizing committee". Thus none of these were "plenary" in a strict sense. Several of the 23 such "addresses" (including one by Tarski of 30 minutes) were even within so-called "conferences", which ran parallel to other specialised sections of the congress. In their later history, the ICMs have returned to plenary sessions in a strict sense.

[45]Among Størmer's papers, there can be found two drafts of the list of plenary speakers at the 1936 ICM and a draft programme, as well as the preliminary programme (separately, in two languages: German and English) circulated with the call for registration (Størmer Archive). The final programme was, of course, printed in the proceedings: (**ICM, 1936**, vol. I, pp. 14–16).

[46]Brevsamling 698B-8: Guldberg & Størmer to Ore, April 1935.

TABLE 2.5. Summary of the deliberations and evolution of the programme for general lectures at the 1936 ICM. *Marked as reserve. **Crossed out. ***High-lighted. The dividing lines are reproduced from the manuscripts, although we are unsure of the meaning of the horizontal dividers. Russian names are given according to the transliteration used in the original document. Note that general lectures were not labelled by section in the final programme; our final column here is therefore inferred from the preceding ones.

	First draft (Spring 1935)	**Second draft: prioritised list (Summer 1935)**	**Preliminary programme**	**Final programme**
Algebra and number theory	Hasse Ore Hardy Siegel Gelfond Landau van der Corput Takagi Artin* Krull* van der Waerden	Hasse Ore Siegel Gelfond van der Corput Takagi Artin* Krull* van der Waerden*	Hasse Ore Siegel Gelfond van der Corput Hecke	Hasse Siegel Ore van der Corput Hecke
Analysis	Ahlfors Ritt M. Riesz** Fréchet Banach Landau Lebesgue Hardy Wiener Tamarkin* Pólya* Fejér* Stone* Karamata Littlewood	Ahlfors Fréchet Banach Lebesgue Landau Wiener Tamarkin* Pólya Fejér Stone Ritt Karamata Littlewood (for Hardy)	Ahlfors Fréchet Banach Landau Wiener	Ahlfors Banach Fréchet Wiener
Geometry and topology	Cartan Nielsen Blaschke*,** Castelnuovo*** Veblen van der Waerden* Severi* Alexandroff* Kuratowski*	Cartan Nielsen Veblen van der Waerden* Severi* Alexandroff* Kuratowski* Blaschke* Lefschetz*	Cartan Nielsen Veblen Severi	Cartan Nielsen Veblen
Probability theory, mathematical statistics, and actuarial mathematics	Khintschine Kolmogoroff*,*** Cramér*	Khintschine Kolmogoroff* Cramér*	Khintschine	
Astronomy	Einstein	Eddington		
Mechanics and mathematical physics	Dirac Weyl Størmer Birkhoff Oseen von Neumann* Levi-Civita* Castelnuovo	Dirac (?) Weyl (?) Størmer Birkhoff Oseen von Neumann	Weyl Størmer Birkhoff Oseen	Størmer Birkhoff Oseen
Philosophy, history and didactics	Skolem Neugebauer Heyting* Gödel*	Skolem Neugebauer	Skolem Neugebauer	Neugebauer

(VII) Philosophy, history and didactics.

This division thus reflects the main disciplinary units within mathematics, at least as perceived by Størmer and Schieldrop; these categories did not subsequently appear in the printed programme for the general lectures, and the sectional lectures were organised according to slightly different divisions (see p. 46). Nevertheless, we organise our discussion of the early preparation of the programme around these broad headings.

2.8.1.1. *Algebra and number theory.* In the first draft, Hasse, Ore, Siegel, and Gel'fond had been singled out as prospective speakers on algebra and number theory (see Table 2.5). Hardy had also been on the list (at number 3), but by connecting Hardy with Littlewood it seemed possible to transfer them to the analysis list where they were put at the top (see Section 2.8.1.2). Among the lower priorities and reserves in algebra and number theory were Landau (with a note that he may also fit elsewhere), van der Corput, Takagi, Artin, Krull, and van der Waerden. Artin and Krull were listed as the reserves. In the second draft, Hasse, Ore, Siegel, Gel'fond, van der Corput, and Takagi were on the list of intended speakers with Artin, Krull, and van der Waerden as reserves.

When the draft programme was prepared, Hecke's name had entered the list without his being mentioned in any of the original drafts. In the end, the plenary lectures were divided into the seven sections, and under algebra and number theory were given by Hasse, Siegel, van der Corput, Ore, and Hecke (see Chapter 9). Thus, compared to the prioritised list, Gel'fond and Takagi had dropped out of the list without being replaced by any of the reserves. Gel'fond had until the very last minute been scheduled to speak on "Théorie des nombres transcendants",[47] but, being part of the Soviet delegation, he was prevented from going to Oslo (see Chapter 4). The reason behind Takagi's being dropped from the programme remains unclear; the Japanese mathematician had achieved international recognition in the late 1920s and early 1930s and had received one of the honorary doctorates bestowed in Oslo in 1929 (see Table 2.3). In addition, he had served as a vice president at the 1932 ICM. However, by 1936, international travel was perhaps no longer feasible for the elderly mathematician who had retired that year and had already made three trips to Europe (**Honda, 1975/76**). Instead, the Japanese delegation consisted of slightly younger mathematicians and mathematics educators.[48]

None of the named reserves (Artin, Krull, van der Waerden) would ultimately attend the congress in Oslo. However, their contributions to new developments in algebra were mentioned in Ore's plenary lecture (on which, see Section 9.20).

2.8.1.2. *Analysis.* In the section on analysis, the brainstorming sessions produced the largest field of candidates to choose from; this is probably both a reflection of the maturity and variety of that discipline and of the interests of the

[47](**ICM, 1936**, vol. I, p. 10); see also Section 9.21.1.

[48]Takagi was born in 1875. The Japanese participants at the Oslo ICM with affiliations in Japan were Matsusaburo Fujiwara (1881–1946), Motoji Kuniyeda (or Kunieda, 1879–1954), Ichiro Shimomura (1893–?), and Masao Tanaka (1895–1976). We note that the list of congress participants features one more Japanese name: Michio Abe, listed as coming from Hamburg (**ICM, 1936**, vol. I, p. 29). Moreover, a report in the Norwegian newspaper *Arbeiderbladet* at the time of the congress (translated here in Appendix A.3) indicated that there were (probably two) Japanese students at the congress who had come from Germany: Michio Abe would appear to be one of them. The other, probably the person interviewed by *Arbeiderbladet* whose name was recorded in the newspaper as "Schur" (see p. 251), is not listed in the congress proceedings.

organisers. The initial draft listed Ahlfors, Joseph Ritt, Marcel Riesz, Fréchet, and Banach, with the possibility of Landau, Lebesgue, Hardy, and Wiener added (see Table 2.5). In the draft, Riesz' name was then crossed out. Tamarkin, Pólya, Fejér, and Stone were listed as reserves, with the further possibility of Wiener, Karamata, and Littlewood added. In the next draft, Ahlfors, Fréchet, Banach, Lebesgue, Landau, and Wiener were check-marked but Hardy had apparently declined the invitation. The list of reserves was now Tamarkin, Pólya, Fejér, Stone, Ritt, Karamata, and "Littlewood (for Hardy)". Between the second draft and the preliminary programme, Lebesgue's name was dropped again. Ultimately, Landau did not speak in Oslo, allegedly because his wife had fallen ill,[49] and it was left to Ahlfors, Banach, Fréchet, and Wiener to deliver the plenary lectures in this section (see Chapter 9).

2.8.1.3. *Geometry and topology.* For the section on geometry and topology, the initial draft included Cartan, Jakob Nielsen, Blaschke, Castelnuovo, and Veblen. Blaschke's name was crossed out and he was moved into the reserves, and Castelnuovo's name was framed. On the list of reserves were van der Waerden, Severi, Aleksandrov, and Kuratowski (who was noted as being a possible stand-in for Banach). In the second draft, Cartan, Nielsen, and Veblen had accepted, but Castelnuovo had declined. Veblen had not yet provided a title though; it is perhaps for this reason that Lefschetz was added to the list of reserves with the annotation "for Veblen". Severi's name was highlighted on the list of reserves, perhaps as a replacement for Castelnuovo. In the section on geometry and topology, it seems that the organisers were content with the list produced in the second draft. It had included some Italians who would eventually not participate (see Chapter 5), but these had already been out of consideration for the preliminary programme.

2.8.1.4. *Probability theory, mathematical statistics and actuarial mathematics.* As we saw in Section 2.6, the inclusion of actuarial mathematics in the programme had worked as an argument to interest banks and insurance companies in supporting the congress financially. And actuarial mathematics had been the primary interest of the original main organiser, Guldberg. However, when the brainstorming for the plenary lectures in the probability section was complete, only two names and one reserve were put forward: Khinchin and Kolmogorov as candidates, and Cramér as a reserve. Yet in that first draft, Kolmogorov's name was framed and moved to the list of reserves below Cramér. In the second draft, Khinchin had accepted, and Kolmogorov had been moved to the top of the list of reserves. However, in the end, neither would speak in Oslo (see Chapter 4). Although Cramér was present, he did not give a plenary lecture, and presided over one of the sections instead.

2.8.1.5. *Astronomy.* Astronomy had been treated in some form among the plenary lectures in Zurich, when Rolin Wavre (1896–1949) from Geneva had lectured on the shapes of planets. And although astrophysics was a topic of great interest in Oslo, the programme committee chose not to include any local candidates. Instead, the first draft just listed Einstein, who declined the invitation when they approached him.[50] Instead, they briefly considered Eddington, but in the end, the

[49]Størmer Archive, Box 10: Landau to Størmer, 26 March 1936. The reason given for Landau's not attending the congress (his wife's illness) does not appear to be truthful, since it was given as an excuse so early, without any attempt to wait for her recovery: see also the comments in Section 9.21.3.

[50]Einstein had been invited successfully to Norway during the 1920s; see (**Johansen, 2006**).

idea of having a plenary lecture on astronomy came to nothing. We note, however, that one of the main speakers at the congress did deliver a talk on a topic that can at least partly be considered mathematical astronomy: Størmer on his magnetic dipole theory, and its application to the study of cosmic rays (see Section 9.2).

In approaching Einstein and having Eddington as their second option, the organisers were considering a branch of astronomy and cosmology that had been very much publicised during the previous 20 years. However, it seems that the organisers had no clear idea of what kind of astronomy to invite and promote, and instead they invited the two biggest celebrities in the field. But, although the two did not come to Oslo, the expansion of the universe continued to capture some public attention in Oslo when Norwegian newspapers briefly interviewed the Belgian Georges Lemaître (1894–1966), who was also centrally involved in this line of research and did attend the congress (see Appendix A.3), though didn't submit a paper for the proceedings.

2.8.1.6. *Mechanics and mathematical physics.* Whereas the proposed section on astronomy would have been peripheral to the ICM, with no plenary talk attached to it, the planned section on mechanics and mathematical physics was much more integral to the mathematics of the 1930s, and the programme committee had a wealth of names upon which to draw. The initial list consisted of Dirac, Weyl, Størmer, Birkhoff, and Oseen, with von Neumann and Levi-Civita listed as reserves, together with the possibility of Castelnuovo. Levi-Civita's name was transferred to the shortlist, but when he was invited, he declined. Thus, the second draft lists Dirac (with a question mark), Weyl (with the annotation "possibly"), Størmer, Birkhoff, Oseen, and Levi-Civita, with the information that he had declined. The only reserve was von Neumann. In the preliminary programme, Weyl's name was listed but without a title; Dirac was not on the preliminary programme. In the end, Weyl would contribute a talk in the section on algebra and number theory (**ICM, 1936**, vol. II, p. 3).

2.8.1.7. *Philosophy, history and didactics.* For the eighth and final section on philosophy, history and didactics, the first draft listed Skolem and Neugebauer, with Heyting and Gödel as reserves. This was repeated in the second draft, when both the short-listed mathematicians had accepted the invitation. However, Skolem ultimately did not deliver a plenary talk, leaving only Neugebauer (see Section 9.9).

Thus, neither Heyting nor Gödel made it into the programme, although they had been considered in the process — neither apparently attended the congress, so they may simply have declined (see Section 8.4). According to the draft program, Skolem's plenary talk would have been "Die Mengenlehre und die Grundlagenforschung". On Skolem's reasons for withdrawing his commitment to give a plenary talk at such a relatively late stage, we find a clue in a letter written to his friend and co-organiser of the congress, Viggo Brun:

> Thank you for your letter! First of all, I have not been very well for a while, and secondly I am very occupied by studies of the foundational literature because I must give a larger lecture at the congress on such matters. It is a very uninviting affair; it is quite different and fun to give a specialised lecture on some topic or another. But as a consequence of these two circumstances, it has been quite difficult for me to devote any real thinking to

> the complicated combinatorial programme that Rowanoff wants solved.[51]

As it turned out, Skolem gave a specialised talk, "Eine Bemerkung zum Entscheidungsproblem", in the session on logic, philosophy and history (**ICM, 1936**, vol. II, p. 268–270).

2.8.2. The proceedings. As we have already noted (p. 34), one of the aims of the Norwegian organisers in preparing the budget for the Oslo ICM was to keep down the cost of production of the proceedings. Thus, the proceedings printed just the full-length versions of the plenary lectures and only one-page abstracts of the contributed papers. Furthermore, publication in the proceedings was restricted to the papers actually presented in Oslo:

> In these proceedings, volume I and II, one finds only the lectures presented during the conference, the committees having refused on this principle, as well as for lack of space, to print other papers that were sent to us.[52]

Indeed, one such attempt to publish in the proceedings without intending to participate in the congress was made by the Russian Dmitriy Grave (1863–1939). A central figure in Soviet mathematics and the director of the Institute of Mathematics of the Ukrainian Academy of Sciences in Kiev, Grave sought, as early as January 1936, to have a paper entitled "Algorithme du calcul des racines des équations algébriques" included in the proceedings, although he was not going to participate in the ICM. At around the same time, Soviet mathematicians invited Størmer to contribute to the celebration of Grave's 50th anniversary as a teacher of mathematics, perhaps as a means of applying some pressure and smoothing the path of Grave's paper into the publication. Nevertheless, the publication committee managed to send a very elegant refusal.[53]

With regard to language, we note that the congress proceedings were published in French, rather than English, most likely because the French were stronger in mathematics. The avoidance of German was not necessarily a political expression but apparently a result of the more practical issue that French was understandable to all.

2.8.3. Summary and analysis of plenary and sectional lectures. In this subsection, we provide a preliminary discussion of the shape of the programme during the planning stages, up to the final version; a more detailed analysis of the plenary lectures of the latter may be found in Chapter 9.

We have seen in some detail how ideas with regard to the plenary lectures were generated in the brainstorming of Størmer and Schieldrop, and then discussed in a larger group before invitations were sent out. But even after the responses to those invitations were integrated into the preliminary programme, a number

---

[51]"Tak for brevet! For det første har jeg været svært lite frisk en tid, og for det andet er jeg nokså optatt med studier i denne grundlag literaturen, fordi jeg jo skal holde et større foredrag på kongressen om slikt noget. Det er en yderst utiltalende affære; det er ganske anderledes greit å holde et specialforedrag over en eller anden bestemt gjenstand. Men som følge av disse to ting har det været mig nokså vanskelig å tænke noe større om det kringlede kombinatoriske problem, som Rowanoff ønsker løst." (Brevsamling 719: Skolem to Brun, 9 January 1936)

[52]Translated from the French in (**ICM, 1936**, vol. I, p. 7). Note, however, that this does not mean that all authors were present: see footnote 34 on p. 16.

[53]Størmer Archive, box 10: Størmer to Gravé, 6 March 1936.

of changes occurred: some invited mathematicians withdrew their commitments to speak, whereas others were prevented from going to Oslo. Thus, Skolem and Weyl both withdrew their plenary lectures in favour of giving sectional talks, while Landau, Severi, Gel'fond, and Khinchin would not make it to Oslo to give plenary lectures, although they had been advertised in the preliminary programme. In the published proceedings, the absence of the Russians is noted laconically:

> The proposed programme became reality except that A. Gel'fond and A. Khinchin were prevented from coming and that as a consequence the announced lectures "Théorie des nombres transcendants" and "Hauptzüge der modernen Wahrscheinlichkeitstheorie" did not take place.[54]

This absence is dealt with more fully in Chapter 4; discussions on the possible content of the proposed lectures by Gel'fond and Khinchin can be found in Sections 9.21.1 and 9.21.4, respectively.

Comparing the final programme in its entirety to the deliberations we have detailed above, one realises that the lectures delivered in plenary sessions by Fueter, Bjerknes, and Mordell had not been considered by the organising committee in the early planning phase of 1935. Fueter and Bjerknes were, however, added in the preliminary programme, but Mordell was not. Mordell at least, therefore, seems to have been a late addition to the plenary programme, perhaps as a reaction to the changes necessarily made to the programme when Landau, the Italians, and the Russians did not show up. Indeed, we can clearly see the effects of the non-appearance of the Italians and Russians if we compare the preliminary programme sent out to registrees (Table 9.1 on p. 219) and the final programme, as presented in the proceedings (Table 8.3 on p. 152). In moving from the former to the latter, Gel'fond's lecture, scheduled for the Wednesday of the congress, disappears, while Neugebauer's, originally planned for the Saturday, takes the slot that should have been occupied by Severi on the Thursday; likewise, Banach's Friday lecture has been moved to what should have been Khinchin's slot on the Saturday, apparently becoming longer in the process. Mordell spoke at the time originally to have been divided between Banach and Landau.

Some of those mathematicians considered by the organising committee and put on the lists of reserves would eventually take part in the parallel sessions: for example, M. Riesz (multiple appearances), Stone, and Cramér. A few of those invited to give plenary lectures also contributed in the parallel sessions: for example, Fréchet and Mordell.

In the descriptions above, Størmer's original list of seven sections for the plenary lectures, as formulated in the earliest draft, was retained as an organising principle for the sections. However, in the preliminary programme, the titles of the sections were modified slightly so that there were now eight sections:

(I) Algebra and number theory;
(II) Analysis;
(III) Geometry and topology;
(IV) Probability, mathematical statistics, actuarial theory and econometrics;
(V) Mathematical physics, astronomy and geophysics;
(VI) Rational and applied mechanics;

[54]Translated from the French in (**ICM, 1936**, vol. I, p. 10).

(VII) Logic, philosophy and history;
(VIII) Pedagogy.

Thus, the last four of the original seven sections were slightly reorganised and renamed; econometrics was added to section IV, mathematical physics was split from mechanics and merged with astronomy into section V, to which geophysics was also added, section VI was recast as mechanics and engineering (rational and applied mechanics), logic was added to section VII, and pedagogy was split from section VII into its own section VIII. This structure was then kept in the sections and in the proceedings (volume II). The parallel sessions in sections were of quite uneven sizes, with the first three dominating much of the programme.

Compared to Størmer's original list of seven sections, some of the changes were mandated by the problems in setting up the plenary lectures; for instance, the organisers were unsuccessful in attracting a plenary lecture on astronomy and that intended section was then merged with other fields into section V.

The additions made to the original categories (econometrics, geophysics, applied mechanics, and logic) all represent fields on the periphery of a pure conception of mathematics. However, they are fields in which Norwegian scholars had obtained international renown. Perhaps the most visible changes have to do with the additions of "econometrics" (section IV) and "geophysics" (section V), which had not been considered for the plenary lectures. These fields were recent strongholds of Norwegian mathematical science, where the nation could boast of such talents as Ragnar Frisch (1895–1973) in econometrics and Halvor Solberg (1895–1974) (and others) in geophysics.[55] Thus, the promotion of these fields into the titles of the sections also served to highlight Norwegian academic interests and expertise in them.[56]

Given the renaming of the sections, it would perhaps even have been obvious to include Frisch among the plenary lectures at short notice when Khinchin did not show up. However, replacing a lecture in probability theory with one in econometrics would probably have stretched the disciplinary boundaries of the ICM too much. Whatever the reason, Frisch did not present a plenary lecture but served as chair of section IV together with Meidell. Overall, despite the fact that the internationally recognised cosmological and geophysical groups in Norway did not contribute any papers to the section on mathematical physics, astronomy and geophysics (section V), the nation was nevertheless well represented at the congress, with Størmer, Bjerknes, and Ore on the list of plenary speakers, and many more—including Brun, Tambs Lyche, H. L. Selberg, Sven Guldberg, Meidell, Frisch, Gran Olsson, Skolem, and Heegaard — contributing sectional talks (**ICM, 1936**, vol. II).

2.9. The outlook and international ambitions of the organisers

In the previous sections, we have detailed some of the aims and ambitions of the local organisers as they pertained to showcasing Norway for its culture, nature,

---

[55]Both Frisch and Solberg had studied in Oslo (mathematics in Frisch's case, astronomy, chemistry, and mechanics in Solberg's) before going abroad during the 1920s. They were then called back to Norway to academic careers that would eventually make Solberg professor of meteorology (1930) and Frisch professor of economics (1931) and director of the new Economic Institute in Oslo the following year. In 1969, Frisch received one of the first *Sveriges Riksbank* Prizes in Economic Sciences in Memory of Alfred Nobel.

[56]Indeed, see Lietzmann's comments on p. 84.

and academic potential, and to the views on mathematics that went into setting up the programme. A third set of aims relevant here has to do with the international outlook of the organisers and the effects they were hoping to achieve through the 1936 congress. These ambitions are perhaps best found in public acts of rhetoric such as speeches or interviews given during the congress, and such sources will be analysed in the following.

On the evening of the day before the official opening of the congress, Heegaard welcomed the participants to the ICM on behalf of the university. At that time, he served as the dean of the Faculty of Mathematics and Natural Sciences of Oslo University, but on this particular occasion, he acted on behalf of the rector of the university, Sem Sæland, who was travelling abroad. In his address, Heegaard expressed the conviction — in French, English, and German — that

> [y]our work will be of great value not only to the development of mathematics but for all sciences in the civilised world; a brilliant example of the peaceful and productive cooperation between the nations.[57]

Thereby, Heegaard emphasised the advancement of culture as a means to secure and extend peaceful cooperation between nations — a theme that would recur throughout the congress.

The ICM in Oslo was officially opened on 14 July in the presence of the King and members of the Norwegian government (see Figure 2.6). Størmer extended the first formal words of welcome. Referring to the universal nature of mathematical knowledge, he argued that international collaboration should be natural to mathematicians, and he saw the international congresses as prime examples:

> More than in any other domain of culture, the results obtained in the sciences are international. The scientific discoveries made by the men of science in one country can just as well be used all over the world. This applies particularly well to mathematics. The truths of mathematics are actually universal and its ways of expression are international. Therefore, collaboration across borders is more natural to mathematicians than to any other scholars. We also hope that our congress will be successful in furthering the desire for international collaboration among mathematicians.[58]

Størmer went on to praise the usefulness of congresses for promoting informal contacts, easy access to new developments, and personal friendships and encouragement, particularly for young mathematicians. He then continued by discussing the particular role of congresses in times of deep political and social differences between nations:

> In our times, in particular, it is of great importance to try to gather the eminent scholars of the entire world ... in view of the danger of vast and threatening political and social problems. What is most important is that it is precisely in the domain of mathematics that all, independent of race or nationality, have

[57]Reported in *Aftenposten*, 14 July 1936b.

[58]Translated from the French in (**ICM, 1936**, vol. I, p. 40).

FIGURE 2.6. The opening ceremony of the Oslo congress (Source: National Library, Oslo, unprocessed 115 Carl Størmer Archives)

> only one goal: to enlarge our knowledge and to obtain truths of the most absolute kind.[59]

Thus, according to Størmer, the universal — or global — nature of mathematical research held special promise for peaceful collaboration between mathematicians of different nations even during troubled times.

Størmer then described how mathematics, apart from being of considerable use, has a particular trait: it strives for marvellous solutions to (mathematical) problems, seeking beauty rather than usefulness. Thus, Størmer, although having become more a man of applications, drew a connection with the modernist, autonomous view of mathematics as an abstract, non-utilitarian science: mathematics as a cultural transcendent. He summarised his view in an emotional appeal to his audience:

> Always preserve your noble enthusiasm for your science and protect its purity from all the harmful motives that are foreign to it.[60]

Following Størmer's official opening, the Norwegian Minister of Foreign Affairs, Halvdan Koht (see p. 32), delivered a welcome address on behalf of the Norwegian government. Schieldrop had approached Koht in July 1935, inviting him to address the congress.[61] Koht was a member of the Labour Party (*Arbeiderpartiet*), which

[59] Translated from the French in (**ICM, 1936**, vol. I, p. 41).

[60] Translated from the French in (**ICM, 1936**, vol. I, p. 42).

[61] The Minister from KUD, Nils Hjelmtveit (1892–1985) — a school teacher and educationalist from the Labour Party — does not seem to have been approached by the committee or been

had taken power in 1935 under the leadership of Johan Nygaardsvold (1879–1952);[62] including the time of the Second World War when the Labour Party formed a government in exile, it governed with small interruptions until 1965. Whereas other countries reacted to the crisis of capitalism after the Great War through fascism (Italy and Germany) or New Deal Liberalism (the USA), the Scandinavian countries adopted social-democratic approaches, and the Labour Party in Norway implemented an ideological transformation towards welfare capitalism; the key ideological figures in this historic compromise were Ole Colbjørnsen (1897–1973) with his industrial socialism and Koht with his folk socialism (*folkesosialisme*) (**Slagstad, 1998**, p. 191). Koht had an academic background as professor of history at the University of Oslo from 1910;[63] considering himself a socialist, he had been a member of the Norwegian Labour Party since 1911. He became a member of the *Institut international de la paix* in 1913 and he served on the Norwegian Nobel Committee (for the Peace Prize) from 1918 to 1936, before he retired over the Ossietzky affair (see the preface). Thus, it is fair to say that for Koht — as for many other Norwegian intellectuals — peace was on the agenda in 1936.

When Schieldrop had approached Koht before the congress, he had presented an argument that drew both on the status of mathematics and on a concern for Norwegian national pride — a combination that had previously been successful in soliciting political interest and support:

> It is the first time that all mathematicians of the world have gathered for a congress in our country, and it is therefore very important for us to provide them with a worthy welcome.[64]

By arguments such as these, and by drawing on Koht's background, Schieldrop succeeded in interesting Koht in the congress, and eventually Koht would become an honorary member. And despite the extremely busy period for a foreign minister in the summer of 1936, Koht delivered an address at the opening of the congress that was very well received by the participants.[65] In this address, which he presented in English, Koht emphasised the importance of mathematics for Norwegian identity, and at the same time sought to create an image of mathematics as a means peacefully to transcend national borders. It made the headlines of Norwegian newspapers as "Foreign Minister pays homage to mathematics as intellectual superpower".[66] Koht's view of mathematics may have been remarkable for a politician, but it coincided with Størmer's views, as expressed above. Koht stressed how political or "practical" people — when they praised mathematics — often pointed to the usefulness of mathematics for technical innovations. However, he preferred to "lay stress upon your science as leading in the expansion of the human mind".[67]

---

involved in any particular way. The somewhat unusual fact that not the education minister but the foreign minister addressed the participants finds its expression also in Fueter's erroneous attribution of this portfolio to Koht in his congress report, published here in Appendix C.

[62]For a detailed study of how science and scientists were involved in shaping Norwegian national identity, see (**Slagstad, 1998**, pp. 163ff and 191ff in particular).

[63]Koht eventually retired from his professorship in 1937.

[64]"Det er første gang verdens matematikere samles til kongress i vårt land, og det er oss derfor meget maktpåliggende å gi dem en verdig mottagelse. Vi håber at De vil støtte oss i disse bestrebelser ved å imøtekomme den henvendelse som vi herved retter til Dem." (Brevsamling 386: Schieldrop to Koht, 4 July 1935)

[65]Brevsamling 386: Schieldrop to Koht, 25 July 1936.

[66]*Aftenposten*, 14 July 1936a.

[67](**ICM, 1936**, vol. I, p. 44) and *Morgenbladet*, 15 July 1936.

Koht's speech is thus a remarkable combination of the particular attention given to the local Norwegian attitudes towards mathematics and a particular view on the nature of the science: Koht's perception of mathematics allowed him to praise the universality of mathematical science as an international endeavour towards the understanding of the world.

Koht, Størmer, and Heegaard were far from alone in their fears for the future and hopes for the outcome of congresses such as the ICM. Scandinavian mathematicians in particular expressed their concerns. For instance, the Danish mathematician Harald Bohr argued in a Norwegian newspaper interview that:

> The importance of a congress such as this one is closely connected to the general character of our discipline, ... That cannot be emphasised enough in times such as these when people from all sides — how should I express myself to avoid becoming too aggressive — when from all sides the general humanity is forgotten in favour of the narrow-minded nationalistic.[68]

Similarly, in presenting a toast during the reception at City Hall, the American mathematician Luther Eisenhart (1876–1965) proclaimed the internationality of mathematics while accepting that politics and economics adhere to national borders. Thus, to Eisenhart, mathematics was apolitical. Another prominent American mathematician, G. D. Birkhoff, soon reiterated this view when he argued, shortly after the ICM, that the abstract, even useless, nature of mathematics was responsible for allowing mathematicians to remain politically neutral:

> Mathematicians are realizing for the first time today what a great blessing it is to be regarded as innocuous and unimportant.[69]

Thus, in Birkhoff's view, mathematics remained apolitical in 1936, largely because it held little political promise or (military) importance.

Birkhoff, who was closely connected to the private Rockefeller Philanthropies, was known to be suspicious of state interference in science. Politically, he was rather conservative, against Roosevelt's New Deal politics, and particularly opposed to the reception of too many refugees by the United States. Also for Størmer, who was from a wealthy and conservative background, one has no reason to assume much sympathy for the cause of the socially progressive Norwegian workers' party, although he certainly welcomed financial and other state support for the congress. There is of course no reason to doubt Størmer's patriotic feelings for Norway. Heegaard, on the other hand, was *known* to be politically conservative. A few days after the end of the Oslo congress, and in seeming contradiction to his words on the eve of the ICM about "peaceful and productive cooperation between the nations" (p. 48), he addressed another congress in Oslo, the Third International Congress of Nationalists.[70] In his opening speech, which he gave in German, he said among other things:

---

[68]Translated from the Norwegian in *Aftenposten*, 13 July 1936; see also Appendix A.1.

[69]Birkhoff Papers, quoted in (**Siegmund-Schultze, 1994**, p. 322).

[70]Heegaard sent his German friend Engel a manuscript (three typewritten pages) of his address to the Third International Congress of Nationalists (EN: NE 210132, received by Engel 8 August 1936) on 25 July 1936 in the Nobel Institute in Oslo. This was apparently an organisation, in which the French novelist Louis Bertrand (1866–1941) was strongly involved, that was opposed to the Communist International. No more about the involvement of Heegaard in this organisation is presently known (cf. (**Bertrand and Keller, 1936**)). Heegaard mentioned this address also in his letter to Engel of 24 July 1936 (see Appendix B).

> Similar to the Olympic fire being sent to Berlin, one has recently brought a fire from the Tomb of the Unknown Soldier in Paris to Verdun, where veterans from both former enemy camps were peacefully gathered. They must have felt deeply the pointlessness of the World War. This gathering was a vivid picture of the cleavage between state nationalism, responsible for the world war, and the new peaceful nationalism of peoples.[71]

These words seem to express deep frustration about any official politically planned and militarily backed peace movement. It is less clear how Heegaard felt about the "state nationalism" of the more recent dictatorial regimes in Italy and Germany, and whether he had illusions about the peace propaganda uttered by these regimes.[72] Being on the conservative side of the political spectrum, Heegaard was most opposed to "communist internationalism", with its threat to "personal freedom", as he said in his speech. Indeed, this was the main target of the Third International Congress of Nationalists. We do not know how much support this apparently short-lived organisation mustered among scientists. However, Heegaard's words also echo similar remarks by Erhard Schmidt and Gaston Julia (1893–1978), uttered during the congress.[73]

One particular instance where geopolitical tensions influenced and interfered with mathematics in Oslo was connected with the legacies of Norway's two greatest mathematical sons: Abel and Lie. The organisers were rightfully showing off the great mathematical talents fostered in Norway,[74] but German and French mathematicians found subtle and rhetorical ways of aligning themselves and their particular nations to the legacies of the great mathematicians. In a sense, they were fighting over who could be the hosts' best friends.

2.9.1. The hosts' best friends: laying claims to the legacies of Abel and Lie. In his opening address on the first day of the congress, Størmer quoted from Bjørnstjerne Bjørnson's poem from the Abel centennial celebrations in 1902.[75] And throughout the ICM, the international guests were constantly reminded that they were visiting the home of both Abel and Marius Sophus Lie (1842–1899).[76]

[71] "Wie das olympische Feuer nach Berlin gebracht wird, brachte man neulich ein Feuer vom Grabe des unbekannten Soldaten in Paris nach Verdun, wo Kriegsveteranen aus beiden ehemals feindlichen Lagern friedlich versammelt waren, und sicher die Zwecklosigkeit des Weltkrieges tief empfunden haben — ein anschauliches Bild des Gegensatzes zwischen dem für den Weltkrieg verantwortlichen Staatsnationalismus und dem neuen friedlichen Volksnationalismus." (EN: NE 210132)

[72] We have many examples of critical positions adopted by Norwegian scientists against the Nazi regime in Germany. Among them is Størmer's annotation on an article about the German NSDAP representative in Norway, Eberhard Günther Kern, who is mentioned in Lietzmann's report (see p. 272). Indeed, the article itself is rather polemical in tone — an English translation, together with Størmer's remarks, as well as a commentary based on (**Kern, 2013**), can be found in Appendix F.

[73] See pp. 57, 59. Heegaard himself had a complicated and contradictory later biography under German occupation, which reminds us in some respects of Julia's behaviour in France. In both cases, a longing for peace led to problematic gestures of adaptation to the occupiers.

[74] Such had been the tradition since the first congress in Zurich in 1897, when a poster reproduced in the proceedings boasted pictures of three of the Bernoullis, Euler, and Steiner.

[75] For more on the 1902 Abel centennial celebrations, see (**Sørensen, 2006b**). For a brief account from the point of view of J. C. Fields, see (**Riehm and Hoffman, 2011**, pp. 92–96).

[76] Abel's status as a national icon had previously been touched upon in the newspapers in connection with the ICM: see *Aftenposten*, 13 July 1936.

FIGURE 2.7. Photograph of the bust of Sophus Lie given to congress delegates (Source: Universitätsarchiv Gießen, Nachlass Friedrich Engel)

A particular manifestation of national pride was the unveiling of a new bust of Lie which took place in the morning plenum of the congress on 15 July.[77] Such public events as the opening of the congress and the unveiling of the bust were covered by the Norwegian media, where the role of Norway in the evolution of mathematics was also emphasised for the domestic readership. During the centennial celebrations in 1902, Bjørnson had stressed that "Abel's works belong to the world although Abel,

[77]In their small congress binders, each participant was given a picture of the Lie bust (Figure 2.7) as well as a compact programme and menus for the different official banquets (Størmer Archive).

himself, belongs to Norway"[78] — but in 1936, the legacies of Abel and Lie were no longer only part of Norwegian national self-representation but took on new political aspects. And it all came to a head on 15 July.

2.9.2. A Nazi wreath at the Abel monument. On the same day as Lie's bust was unveiled, one of the leading Oslo newspapers published a short notice reporting that the day before, the German delegation had commemorated Abel in front of his memorial statue, i.e., Vigeland's famous *Abelhaugen* in Oslo; the accompanying photograph showed a wreath with Nazi insignia placed at the base of the monument (see Figure 3.4 on p. 80).[79] The short report, however, only touched the surface of the tensions involved behind the scenes. The leader of the German delegation, Walther Lietzmann, had planned the event and made preparations. Later, he recollected how the German mathematicians were anxious (*gespannt*) about the reactions of their colleagues and the Norwegian population towards National Socialism in general, and themselves in particular (**Lietzmann, 1960**, p. 60ff). Thus, when the German delegation wanted to honour Abel's memory, they contacted the German envoy in Oslo to ask for guidance. The German ambassador to Oslo, Heinrich Sahm (1877–1939), reassured the delegates that they should proceed with the memorial tributes and let the police take care of preventing trouble. Therefore, the wreath was placed at the Abel memorial in the park in front of the royal castle, and Erhard Schmidt gave a short speech while the police stood by at a distance. Afterwards, the police guarded the wreath around the clock for a week to prevent any reaction from Norwegians opposed to Nazism or the desecration of the Norwegian icon (**Lietzmann, 1960**, p. 60ff).[80]

As was and is well known, Abel had died from tuberculosis shortly before a call to Berlin could have reached him with the position that the Norwegian government had "neglected" to provide.[81] It is likely that this conflict between Norwegian neglect and German recognition of genius was evoked for the commemoration at the Abel statue; years later, Lietzmann referred to it in a way that may reflect what was said that day in July 1936:

> Abel's mathematical genius was discovered and developed in Germany; when his call to the Berlin University arrived, he had died two days before from an insidious illness.[82]

We will revisit the wreath-laying from the German perspective in Section 3.2.

2.9.3. French and German interests in the legacy of Lie. Whereas the issue of the German attempt to affiliate with Abel's legacy had been a matter of balancing only German and Norwegian national interests, and had taken place outside the congress itself, the unveiling of the bust of Lie provided a larger scene for comparable rhetoric within the congress. By 1936, the printing of the collected works of Lie was at last reaching its final stages. The editorial work had been undertaken by Heegaard and Engel, and at the ICM in Oslo, Heegaard could proudly announce

---

[78] This is a free translation of Bjørnson's words, which run: "En vestlands-gut var han; paa vel en snes aar. Nu verden ham eier; men gutten var vor" (**Holst *et al.*, 1902**).

[79] *Aftenposten*, 15 July 1936a.

[80] See also p. 385 of Lietzmann's report in Appendix D.

[81] On Abel's legacy in Norway, see (**Sørensen, 2006b**), and also the findings of a metabiographical project in (**Sørensen, 2012**). The alleged "neglect" of Abel by the Norwegians is there discussed critically as a historical construction.

[82] Translated from the German in (**Lietzmann, 1960**, p. 62).

that only the final volume was missing before this enterprise could be concluded.[83] To honour Lie's memory, a bust had been commissioned by a group of well-to-do Norwegians; the artistic task was given to Dyre Vaa (1903–1980), who, together with Vigeland, dominated Norwegian sculpture in the 1930s, and produced many works of art still decorating the public sphere in Oslo. The bust of Lie was unveiled in the great theatre of the university on the morning of 15 July, and the event was reported extensively in Norwegian newspapers (Figure 2.8).[84]

After the unveiling of the bust, Élie Cartan delivered an official lecture on Lie's mathematics (**Cartan, 1936/7**).[85] In that lecture, Cartan explained the central aspects of Lie's mathematics in a general way and contextualised Lie's life and work within the traditions of both French and German mathematics (see Section 9.4). But Cartan was not alone in paying tribute to Lie. In his later reminiscences, Lietzmann noted a more informal event involving Lie's family, who were present in Oslo (see below):

> At breakfast, to which the family invited a closer circle, there was an opportunity to celebrate the importance of mathematical science. The Frenchman Cartan took up Lie's visit (*Aufenthalt*) to Paris, I talked about his attachment to Germany, the Dutchman Schouten about his international importance, the Leipzig mathematician Koebe for his university.[86]

Thus, at least from the German point of view, it was important to emphasise the ties that Lie had with Germany and thus to counter what was perhaps seen as a French attempt to appropriate his legacy. Such a line of argument was to appear again when the German delegation wanted to commemorate Lie by placing a second wreath on his grave (Figure 2.9).[87] The German diplomatic delegation in Norway had invited Lie's family — his two daughters had both married in Germany — to a small celebration at the grave together with three delegates of the congress. After the wreath had been placed, all those present had joined in the Nazi salute.[88] Although the Cemetery of Our Saviour (*Vår Frelsers Gravlund*), where Lie's grave is located, was placed centrally in Oslo, it was, nevertheless, not such an open, exposed place as the Abel monument in front of the castle, and the ceremony did not attract the same kind of public attention as the wreath placed at the Abel monument.

2.9.4. Collaboration or Olympic competition? A third occasion for foreign guests to align themselves rhetorically with the Norwegian hosts arose on the evening of 15 July, when the Municipality of Oslo officially welcomed congress participants at the City Hall. There, delegates from the three major language groups presented toasts and short speeches extending their thanks and praise to their Norwegian hosts.

---

[83] Although Heegaard could celebrate the near-completion of Lie's collected works in Oslo, Engel was not there to join him (see p. 73). On Engel's priority over Heegaard as an editor of Lie, see footnote 54 on p. 73.

[84] *Aftenposten*, 15 July 1936b; *Dagbladet*, 15 July 1936; *Morgenbladet*, 16 July 1936; *Tidens Tegn*, 16 July 1936.

[85] See also (**Stubhaug, 2002**, pp. 376, 494–495).

[86] Translated from the German in (**Lietzmann, 1960**, p. 62).

[87] This incident is mentioned briefly in (**Siegmund-Schultze, 2002**, p. 342).

[88] See p. 385 of Lietzmann's report in Appendix D. The salute is not mentioned, however, in the relevant passage of Fladt's edition of Lietzmann's memoirs (**Lietzmann, 1960**, pp. 61–62).

Sophus Lies byste efter avsløringen. På talerstolen professor Heegaard.

Matematikerkongressen hylder geniet Sophus Lie.

Dagen avsluttedes med Oslo kommunes belivede middag på Bristol.

skere, som kaster like stor glans over sitt land og sin videnskap som det veldige norske geni Sophus Lie.

Efter at dekket var falt fra den imponerende byste besteg det matematisk-naturvidenskapelige fakultets dekanus, professor Heegaard

FIGURE 2.8. *Tidens Tegn*, 14 July 1936, p. 6 (Cartan lecturing on Lie)

Shortly after the congress, Lietzmann published a short report in which he said, among other things:

> Let us be frank: In some quarters, one saw the appearance of German mathematicians regarded with excitement, and — as some coincidentally overheard remarks made clear — with some scepticism.[89]

This remark by Lietzmann was of course related to Nazi politics since 1933 and the feelings about it abroad.

[89]Translated from the German in (**Lietzmann, 1936**, p. 304).

FIGURE 2.9. Sophus Lie's grave in Oslo (Picture by the second author)

Let us look for a moment at how the senior Berlin mathematician Erhard Schmidt, then president of the DMV, met these expectations, when he addressed the congress in Oslo with a speech to thank his Norwegian hosts (**ICM, 1936**, vol. I, pp. 52–53). Schmidt, a student of Hilbert, had always been very German-nationalistic. This had been revealed, for instance, by his staying away from the Bologna ICM in 1928 (p. 12) because he felt that the honour of the Germans had not been sufficiently restored after their exclusion.

In his speech in Oslo, Schmidt apparently felt he could not refrain from alluding to past and present competition between nations,[90] while the organisers in their addresses had mainly stressed the collaborative efforts of scientists of various nations to advance the global state of mathematics. But Schmidt tried to emphasise the peaceful side of competition when alluding to the impending Olympic Games that would begin in his home town on 1 August 1936:

---

[90]What he described as "peoples" ("Völker") was in good agreement with the terminology preferred by the Nazis.

> When I left Berlin, everything was resounding with the joyful preparations for the Olympic Games. Everybody was prepared to welcome the gathering peoples with honour and to celebrate the spectacle of their peaceful competition.[91]

Schmidt called the congress in Oslo an "Olympia of the Intellect" and continued:

> Here [at the mathematical congress] everybody is thrilled by the successes of the others. Because in the bright and pure atmosphere of the intellect one more easily understands that the wealth and the rise of one people is no disadvantage, but only a great fortune for all other peoples.[92]

There seems no doubt that "one more easily understands" can only mean that the "rise of one people" also in other, non-intellectual respects had to be honoured, and that peaceful competition was possible and worthwhile in sports, in mathematics, and in other domains.[93]

If we look at the concrete historical situation in 1936, there was only one "people" that in the opinion of some foreigners, and certainly in the opinion of many Germans inside the country, was "on the rise": Nazi Germany. It is well known that Schmidt was no anti-Semite or member of the Nazi party. But he was impressed by the recent "successes" of Nazi foreign policies, such as the recovery of the Rheinland and the Saar region, and rearmament.[94] One might imagine how those mathematicians such as Gumbel, Courant, Neugebauer, Weyl, and others, who had been forced out of Nazi Germany, and had thus been excluded from this kind of "rise", must have felt about this phrasing in Schmidt's talk. Schmidt was not alone at the time in being naive with respect to the Games in Berlin,[95] although one participant in the congress, Emil Julius Gumbel (1891–1966), an early emigrant to France, remarked mockingly in an anti-Nazi book two years later that Schmidt gave a propaganda speech for the Olympic Games which was not quite appropriate for the mostly elderly professors gathered in Oslo (**Gumbel, 1938**, p. 253). But the decisive fact here is that Schmidt, by invoking the Games, willingly and in contrast to previous speakers, raised the question of competition between "peoples" and took a clear position over the "rise" of Nazi Germany. In addition, Schmidt stressed the relations of Abel and Lie to Germany and thus construed a special bond between Norwegian and German mathematics without unduly claiming strong German merits in the developments of the two Norwegian heroes. Schmidt's German nationalism remained tactful but was palpable nevertheless.

---

[91]Translated from the German in (**ICM, 1936**, vol. I, p. 52). The Olympic Games took place in Berlin from 1 to 16 August 1936, i.e., after the Oslo congress. The regime took measures to hide its true character during these days, for instance by moving Roma people to camps outside Berlin. Of course, just as at scientific congresses, the Nazis had to accept "non-Aryan" foreigners, such as the US sprinter Jesse Owens, as participants, and the Nazis even accepted two "alibi Jews" as German participants, such as the "half-Jewish" fencer Helene Mayer. For more details, see for instance: `http://www.s-port.de/david/ns/hist%20about.html` (accessed 9 December 2018).

[92]Translated from the German in (**ICM, 1936**, vol. I, p. 52).

[93]One may consider this comparison of competition in various fields as an expression of "Olympic internationalism", as described by (**Somsen, 2008**, pp. 365–367).

[94]See (**Siegmund-Schultze, 2009**, pp. 86–87).

[95]Many foreigners, in particular Americans, were deceived by the Nazis during the Olympic Games. For example, otherwise-banned American dance music was permitted in order to please the guests. See also footnote 91 above.

The metaphor of competitions entered tacitly into another of the addresses given at the City Hall by a representative from the other side of the First World War, the French mathematician Gaston Julia (**ICM, 1936**, vol. I, pp. 54–56).[96] Julia, who wore a mask to cover the loss of his nose to shrapnel during the First World War, laid a different claim to the hearts of the Norwegian people. When he had suffered the wound, a Norwegian nurse who had volunteered for war service had treated him. By referring to this anecdote in this particular setting, Julia was expressing his thanks to all Norwegians for their efforts for humanity but also, implicitly, siding (all) Norwegians with the French and linking the legacies of Abel and Lie to France, French institutions, and French mathematics:

> Your Abel and our Galois are two brothers both as geniuses and in their sufferings.[97]

Thus, in a nutshell, the complex situation for neutral states and individuals in the polarised political spectrum was implicitly raised by Julia's well-received words of thanks. Indeed, Julia's speech is worth quoting for its emotional tone:

> Ladies and Gentlemen — this generous woman, this strong woman, was a daughter of Norway. You can imagine that I feel bound to this land by a particular debt of gratitude.[98]

Together, the two acts of commemorating great Norwegian mathematicians and the speeches by Schmidt and Julia present a complex picture of the rhetoric involved in situating certain national interests within a field where universalism and international collaboration were the dominant doctrines of the organisers. In order to promote national interests at an international congress and with the Norwegian hosts, the delegates of Europe's most powerful nations tried to position themselves in the light emanating from great Norwegian heroes.

2.10. How Oslo superseded Princeton for one week

As we have already noted (p. 35), although the organisers had originally budgeted for a number of participants and accompanying persons around 600, the actual numbers were lower; exactly how much lower is difficult to ascertain because of the late cancellations of the Russians and Italians. Thus, the official list of delegates in the proceedings differs from the list given in the report by (**Dunnington, 1936a**) — see Table 2.6, which includes all delegations of 10 or more participants (according to the list in (**ICM, 1936**), vol. I, p. 39), together with Romania, Italy, and the Nordic countries. The columns under "Proceedings" provide information extracted from the list of delegates in the congress proceedings; those under "Dunnington" are reproduced from (**Dunnington, 1936a**). The numbers shown in the second column indicate the number of delegates and the number of accompanying persons. The numbers in the third column indicate the number of plenary and sectional lectures, respectively (counted with multiplicity), and the numbers in the fourth column indicate the number of presidencies. The fifth column lists the number of delegates, whereas the sixth column lists the numbers of plenary and sectional lectures, this time according to (**Dunnington, 1936a**) and thus a more

[96]There is an interesting parallel to be drawn with the speech that Julia delivered in Göttingen in 1937 at the bicentennial of the University of Göttingen, which was turned into an extensive Nazi propaganda event. However, such a comparison must wait for another occasion.

[97]Translated from the French in (**ICM, 1936**, vol. I, p. 54).

[98]Translated from the French in (**ICM, 1936**, vol. I, p. 55).

TABLE 2.6. Participation in the 1936 ICM based on nationality

Nation	Proceedings			Dunnington	
	Participants	Lectures	Presidents	Participants	Lectures
USA	86+56	5/26	3	81	4/28
Norway	59+25	2/16	1	59	2/13
UK	48+18	1/23	7	41	1/21
Germany	35+2	3/15	3	20	3/13
France	28+13	2/19	5	28	2/15
Sweden	26+10	1/16	4	28	1/13
Poland	25+4	1/15	2	24	1/16
Denmark	22+5	2/2	4	23	2/2
Switzerland	20+11	1/10	6	20	1/11
The Netherlands	15+4	1/7	2	16	1/9
USSR	11+0	0/2	0	0	0/0
Austria	10+2	0/9	2	10	0/8
Czechoslovakia	10+3	0/7	1	11	0/8
Romania	9+6	0/8	1	10	0/10
Finland	8+1	0/3	1	9	1/5
Italy	5+2	0/0	0	0	0/0
Iceland	1+0	0/1	0	1	0/1
Total	**487+182**	**19/204**	**50**	**424**	**19/227**

immediate testimony by a witness. All identifications of nationality for the column "Proceedings" are based on listed affiliations; many of these people were, of course, emigrants (see below). The mathematicians from the USSR and from Italy did not appear—contrary to the Proceedings.

It was not merely the nations with large delegations that gained visibility in the programme. For instance, despite only sending nine participants, Romanian mathematicians were responsible for an impressive eight lectures in the sections.[99] Similarly, the three Egyptian representatives managed to deliver five lectures in the sections. Of course, these numbers indicate a trend that transcends these small delegations:[100] many participants contributed more than one paper. For instance, most prolifically, Marcel Riesz (1886–1969) from Sweden (though of Hungarian origin and brother of Frederic Riesz) gave four papers during the congress.

On the other hand, the Danish delegation, although consisting of 22 mathematicians, contributed only four lectures, of which two were plenaries, thereby providing an example of a nation mainly going to Oslo to listen and to socialise and less to present its own research. However, the Danish delegation was also a curious mixture of established researchers and emigrants who had arrived in Denmark after 1933. Among the Danish delegates were Werner and Käte Fenchel (1905–1988; 1905–1983), Herbert Busemann (1905–1994), Paul Nemenyi (1895–1952), and Otto Neugebauer (1899–1990), who were all German emigrants. Moreover, the Swedish delegation included Willy Feller (1906–1970), another refugee from Germany, who had originally fled to Denmark in 1933; the Oslo congress therefore reunited the group that had so influenced the Copenhagen mathematical milieu since 1934.[101]

Although the German language dominated in the plenary lectures (see p. 149 and Chapter 3), the American and Scandinavian delegations nevertheless had a large impact on the congress. The Scandinavians were slightly over-represented, not least with respect to the sizes of their delegations. However, with the largest delegation having more than 80 delegates and more than 50 accompanying persons, as well as a handful of plenary lectures, the American mathematicians left their strong mark on the congress. We will pick up on this point again in Chapter 7.

[99]The Romanian delegation was featured in a newspaper article: *Aftenposten*, 18 July 1936.

[100]Following the Second World War, the policy governing the selection of talks at the ICMs had to change, along with the notion of "invited address", in light of the greater regulation required at the larger post-war congresses.

[101]See (**Jessen, 1993**), as well as (**Sørensen, 2014**).

CHAPTER 3

The German Delegation: Swaying between Expansionism and Isolationism

Having established the circumstances surrounding the preparation of the congress by the Norwegian organisers, we now turn our attention to the assembly of one particular delegation: the Germans. As described in Chapter 2, German mathematicians had had a somewhat difficult relationship with the ICMs, but the 1936 congress provided an opportunity for them to reappear in numbers similar to those at the congresses before the First World War. However, membership of the official German delegation was strictly controlled, with racial and political considerations dictating its make-up. There are indications — both in the official guidelines and in particular in Lietzmann's report quoted below — that the German delegation in Oslo propagandised German mathematics and German politics and tried to gain more influence over Eastern European mathematicians. In doing so, the German participants in the congress exhibited the "expansionist" strategy identified in Chapter 1; nevertheless, as we will indicate in the present chapter, we also see traces of the legacy of Germany's relationship with the ICMs in hints of an "isolationist" approach, which was increased under the conditions of state-supported anti-Semitism after 1933 and the politics of cleansing the universities.

3.1. Preparation for the congress

The "Deutsche Kongreßzentrale" (DKZ) was founded in 1934 as a formally independent political organisation. From 13 November 1936, however, it was subordinated to Goebbels' Ministry of Public Enlightenment and Propaganda. On the same date, Hitler declared that all international congresses and national meetings — not only in the sciences — had to be registered (DKZ 2, p. 8). Although sending information about impending congresses to the DKZ had been voluntary before that date, the DKZ had already (in 1935 and 1936) acquired considerable influence over the preparation of international congresses both at home and abroad.[1] In a confidential (and therefore unpublished) 17-page document of 1936, entitled "Richtlinien für die Leiter deutscher Abordnungen zu Kongressen im Ausland" ("Guidelines for the leaders of German delegations to congresses abroad": DKZ 1), one reads, among other things, the following passage, which comes after an allusion to the boycott policies of the *Entente* during the 1920s:

> Nothing is further from our intention than retaliating and using science as a political weapon now from our side. But we have to demand that the German scholar, who — without any doubt — is leading in many fields, finally gets the influence in

[1] This is clear also from the extensive correspondence and reports by scientists in the files of the DKZ prior to 1935.

> the organisations of international exchange of experiences which corresponds to his expertise. [underlining in original][2]

Detailed instructions were given in the "Guidelines" regarding contacts between the leaders of delegations and local German diplomatic representatives at the place of the congress, relations with the press, the preferential use of the German language, and even the dress code for the delegation.

The invitation of future congresses to Germany was declared to be "in principle always very desirable",[3] although it was supposed to be discussed with German political authorities in detail before departure for the congress. In this connection, "the Jewish question" ("die Judenfrage") was expressly mentioned as a potential obstacle (DKZ 1, p. 17), but the leaders of delegations were entitled to declare that "the participation of foreigners of the Jewish race ... [be] left up to interested foreign organisations".[4]

The main instrument at the disposal of the DKZ in pursuing its political goals was mentioned in more detail in another equally confidential document: its 8-page "Jahresbericht" ("Annual Report") for 1936 (DKZ 2). Here one reads:

> The registration of all participants traveling to the congress is guaranteed by making the DKZ responsible for the control of the necessary foreign currency. By decree of the Reich Foreign Currency Authority, no foreign travel to congresses etc. is possible without permission from the DKZ. ... The incoming foreign currency exchange applications form the basis for the personal files of the archives. The reports demanded from the leaders of delegations are essential for various authorities.[5]

To understand this, one has to be aware that Hitler's Germany pursued a policy of economic autarky, i.e., independence from import of strategically important raw materials and restriction of trade altogether. The resulting shortage in foreign currency heavily influenced many realms of public life, including science. There were problems in acquiring foreign literature, and impediments to foreign travel, because scholars had no possibility of changing even their own money into foreign currency. But it is most important to see that these economic conditions were used as a political instrument. This concerns also the policies of the DKZ, which was so close to Goebbels' ministry. To be sure, the DKZ expressly declared in its report:

[2] "Nichts liegt uns ferner, als nun auch unsererseits die Wissenschaft zur politischen Waffe zu machen. Wir müssen aber fordern, dass der deutsche Forscher, der auf zahllosen Fachgebieten unbestritten führend ist, in den Organisationen des internationalen Erfahrungsaustausches endlich den Einfluss erhält, der seiner Leistung entspricht." (DKZ 1, p. 3)

[3] "Die Einladung von späteren Kongreßtagungen nach Deutschland ist grundsätzlich stets sehr erwünscht und soll — wo immer möglich — versucht werden." (DKZ 1, p. 16)

[4] "... die Beteiligung von Ausländern jüdischer Rasse ... an dem Kongreß interessierten ausländischen Organisationen überlassen bleibt ..." (DKZ 1, p. 17)

[5] "Die Erfassung aller ins Ausland reisender Kongressbesucher ist gewährleistet durch die Übertragung der Devisenkontrolle auf die Deutsche Kongress-Zentrale. Durch Verfügung der Reichsdevisenstelle ist ohne Prüfung und Genehmigung der Deutschen Kongress-Zentrale keine Auslandsreise zu Kongressen, Vorträgen und Studienreisen möglich. ... Die eingehenden Devisenanträge bilden das Rohmaterial für den Aufbau der Personenkartei des Archivs. Die von den Delegationsführern ... eingeforderten Berichte ... bilden oft wertvollste Anhaltspunkte für ... die verschiedensten Dienststellen." (DKZ 2, pp. 4–5)

> Interference in the technical content of the congresses is not the responsibility of the DKZ![6]

But at the same time the DKZ aimed at "propagandistically effective actions"[7] that would

> inform participants at the congresses, above all foreigners, about the accomplishments of the new Germany.[8]

A further goal was to use

> personal relationships with scholars ... and other congress participants to support German interests, ... for instance to draw over an important meeting to Germany.[9]

Thus the DKZ pursued political purposes, in particular cultural-expansionist ones, which later, during the Second World War and the occupation of almost the entire continent, were continued in strategies for a re-ordering (*Neuordnung*) of European science, including mathematics.[10]

The DKZ was directly involved in preparations for the Oslo mathematical congress, not least because of the question of money. In a letter of 18 December 1935, the DKZ informed the Association of the German Trade Fair Industry (Ausstellungs- und Messe-Ausschuss der Deutschen Wirtschaft: AUMA) about the upcoming "First International Mathematical Congress" in Oslo.[11] The DKZ, in the person of a Dr. Knapp, explained the curious numbering of the imminent congress by claiming that "the previous postwar congresses in part did not admit the Germans on an equal footing".[12] Although there had been discussions in Germany and even abroad about whether the Strasbourg (1920) and Toronto (1924) congresses should be counted among the international congresses (see footnote 17 on p. 9), nobody so far had come up with the idea of not counting the congresses of Bologna (1928) and Zurich (1932) — not even the nationalistic Germans around Bieberbach, who had abstained from participating in Bologna (see p. 12). It was thus left to the DKZ to stress the unique character of the Oslo congress under the new auspices of Nazi rule in Germany.

As early as 22 June 1935, the leading mathematical educator in Germany, high-school teacher and extraordinary professor for the didactics of mathematics at the University of Göttingen, Walther Lietzmann (1880–1959; see Figure 3.1), had written to the "leader" ("Führer") of the Mathematischer Reichsverband (the association of German mathematical societies), Georg Hamel, about the meeting

[6] "Einmischung in den fachlichen Inhalt des Kongresses ist nicht Aufgabe der DKZ!" (DKZ 2, p. 5)

[7] "propagandistisch wirksame Aktionen" (DKZ 2, p. 6)

[8] "... den Kongressteilnehmern, insbesondere den ausländischen, ein Bild von den Leistungen des neuen Deutschlands zu geben." (*Ibid.*)

[9] "Besonderer Wert wird darauf gelegt, alle durch gemeinsame Arbeit im Kongressleben entstandenen persönlichen Beziehungen zu Gelehrten, Forschern, Kaufleuten, Technikern und sonstigen Kongressteilnehmern sorgfältig zu erfassen und zu registrieren um sie in geeigneten Fällen für deutsche Belange dienstbar zu machen. ... z.B. das Hinüberziehen einer wertvollen Tagung nach Deutschland ..." (*Ibid.*)

[10] See (**Siegmund-Schultze, 2002**); see also the remarks in Chapter 7.

[11] "Wir erhalten die Nachricht, daß im kommenden Jahr in Oslo der I. internationale Mathematiker-Kongreß stattfindet." (DKZ box 203, f. Mathematikervereinigung)

[12] "Er wird als I. Kongress bezeichnet, weil bei den früheren Nachkriegskongressen zum Teil wenigstens die Deutschen noch nicht als gleichberechtigt anerkannt waren." (*Ibid.*)

FIGURE 3.1. Walther Lietzmann (1880–1959) (Source: (**Lietzmann, 1960**))

of the International Commission on Mathematical Instruction (ICMI), which was planned in connection with the Oslo congress. ICMI had been founded by Felix Klein around 1908, and Lietzmann was strongly involved with it from the German side (WLP). Lietzmann's initiative was apparently rewarded by the Reich Ministry of Education (Reichserziehungsministerium: REM), which made Lietzmann the head of the German delegation for the congress. After the congress, Lietzmann was honoured by German mathematicians by being elected chairman (*Vorsitzender*) of the DMV for the following academic year 1936–7.[13]

Prior to November 1935, several applications for participation in Oslo had been sent to REM by established mathematicians, such as Otto Haupt and Heinrich Behnke (who applied both for himself and also on behalf of his younger collaborators in Münster, Gottfried Köthe and Helmut Ulm). In an inner-ministerial note, dating from about that time, the responsible official commented on a list of eleven mathematicians whose names had come up so far as possible participants in Oslo: Behnke, Wilhelm Blaschke, Gustav Doetsch, Georg Feigl, Hamel, Haupt, Erich Hecke, Walther Lietzmann, Friedrich K. Schmidt, Egon Ullrich, and Otto Volk. The official wrote:

> There are no old party comrades among the above. Lietzmann appears to me as the ablest. I should propose him as Führer of the delegation, although he is merely a high school teacher.[14]

In its above-quoted letter of 18 December 1935 to AUMA, the DKZ had also mentioned that Erhard Schmidt, professor of mathematics at the University of Berlin and president of the DMV (see Figure 3.2), was responsible for preparing for the congress. Schmidt would indeed represent German mathematical research at the Oslo congress in various ways,[15] but he was not an organiser or a worker on details, and so he left this to others. In early 1936, Schmidt had been involved by the ministry, which, however, still intended to send no more than about twelve mathematicians to Oslo, as one of the planned general speakers, Helmut Hasse, related to another, Hecke.[16]

On 7 March 1936, REM officially appointed Lietzmann "Führer" of the German delegation (WLP). By this date, the list of delegates had almost doubled in size and now included 22 participants, among them two representatives from REM. One of these was Theodor Vahlen, an old Nazi and a mathematics professor who was at that time an official at the ministry.[17]

Lietzmann immediately began negotiations for foreign currency with the DKZ, and further applications for participation in Oslo came in during the following

---

[13] *Jahresbericht der Deutschen Mathematiker-Vereinigung* **46** (1936), 2. Abt., p. 88.

[14] "Alte Nationalsozialisten sind unter den Genannten nicht. Am gewandtesten erscheint mir Lietzmann. Dieser ist aber Gymnasiallehrer. Trotzdem würde ich ihn zum Führer der Delegation vorschlagen." (BAB-REM 2905, fol. 54) The note is signed "Me", which could be the high official Rudolf Mentzel or Metzner (see note 17 below). Note that what we have translated here as "ablest" appears in the original as "gewandtesten", which is something like clever or politically experienced, and is clearly not meant in a mathematical respect.

[15] Schmidt was elected one of the vice presidents of the congress (**ICM, 1936**, vol. 1, p. 44).

[16] Hecke Papers, Hasse to Hecke, Göttingen 8 January 1936: "Ich hörte in Berlin noch, dass für Oslo höchstwahrscheinlich nur 12 Ausreisegenehmigungen erteilt werden, darunter selbstverständlich wir zum Vortrag aufgeforderten. Auch für diese wenigen Bevorzugten würden wahrscheinlich die Mittel nur in sehr beschränktem Umfange zur Verfügung gestellt werden."

[17] In addition to Vahlen, the other ten new members of the planned delegation were one professor Metzner (REM, a non-mathematician), and the mathematicians Hasse, Feigl, Bessel-Hagen, Geppert, Engel, Haenzel, Artin, E. Schmidt, Timpe, and Kowalewski. The same names were also reported to Richard Courant in a letter from Ulm, dated Münster, 14 March 1936, in which he stated that his own application had been turned down for financial reasons (RCP, Series I, box 8, folder 8).

FIGURE 3.2. Erhard Schmidt (1876–1959) (Author: Konrad Jacobs. Source: Archives of the Mathematisches Forschungsinstitut Oberwolfach)

months. As a matter of course, they had to be accompanied by letters of support (or otherwise) from local political authorities, in particular from the local Nazi faculty organisations, the "Dozentenschaften".

The most clear-cut case of a politically motivated rejection of an application was that of Erich Kamke from Tübingen. Kamke was an acknowledged analyst and editor of the *Mathematische Zeitschrift*, published by Springer Verlag. He mentioned this responsibility as his main reason for wishing to participate in Oslo, where he wanted to contact foreign colleagues, because "German mathematical journals must today fight for their position in the scientific world".[18] However, the

[18] "... die deutschen mathematischen Zeitschriften heute um ihre Behauptung in der wissenschaftlichen Welt kämpfen müssen ..." (REM 2905, fol. 123, Kamke to REM, 24 March 1936).

accompanying statement by the Tübingen "Dozentenschaft", signed by its head, the mathematician (!) Erich Schönhardt, was outright negative:

> I cannot accept that the editor's discussions with foreign colleagues are so important in maintaining the level of a German journal, unless one intends to maintain that level not by the contributions of German scholars, as it would suit a German scientific journal, but by those of foreign and Jewish colleagues.[19]

But Schönhardt went even further, accusing Kamke of being an "opponent of the National Socialist Weltanschauung"[20] and referring to Springer as a "non-Aryan" publisher. The letter also mentioned that Kamke was married to a "non-Aryan". This last point alone could have been sufficient to lead to a rejection by the ministry of Kamke's application. The Nuremberg laws of September 1935 had expressly forbidden new marriages between Germans and so-called "non-Aryans", and, from 1937 onwards, those university professors who were in the "Jewish clan" ("jüdisch versippt") were dismissed. One should assume also that Otto Haupt's ultimate withdrawal from the congress (see p. 76) was in some way related to the fact that he was married to a Jew.[21]

Indeed, German mathematicians were fully aware that their participation was decided by REM on political grounds, and that they needed both foreign currency and political clearance in order to travel to Oslo. And they also knew that in an atmosphere of continuous and systematic mutual political mistrust in Nazi Germany,[22] decisions could easily be taken against them without their knowledge and without any possibility of appeal — this atmosphere in turn induced some Germans to show their engagement with the regime most clearly and unambiguously.[23]

In a draft of a letter that the emigrant Richard Courant planned to send on 15 April 1936 to Harald Bohr in Copenhagen, one finds:

---

[19] "Dass zur Erhaltung des Niveaus einer deutschen Zeitschrift ... die Aussprache der Herausgeber mit ausländischen Fachgenossen so besonders wichtig sein soll, kann ich nicht einsehen, es sei denn, dass man eben jenes Niveau nicht durch Beiträge deutscher Gelehrter hochzuhalten versuchen will, wie es sich für eine deutsche wissenschaftliche Zeitschrift geziehmen würde, sondern durch Beiträge ausländischer oder jüdischer Fachgenossen." (REM 2905, fol. 124)

[20] "... dass er zu den Gegnern der nationalsozialistischen Weltanschauung gehörte und noch gehört." (*Ibid.*)

[21] Although Haupt's initial application of 30 October 1935 had been accompanied by a formal supporting statement from the local "Dozentenschaft", this statement also noted that Haupt was married to a "non-Aryan" (REM 2905, fol. 26). However, at the time the discrimination against German citizens in the "Jewish clan" was not yet fully written into Nazi law, as the temporary nomination for the congress of Artin shows, who was also married to a "non-Aryan".

[22] The general political atmosphere of mutual distrust and denunciation is vividly described in the recollections of one participant in the Oslo congress: (**Behnke, 1978**).

[23] The ministerial files for the preparation for the Oslo congress provide good examples of this. Even mathematicians who had shown their political reliability (in the Nazi sense), such as Blaschke, Paul Koebe, and Doetsch, were labelled in a political inner-ministerial note as potentially unreliable (REM 2905, fol. 254, undated). Doetsch was alleged to have been a political sympathiser in the 1920s of the communist and anarchist Max Hoelz (1889–1933), an accusation that is hardly likely. Doetsch had held pacifist convictions during the Weimar Republic and felt that he needed to overcompensate for these by adapting to the Nazis. All three of these mathematicians eventually went to Oslo. In Behnke's case, his previous marriage to a Jew, meanwhile deceased, was mentioned in the same inner-ministerial note. Similar concerns may have contributed to his temporary removal from the list (see Figure 3.3, p. 75).

> I hear from many people, such as Rellich, Ulm, Friedrichs, v.d.Waerden, Köthe that they cannot come or are not allowed to come because they lack the general permission or foreign money or both or because they are even afraid to apply.[24]

One may safely assume that quite a few German mathematicians did not apply because they loathed the promise of a "politically streamlined" delegation in Oslo.[25] The able and promising Kurt Friedrichs of Braunschweig, who was at the time collaborating closely (though usually only by mail) with Courant in New York City on the second volume of the "Courant–Hilbert" book on mathematical physics (which finally came out with Springer in 1937) ruled out going to Oslo, in spite of what Courant had hoped. He wrote to the latter on 22 April 1936:

> Very friendly your proposal with respect to Oslo. I am afraid though I won't go there. Visits to congresses are now regulated in such a way that one is no longer an individual [literally, a loner: *Einzelgänger*] but part of a delegation, whose behaviour is ruled by a leader. But hopefully it is topologically possible to meet in Europe anyway.[26]

In addition, a reaction by Otto Toeplitz in a letter to Courant, dated 11 March 1936, shows the predicament for prospective participants, in view of Nazi censorship:

> My wish to see you in Oslo is great. N. [probably Otto Neugebauer] has indicated to me how he would take care of my upkeep there — I would only have to purchase the train ticket here. However, my friends here have doubts about going. But their situation is different from mine. One of them still has emeritus professor status and can only travel with permission. The other is on a pension, as I am, but particularly exposed [politically]. I find, however, that it will be very difficult to adopt the proper relationship to the various groups attending the congress, and I feel therefore that I have to think about it again. Maybe I was too naive. But at least I can be sure to have enough control over what I say at the congress and so I could take the risk and go into this tricky situation.[27]

---

[24] "Im übrigen höre ich, dass sehr viele Leute, wie Rellich, Ulm, Friedrichs, v.d.Waerden, Köthe nicht kommen können bzw. dürfen, weil ihnen entweder die allgemeine Erlaubnis oder die Devisen oder beides verweigert wird oder sie sich sogar fürchten, entsprechende Anträge zu stellen." (RCP, Series I, box 2, folder 4) The draft is marked as "not sent" ("nicht abgesandt").

[25] Behnke wrote in 1978: "Most German colleagues gave up from the outset on trying to obtain permission to attend the congress." (Translated from the German in (**Behnke, 1978**, p. 247).)

[26] "Sehr freundlich Dein Vorschlag bezüglich Oslo. Ich fürchte allerdings, dass ich doch nicht dorthin reisen werde. Kongressbesuche werden jetzt so geregelt, dass man nicht mehr als Einzelgänger erscheint, sondern als Glied einer Abordnung, deren Auftreten von einem Leiter bestimmt wird. Vielleicht ist es doch topologisch möglich, sich schon in Europa zu sehen." (RCP, Series I, box 4, folder 7)

[27] "Mein Wunsch Sie in Oslo zu sehen ist (entsprechend) groß. N. hat mir einen Weg angedeutet, wie er mir für meinen Unterhalt dort sorgen würde — das Billet müsste ich hier aufbringen. Aber meine hiesigen Freunde haben Bedenken hinzugehen. Ihr Fall liegt in mancher Beziehung anders. Der eine ist emeritiert und darf nur mit Genehmigung fahren. Der andere ist wie ich pensioniert, aber besonders exponiert. Ich finde aber dass es recht schwierig sein wird, zwischen den dort befindlichen Gruppen die richtige Haltung einzunehmen und muss mir die Sache wirklich noch überlegen. Ich habs [*sic*] vielleicht zu naiv angesehen gehabt. Immerhin weiß ich,

Toeplitz did not, in the end "take the risk", but met Courant in Copenhagen after the congress,[28] as did others who had not gone to Oslo, including Friedrichs.[29]

Toeplitz' undecided letter to Courant also made it clear that even retired professors, if they were in the otherwise privileged position of being "emeritus", had to apply for permission from the ministry, regardless of whether funds were provided for them or not.[30]

Thus, shortage of money did influence ministerial decisions but it was not by itself decisive and probably was not the primary reason for final decisions. However, that shortage had the effect of restricting the number of participants from the outset,[31] which was particularly damaging for the chances of young mathematicians who wanted to participate in addition to the prominent ones.[32] To a certain extent, this problem was acknowledged by REM when it refused permission for the participation of the older Bonn mathematician Hans Beck (1876–1942) on the grounds that "given the restricted financial means at our disposal the support for younger, less salaried docents is more important".[33]

But the reality looked different: one of the few younger mathematicians and non-professors who finally took part in Oslo, Gottfried Köthe (1905–1989), could only go because he was born an Austrian and his mother provided the foreign currency for him. Another young participant, Emanuel Sperner (1905–1980), had the political recommendation of being an "old comrade" of the Nazi Party, which he had joined in 1933.[34] An even younger promising mathematician, Theodor Schneider (1911–1988), had planned to go to Oslo too, but had apparently failed to acquire the money.[35] Politically unreliable young mathematicians such as Hans Zassenhaus (1912–1991) obviously had, from the outset, no chance at all of attending.[36]

---

dass ich auf dem Kongress das, was ich sage, einigermaßen in der Gewalt habe und vielleicht eine heikle Situation wagen kann." (RCP, Series I, box 8, folder 7)

[28]RCP, Series I, box 8, folder 7: Toeplitz to Courant, Bonn, 17 August 1936.

[29](**Reid, 1976**, p. 186) relates that Courant met Friedrichs in Copenhagen around the time of the congress and conjectures that Friedrichs preferred "meeting Courant in Copenhagen rather than at the congress in Oslo, where their meeting would be observed".

[30]This was particularly clear in the case of the Jewish mathematician Artur Rosenthal, who had been an "emeritus" in Heidelberg since 1935 (see p. 71).

[31]Even the ministry itself showed some restraint. Metzner, who had been on the original list in March 1936, resigned in a letter to Lietzmann, dated 29 April 1936, with reference to the currency situation (WLP).

[32]There is a clear parallel in this respect to the situation of young scholars in the German Democratic Republic decades later, where political criteria also played a decisive role with respect to participation in international congresses. Størmer, in his opening address to the Oslo congress, stressed the importance that congresses have in particular for young mathematicians (**ICM, 1936**, vol. I, p. 41).

[33]"Bei den geringen Mitteln, die zur Verfügung stehen, scheint mir die Unterstützung minder besoldeter jüngerer Dozenten wichtiger." (REM, 2905, fol. 142: undated inner-ministerial note)

[34]REM, 2905, fol. 254. Undated inner-ministerial note.

[35]See the draft of Courant's letter to Harald Bohr of 15 April 1936, quoted above (p. 67), where Courant reported on a recent letter he had received from C. L. Siegel, mentioning Max Dehn and Theodor Schneider from Frankfurt as prospective participants for Oslo. Schneider had already written to Størmer in Oslo, offering a talk (Størmer Papers, 14 March 1936).

[36]The application on behalf of Zassenhaus by his boss in Rostock, Robert Furch, was dated 4 May 1936 (WLP). Furch sent his application, which included an application for his own participation as well, to Lietzmann, but it was apparently not even discussed in the ministry.

Shortage of foreign money even affected the chances of "politically reliable" mathematicians, particularly if the applications came in too late.[37] The shortage of Norwegian crowns led to a reduction both in the size of the delegation and in the funds available for each participant (the equivalent of 185 Reichsmarks instead of the 300 originally planned).[38]

But in many cases the currency problem was used as a welcome excuse to refuse participation for "unreliable" or otherwise unwanted applicants. On 23 June 1936, the Göttingen aerodynamicist Ludwig Prandtl (1875–1953) protested, in a letter to the ministry, against the negative response to his proposal to send the younger applied mathematicians Walter Tollmien (1900–1968) and Irmgard Lotz (1903–1974) to the congress:

> I add that from my standpoint as head of the biggest German research institute for hydrodynamics and aerodynamics I deem it unacceptable that the institute be excluded from a congress where among the main talks is one by Bjerknes entitled "New Lines in Hydrodynamics".[39]

Prandtl therefore renewed his application, at least for Tollmien. The ministerial official wrote on that letter with respect to Tollmien:

> Because he is likely not to represent Germany in a worthy manner we should stick to the original decision.[40]

Even an appeal by the famous physicist and president of the Kaiser Wilhelm Gesellschaft (to which Prandtl's institute belonged), Max Planck, did not change the ministry's negative decision, which was sent on 6 July, but only mentioned currency problems.[41]

In this case, we cannot rule out the role played by the fact that applied mathematics was not a major focus at the Oslo congress (see Chapters 8 and 9), and that separate international congresses for applied mechanics had in the meantime been instituted.[42] This supposition seems to be confirmed by the treatment given to the application from Alwin Walther, who held the chair of practical mathematics at the Technical University of Darmstadt. In his comparatively early application of February 1936, Walther could refer to special connections to applied mathematics in Scandinavia (N. E. Nørlund, E. J. Nyström) and, in particular, to the new developments in (analogue) calculating machines that were being cultivated in the institute of one of the Norwegian organisers of the congress, Størmer:

[37]As in the case of Helmut Kneser of Greifswald, who applied only on 27 June, two weeks before the congress began. See REM 2905, fol. 222 (undated inner-ministerial note): "In principle, no reservations against Kneser, however no money" ("Gegen Kneser an sich keine Bedenken. Jedoch aus Devisengründen für Ablehnung").

[38]WLR 1936, fol. 384. Using the same online resource as in footnote 29 on p. 35, we note that 1 RM at 1936 rates is equivalent to US$5.70 or 5,13 € at 2015 rates.

[39]Prandtl to REM, 23 June 1936 (REM 2905, fols. 224–225).

[40]"Da er m.E. Deutschland nicht in würdiger Form vertreten wird sollte man die urspr. Entscheidung aufrecht erhalten." (Signed: Me.) (*Ibid*)

[41]REM to Prandtl (REM 2905, fol. 227). Planck's intervention is fol. 226 (26 June 1936).

[42]Bjerknes' talk, mentioned by Prandtl, was the shortest of all the general talks given at the congress, at least as it appears in the proceedings (see Section 9.11). In his published report, Lietzmann mentioned three further general talks on applied mathematics (out of a total of 19) by C. Størmer (on electron orbits), C. W. Oseen (geometrical optics), and G. D. Birkhoff (foundations of quantum mechanics), but added: "Applied mathematics ... was clearly in the background." (Translated from the German in (**Lietzmann, 1936**, p. 301).)

> It is necessary for me to see the machine for the instrumental solution of differential equations currently under construction with Prof. Störmer in Oslo. Such machines have so far existed in a usable form only with Prof. Bush in Cambridge (Mass., USA). ... In my institute in Darmstadt this process is now being started for Germany for the first time. We will save much unnecessary effort if I can see Prof. Störmer's machine and can talk to him.[43]

Walther's application was granted in principle but he apparently did not receive the necessary foreign currency. He therefore returned his permission to Lietzmann on 12 July 1936, one day before the congress started, adding the frustrated remark that "his problem had unfortunately not been solved by the silence [Stillschweigen]" of the ministry.[44] Ironically, the Germans would be very keen to see the Rosseland–Størmer machine after 1940, when they occupied Norway during the war.[45]

In some instances, the ministry's tactic of hiding political decisions behind the currency problem led to unforeseen results. Rather late, on 9 July 1936 (i.e., four days before the congress was due to start), the ministry had to agree to the participation of William Threlfall and Artur Rosenthal, although they both had various political "defects" counting against them,[46] Rosenthal even being Jewish (by the Nazi definition) and having been dismissed from Heidelberg in 1935. Originally, Threlfall and Rosenthal had been denied participation, with the foreign currency problem cited as the reason. But after they had declared that they no longer needed money from the government, the ministry felt:

> With regard to the effect which such a measure would have abroad, we cannot now refuse participation without an explanation or with a different one.[47]

The problem of foreign mathematicians working in Germany and applying to participate in Oslo also seems to have been a difficult one. This is most visible in the case of B. L. van der Waerden (1903–1996), the Dutch algebraist in Leipzig. As early as 28 March 1936, even before he officially applied to REM on 10 April, van der Waerden wrote to Courant:

---

[43] "Ferner ist es für mich notwendig, die bei Prof. Störmer in Oslo im Bau befindliche Maschine zur instrumentellen Auflösung von Differentialgleichungen zu sehen. Derartige Maschinen waren bisher nur bei Prof. Bush in Cambridge (Mass., USA) in gebrauchsfähiger Form zu finden. ... In meinem Darmstädter Institut kommt diese Angelegenheit jetzt zum ersten Mal für Deutschland in Fluss. Es wird sehr viel unnötige Mühe sparen, wenn ich die Maschine von Prof. Störmer sehen und mit Prof. Störmer selbst sprechen kann." (REM 2905, fol. 85, Walther to the rector of the university [who sent it on to the ministry with his support], 18 February 1936)

[44] "durch Stillschweigen des Reichserziehungsministeriums leider nicht erledigt" (WLP)

[45] Svein Rosseland, who was the real initiator of the Oslo machine, would then make it inoperative for the Germans: see (**Holst, 1996**). On the machine, see also Section 9.2.

[46] REM, 2905, fol. 244. Threlfall was the son of an English father and known to be critical of the regime. According to (**Volkert, 2018**, p. 2), Threlfall had money in England and thus could finance the trip for himself and Seifert. On Rosenthal, see (**Siegmund-Schultze, 2009**), where the Nazi student boycott against his and Heinrich Liebmann's lectures in Heidelberg is also documented.

[47] "... ist es mit Rücksicht auf die Wirkung einer derartigen Maßnahme im Ausland unmöglich, die Genehmigung jetzt ohne Begründung oder mit einer anderen Begründung zu versagen." (REM, 2905, fol. 244) See also REM to Lietzmann, 9 July 1936, in which the ministry explained that Rosenthal had received money from a Norwegian colleague.

> For Oslo I will probably receive neither confirmation nor foreign currency. There will only be a restricted number of Germans who are allowed to go there.[48]

The stress in this quotation seems to lie on "Germans", meaning that the latter would receive preferential treatment. There was of course no doubt about van der Waerden's international reputation and the influence of his recent book *Moderne Algebra* (**van der Waerden, 1930**). In addition, van der Waerden was heavily involved in various of the internationally leading mathematical enterprises of Springer Verlag, in particular the *Mathematische Annalen* and the *Grundlehren der mathematischen Wissenschaften* series. Van der Waerden did not fail to mention these activities in his application letter to the ministry, dated 10 April 1936. He also mentioned the fact that Springer provided — through their international book sales — considerable amounts of foreign currency to the German Reich.[49] But as the case of Kamke has shown (p. 67), engagement with the internationalist Springer publishers was not an activity that endeared a mathematician to the Nazis. Moreover, the rather young van der Waerden (born 1902), who had been a student of the Jewish mathematician Emmy Noether and had recently written a courageous obituary of her in the *Mathematische Annalen* (**van der Waerden, 1935**), was probably not the kind of mathematician the Nazis would trust very much. In addition, there was the general problem of how to include foreigners in a German delegation.[50] Even Ernst Foradori from Austria (in 1936 still not occupied by the Germans), who had received much political support from Nazis such as Bieberbach in Berlin, failed in his application, with the usual financial argument being given.[51]

One would assume that a strong argument in favour of participation in Oslo would be to cite special and close mathematical relationships with Norwegian mathematics. Two German mathematicians in particular could claim such close relationships: Friedrich Engel and Wilhelm Lorey. The former, already 75 years of age and an "emeritus" in Giessen since 1930, had devoted much of his career to an edition, together with Poul Heegaard, of the works of his teacher, the Norwegian mathematician Sophus Lie. Engel spoke and wrote Norwegian, and he was on the original list of 22 March 1936. But something seems to have got in the way of his trip, because Engel wrote to Lietzmann towards the end of April:

> None of the attending mathematicians, neither the Germans nor the foreign ones, has such long and close relationships to Norway and to Norwegian mathematics as I have. This should entitle me to give a talk there.[52]

---

[48] "Für Oslo werde ich voraussichtlich weder die Bewilligung noch die Devisen bekommen. Es wird nur eine beschränkte Anzahl von Deutschen dorthin können." (RCP, Series I, box 9, folder 5)

[49] WLP, van der Waerden to Lietzmann, 7 June 1936, in which he quoted indirectly from his earlier application to the ministry. A copy of the latter application is also in van der Waerden's personal file at the University of Leipzig Archives: PA 70, fol. 35. The file shows that van der Waerden had serious political problems with the Nazi administration both in Leipzig and in Berlin.

[50] Things were apparently different with respect to the older Constantin Carathéodory of Munich, who had long been seen as a representative of the German mathematical community, despite having been brought up in Belgium by Greek parents (see p. 76).

[51] REM 2905, refusal by ministry dated 7 June 1936 for "Devisengründe".

[52] "Keiner der anwesenden Mathematiker, weder der Deutschen noch der ausländischen hat so langjährige und enge Beziehungen zu Norwegen und zur norwegischen Mathematik aufzuweisen

This could mean that Engel had been crossed off the list by the Germans, but it seems more likely that the organisers in Oslo had not given the old man the opportunity for a longer talk, maybe on the occasion of the celebration for Lie. In any case, Engel finally appeared on the list of delegates in June 1936, but he withdrew at short notice, officially due to the health problems of his wife.[53] However, Engel's edition of Lie's works was mentioned both in Lietzmann's report and in the general talk by Élie Cartan on the occasion of unveiling the bust of Lie (see Section 2.9.3). Cartan — who was certainly a much more prominent researcher than Engel — spoke on "The role of Lie groups in the development of modern geometry" (see Section 9.4).[54] In addition, the congress sent a telegram of gratitude to Engel, maybe partly out of bad conscience on the part of the organisers.[55]

Engel had also — together with Wilhelm Lorey — proposed to lay a wreath at the Abel monument in Oslo (see Figure 2.5 on p. 39). The proposal had been communicated to E. Schmidt by Lorey.[56] The latter, who had close relations with Scandinavian and Norwegian insurance mathematics,[57] had originally not applied for official participation in Oslo, maybe because he had never held a position at a German university. But the Norwegian insurance mathematician, Birger Meidell, who was involved in the preparation of the congress, asked Lorey, in April 1936, to give a talk on standardised terminology in mathematical statistics and offered to accommodate him at his home in Oslo.[58] Lorey finally applied to the ministry in June: the ministry replied that there were no objections to him travelling to Oslo, but he could not be an official delegate due to his late application.[59] Lorey,

wie ich. Ich habe daher wohl einen Anspruch, zu Worte zu kommen." (WLP, Engel to Lietzmann, 30 April 1936)

[53]In a letter to Lietzmann after the congress, Hasse quoted indirectly from a letter from Engel, saying the following: "I was not in Oslo after all because my wife was sick. I would have travelled alone, but the executive committee of the congress had so annoyed me that I gave up the whole thing. How much it annoyed me you can infer from the fact that I already had foreign exchange approval. Moreover, Lie's son had provided me with foreign currency!" ("Ich war schliesslich doch nicht in Oslo, da meine Frau krank wurde, wäre ich [*sic*] auch allein gereist, aber das Exekutivkomitee des Kongresses hat mich so geärgert, dass ich auf die ganze Sache verzichtete. Wie sehr es mich geärgert hat, können Sie daraus entnehmen, dass ich schon die Devisengenehmigung hatte. Ausserdem hatte mir der Sohn von Lie Devisen zur Verfügung gestellt!" — Hasse Papers: 1: 996, Hasse to Lietzmann, 18 August 1936.) See also the letter that Heegaard sent to Engel after the congress, reproduced in our Appendix B.

[54]It should be stressed that Engel, not Heegaard and not Cartan, was the real expert on Lie's mathematics. The fact that Cartan had developed a major theory inspired by Lie's ideas (**Hawkins, 2000**) hardly means that he had a deep knowledge of Lie's work. Engel obviously did, if only Lie's more mature ideas. The earlier work was totally foreign to him, so he began work on the edition of Lie's papers in wartime with volumes 3–6. Heegard was brought in much later, perhaps after 1930, because Engel needed a geometer who could try to understand Lie's earliest works: those written before his group-theoretic programme took hold. Heegaard never really completed the commentaries on this early work, which he struggled with mightily. (We thank David Rowe, who kindly supplied us with this commentary.)

[55]See (**ICM, 1936**, vol. I, p. 46); see also Cartan's subsequent letter to Engel, Dolomieu-Isère, 13 August 1936 (EN: NE180451).

[56]WLP, Lorey to Erhard Schmidt, 30 June 1936 (copy).

[57]That inspiration for German statistics and insurance mathematics had come from Scandinavian and English colleagues was recognised by Lietzmann in his official report (**Lietzmann, 1936**, p. 302).

[58]WLP, Meidell to Lorey, Oslo, 16 April 1936 (copy).

[59]WLP, Lorey to REM, 11 June 1936 (copy), and WLP, REM to Lorey, 27 June 1936 (copy).

who already felt particular pressure on him in Nazi Germany for other reasons,[60] evidently chose not to go to Oslo under these circumstances. In his report on the congress, Dunnington made his only critical remark in the direction of the German delegation with the following words:

> Section IV [Calculus of Probabilities, Mathematical Statistics, Insurance Mathematics and Econometrics] was deprived of a valuable paper "On Unified Nomenclature in Mathematical Statistics" by Prof. Wilhelm Lorey (Frankfort a.M.), well known historian of mathematics, who was prevented from attending.[61]

On 26 June 1936, the Foreign Office of the German Reich sent what was supposed to be the official and final list of 30 German delegates to the Norwegian Embassy in Berlin (Figure 3.3).[62] The following nine mathematicians on this list would ultimately *not* show up in Oslo, for various, not in all cases verifiable, reasons: Vahlen, Feigl, Engel, Artin, Kowalewski, Reidemeister, Lettenmeyer, Salkowski, Graf. Because Behnke managed to reappear on the list, the official delegation finally consisted of 22 men.

Among the nine on the list of 26 June who did not go to Oslo, the last-mentioned three, Lettenmeyer, Salkowski and Graf, were not first-class mathematicians but persons who were reliable in the sense of the regime. Lettenmeyer even wrote to Lietzmann that he was surprised to be nominated even though he had not applied, but he felt unable to come, because — due to the late invitation — he had no talk.[63] Salkowski wrote to Lietzmann that the late decision in his favour prevented him from booking himself onto the ship to Norway, which was the only way to travel with domestic German currency.[64]

Arbitrary decisions, which may also have been taken temporarily in the case of Behnke and which were usually explained by the currency situation, led to late cancellations of permissions, even after the list of 26 June had been issued:[65] Volk, Vogel, and Haenzel were crossed off the list at the last minute. However, the latter two had already departed for Oslo and had therefore to be accepted there.[66] Following his temporary removal from the list, Behnke had protested against this by pointing to his many international connections,[67] and was reinstated just four

[60]Namely for his sexual orientation: see (**Siegmund-Schultze, 2009**, p. 25).

[61](**Dunnington, 1936a**, p. 91). Moreover, Lorey's paper was not printed in the proceedings.

[62]KUD. The same list had been sent to Lietzmann on 20 June 1936, according to WLP.

[63]Lettenmeyer to Lietzmann, 2 July 1936 (WLP).

[64]Salkowski to Lietzmann, 28 June 1936 (WLP).

[65]Erich Bessel-Hagen had already realised that he was no longer on the list when he wrote a frustrated letter to Lietzmann on 26 June 1936 (WLP). Three days earlier, he still had hope, as expressed in a letter to Courant, dated Bonn, 23 June: "I still do not know whether I can come to Oslo. ... Lietzmann ... has just announced that in view of the large number of congresses taking place over the summer, attendance is likely to be even more restricted. I must therefore be patient and continue to wait." (RCP, Series I, box 2, folder 3: "Ob ich nach Oslo kommen kann, weiss ich bis heute noch nicht. ... Lietzmann ... hat gerade mitgeteilt, dass in Anbetracht der grossen Zahl der im Sommer stattfindenden Congresse die Teilnehmerzahl wahrscheinlich noch stärker beschränkt werden wird. Ich muss mich also gedulden und weiter abwarten.")

[66]REM 2905, fol. 268 (undated). Volk, who had not yet left, protested in vain against this decision in a letter to Lietzmann, dated 10 July (WLP). In his recollections of 1942, Lietzmann mentions one mathematician who successfully intervened in Berlin and later came to Oslo by plane (**Lietzmann, 1960**, p. 60). It is unclear who is meant, but it cannot have been Behnke, who came by boat (**Behnke, 1978**, p. 248).

[67]Behnke to Lietzmann, 23 June 1936 (WLP).

Das Auswärtige Amt beehrt sich, der Königlich Norwegischen Gesandtschaft auf die Verbalnote vom 17. März 1936 — Nr. 35 — und im Anschluß an seine Verbalnote vom 6. April 1936 — VI W 3090 — mitzuteilen, daß die Deutsche Regierung der liebenswürdigen Einladung zur Teilnahme an dem vom 13. bis 18. Juli 1936 in Oslo stattfindenden Internationalen Mathematiker-Kongreß gern Folge leisten wird. Die Deutsche Regierung wird auf dem Kongreß durch folgende Herren vertreten werden:

1)	Professor Dr. Lietzmann, Universität Göttingen, Führer der Delegation,		
2)	Ministerialdirektor Professor Dr. Vahlen, Berlin, Reichserziehungsministerium,		
3)	Professor Dr. Hasse	Universität	Göttingen,
4)	Professor Dr. Hamel	Technische Hochschule	Berlin,
5)	Professor Dr. Feigl	Universität	Breslau,
6)	Professor Dr. Hecke	Universität	Hamburg,
7)	Professor Dr. Blaschke	“	“
8)	Professor Dr. Engel	“	Gießen
9)	Professor Dr. Geppert	“	“
10)	Dr. Artin	Universität	Hamburg
11)	Dr. Haenzel	Techn. Hochschule	Karlsruhe
12)	Dr. Erhard Schmidt	Universität	Berlin
13)	Dr. Kowalewsky	Rektor der Technischen Hochschule Dresden	
14)	Dozent Dr. Köthe	Universität	Münster
15)	Dozent Dr. Ullrich	Universität	Göttingen
16)	Dr.habil. E. Peschl	Universität	Jena
17)	Professor Dr. Reidemeister	Universität	Marburg
18)	Professor Dr. Siegel	“	Frankfurt a.M.
19)	Dr. Riebesell	“	Berlin
20)	Dr. Sperner	“	Königsberg
21)	Dr. P. Koebe	“	Leipzig
22)	Dr. Reinhardt	“	Greifswald
23)	Dozent Dr. Seiffert	“	Heidelberg
24)	Dr. H. Grunsky in Berlin-Charlottenburg, Kaiser Friedrichstr. 59, Schriftleitung des Jahrbuches über die Fortschritte der Mathematik der Preußischen Akademie der Wissenschaften		
25)	Professor Dr. Caratheodory	“	München
26)	Professor Dr. Lettenmeyer	“	München
27)	Dozent Dr. Vogel	“	München
28)	Professor Dr. Timpe	Techn. Hochschule Berlin	
29)	Dr. Salkowski	“ “ “	
30)	Dozent Dr. Graf	Universität	Berlin

Berlin, den 26. Juni 1936.

FIGURE 3.3. Transcription of the official and final list of German delegates, sent by the Foreign Ministry of the German Reich to the Norwegian Embassy in Berlin, 26 June 1936 (KUD)

days prior to the congress, apparently because the Norwegian organisers took over his hotel costs.[68] Several mathematicians withdrew their applications in letters

[68] REM, 2905, fol. 244, 9 July 1936. Here the ministry officially granted participation on the grounds that Behnke was supposed to give a talk. See also Behnke's letter of thanks to Størmer after the congress, dated 19 July 1936. In this letter Behnke mentioned the “humiliating” material conditions (“entwürdigende Bedingungen”) for Germans in Oslo and thanked Størmer for material support (Størmer Brevarkiv, box 10, Nasjonalbibliotek Oslo).

to Lietzmann, mostly without giving reasons.[69] In several of these cases, political motives may have been involved.

3.2. German attendance and Lietzmann's reports

The following 22 mathematicians were finally listed as official members of the German delegation when Lietzmann sent his report to the ministry after the congress: Behnke, Blaschke, Carathéodory, Harald Geppert, Helmut Grunsky, Haenzel, Hamel, Hasse, Hecke, Paul Koebe, Köthe, Lietzmann, Ernst Peschl, Karl Reinhardt, Paul Riebesell, E. Schmidt, Herbert Seifert, C. L. Siegel, Sperner, Aloys Timpe, Egon Ullrich, and Kurt Vogel. Among the 22 official German delegates there were two applied mathematicians (Riebesell, Timpe), one didactician (Lietzmann), and one historian of mathematics (Vogel), while the clear majority — in tune with the topics of the congress — represented purer areas of mathematical research.

Carathéodory and Siegel[70] had obviously been added to the original list of March 1936 because they were among the internationally most-established mathematicians, and the latter had been invited to give a plenary talk. Carathéodory, who, as late as June, had not intended to go to Oslo,[71] would eventually represent Germany at the congress in various capacities, including his laudatory speech for the two Fields Medal laureates Lars Ahlfors and Jesse Douglas (**Carathéodory, 1936/7**). With Hasse, Hecke, and Siegel speaking, the German delegation was responsible for three out of a total of 19 plenary mathematical talks, which means that they were almost on a par with the Americans (Birkhoff, Ore, Veblen, and Wiener).

Probably in order to minimise the danger of politically precarious encounters with foreign mathematicians and, in particular, German expatriates, there were no official Nazi functionaries or outspoken Nazi propagandists among the German participants in Oslo.[72] Bieberbach, the creator of the racist theories of the "Deutsche Mathematik", who had for some years been in a quarrel with several members of the DMV, and had the most to fear from confrontations with émigrés in Oslo (to whose dismissal he had at least indirectly contributed), apparently had preferred not to apply. His companion Vahlen in the mathematical journal with racist overtones, *Deutsche Mathematik*, stepped back from participation one week before the congress.[73] Neither Bieberbach's name, nor those of two other exposed Nazis Erhard

[69]The letters of withdrawal are in WLP: Doetsch, 28 March; Haupt, 5 April; F. K. Schmidt, 15 May; Artin, 6 July; Feigl, 7 July; Vahlen, by telegram, 6 July; Engel, 8 July.

[70]Siegel only indicated his interest as late as 22 June 1936 (in a letter to Lietzmann), by which stage he had probably already been on the list of general speakers for some time (WLP). He then appeared on the official list of 26 June.

[71]This is according to a letter to Erich Hecke, dated 7 June 1936, in which Carathéodory wrote: "Unfortunately, I will not go to Oslo. ... In the coming weeks I must write a report on the two winners of the Fields Medals that are to be awarded in Oslo, a not very rewarding task." (Hecke Papers: "Nach Oslo werde ich leider nicht hingehen [*sic*]. ... Ich muss in den nächsten Wochen den Bericht für die beiden Preisträger der Fieldsmedaillen schreiben, die in Oslo vergeben werden, eine nicht sehr dankbare Aufgabe.")

[72]Behnke later wrote: "There is no one among the lecturers from 'Die Deutsche Mathematik'." (Translated from the German in (**Behnke, 1978**, p. 249).)

[73]Telegram from Vahlen to Lietzmann, 6 July 1936: "I cannot go to Oslo." ("Ich kann nicht nach Oslo fahren. Vahlen" — WLP)

Tornier and Oswald Teichmüller, the first an able the second an excellent mathematician, show up at all in the various files kept in preparation for the congress.[74]

Nevertheless, and as a result of the prior political steering of the arrangements by REM, the DKZ and the local "Dozentenschaften", Lietzmann could more or less guarantee the appropriate (in the Nazi sense) political behaviour of the German delegates. On 7 July 1936, a week before the congress, Lietzmann sent a letter to all members of the delegation, containing an excerpt from the DKZ guidelines for the leaders of delegations (DKZ 1).[75] Lietzmann urged his delegation to attend at least all the plenary talks of German mathematicians, and as many in the sections as possible. During votes he urged them to act "unanimously"[76] and he added:

> Of course all members use the German language, also in discussions.[77]

Immediately after the congress, Lietzmann sent a 15-page report to the ministry (WLR 1936). It consisted of two parts, a more political one (pages 1–5 = fols. 384–388) and a more scientific one (pages 6–15 = fols. 389–398); the entire document is reproduced in Appendix D, both in transcription and in English translation. Lietzmann had asked several members of the delegation to prepare drafts for the special sections[78] to be included in the general report. He sent the entire report to REM, but to the DKZ he sent only the first part, in which he wrote in detail about the political strategies which he and his delegation had observed (WLR 1936). Nevertheless, one also finds political remarks in the second part of the report, most of which (though not all)[79] were probably added by Lietzmann himself.

In the first, purely political, part of the report, Lietzmann carefully drew a variety of lines between the members of the delegation — of whom the majority, but not all (e.g., Köthe, as mentioned above), had received support from the DKZ in the form of Norwegian crowns — and the various other German mathematicians whom he called merely "German participants in the congress". As someone who, along with all of his compatriots, had been accustomed for several years to the infamous Nazi policies that forced Germans to prove their "Aryan descent", Lietzmann was of course used to drawing such lines. There were, first of all, the two German mathematicians who had been "named by the Minister as participants in the congress",[80] namely Threlfall and Rosenthal, who, as we have seen (p. 71), had not received foreign currency, and were in fact unwelcome to figure among the official delegation. But it is interesting that at a time when the recent Nuremberg laws of

[74] WLP and REM 2905. Encounters between Tornier with his former friend Willy Feller, now living in Sweden, would have been embarrassing indeed.

[75] REM, 2905, fols. 251–2, Lietzmann to "Herren Kollegen", 7 July 1936.

[76] *Ibid.* The German original "gemeinsam zu stimmen" is somewhat ambiguous because it could also mean to "vote together".

[77] "Selbstverständlich bedienen sich alle Mitglieder, auch in den Diskussionen, der deutschen Sprache." (*Ibid.*, fol. 251)

[78] According to WLP: Köthe for the general talks, Hasse for algebra and number theory, Ullrich for analysis, and Timpe for applied mathematics. The reports themselves are not preserved in the Lietzmann Papers in Göttingen. Köthe's report must have been very cursory and general: see Köthe's letter to Lietzmann, dated Graz, 28 July 1936 in WLP. Hasse's very short report survives in a copy from 14 August 1936 in the Hasse Papers (1: 996).

[79] Lietzmann quotes verbatim the remarks from Hasse's short report (Hasse Papers: 1: 996), including the positive comments about Weyl and Banach (p. 83). The latter remark had clear political overtones, not conforming to official Nazi opinions.

[80] WLR 1936, fol. 384: see the remarks above and Appendix D, p. 271.

September 1935 had introduced a difference between "Reichsbürger" (German citizens) and mere "Staatsangehöriger" (members of the Reich, meaning Jews, among others), Lietzmann still dared to call Rosenthal a "German mathematician". At any rate, all 22 + 2 mathematicians who were mentioned in the report were connected to the German university system. Another line was drawn between these 24 and those German mathematicians attending Oslo who had no current university affiliation. Lietzmann mentioned first the statistician Karl Boehm of the Institut für Konjunkturforschung Berlin (Institute for Business Cycle Research), who had applied for official participation in a letter to Lietzmann in March 1936,[81] and who now in Oslo, according to Lietzmann, "asked for a connection to the delegation".[82] Three more Germans without official attachment to the German university system, who were also in Oslo, but who apparently did not "ask for a connection", were mentioned in Lietzmann's report also. These were the representative of the Teubner publishers in Leipzig, Dr. Ing. Herbert Heisig,[83] the Frankfurt topologist, Max Dehn, and the young applied mathematician from Berlin, Erich Reissner, the latter two being of Jewish descent. Dehn had been ousted from his professorship in 1935 even before the Nuremberg laws would have forced him to go, with the rationale that his chair was superfluous "for the sake of a streamlined administration" ("Vereinfachung der Verwaltung").[84] Most remarkably, in both his published and unpublished reports, Lietzmann mentioned Reissner's talk in the section (VI) "Rational and Applied Mechanics", together with one by a member of the delegation, Georg Hamel. In the published report he even referred only to these talks, without mentioning others (**Lietzmann, 1936**, p. 302). Apparently, Lietzmann took a special interest in the now-threatened career of the son of the well-known applied mathematician Hans Reissner (1874–1967); later in the decade, father and son would emigrate to the United States.[85]

While Dehn and Reissner were then still living in Germany, Lietzmann felt compelled to draw an even clearer line between "German mathematicians" and "emigrants from Germany", of whom the majority were of course "non-Aryan" in the Nazi sense:

> The "emigrants" at the congress were, as far as I know, the professors [Richard] Courant–New York, [Hermann] Weyl–Princeton, [Otto] Neugebauer–Copenhagen, [Fritz] Noether–Tomsk, [Adolf Abraham] Fränkel–Jerusalem, and finally also the infamous Herr [Emil Julius] Gumbel–Lyon. There were no disturbances or unwanted discussions caused by these gentlemen.[86]

---

[81] WLP, Boehm to Lietzmann, 30 March 1936.

[82] WLR 1936, fol. 384: see Appendix D, p. 271.

[83] Heisig would negotiate in Oslo with Hasse, Hecke, and other prospective collaborators in the second edition of the *Enzyklopädie der Mathematik*: see Hecke Papers Hamburg.

[84] There was a special §6 in the respective Nazi law of 7 April 1933 to do this.

[85] On 29 September 1936, REM asked Lietzmann whether the Reissner at the congress was identical with Professor Hans Reissner at the Technical University in Berlin (WLP).

[86] WLR 1936, fol. 385: see Appendix D, p. 271. Gumbel was like a red rag to a bull for the German officials and was therefore mentioned no less than three times in Lietzmann's report. There is a remark in Lietzmann's report about section IV (Calculus of Probabilities, Mathematical Statistics, Insurance Mathematics and Econometrics) of the congress, in which Gumbel spoke: Lietzmann says that the German delegates "reacted passively" ("verhielten sich passiv") to the emigrant in the section. On Gumbel, more below, in particular in Appendix A.7 (see also the commentary in Chapter 6).

Lietzmann dared to give favourable mention to at least one of the two non-Jewish delegates among these emigrants — he was again quoting indirectly from Hasse's partial report but obviously taking responsibility for it — when he discussed the work of the section (I) "Algebra and Number Theory":

> Among a total of 27 lectures, only the somewhat longer one by H. Weyl–Princeton was of general scientific interest and an outstanding achievement.[87]

Looking at the various and scattered historical documents for the congress one finds at least the following 13 additional German expatriates who took part: (probably) Felix Behrend, (probably) Herbert Busemann, Willy Feller, Werner Fenchel, (probably) Hans Heilbronn, Kurt Hirsch, Boris Kaufmann, Paul Nemenyi, Bernhard Neumann, Richard Rado, Wolfgang Sternberg, (probably) Otto Szász, and Alexander Weinstein. All of these are listed as "members of the congress" (**ICM, 1936**, vol. I, pp. 29–38), and all but Behrend, Busemann, Heilbronn, and Szász also have printed contributions in volume II of the proceedings.

Finally, it is worth noting that there were other mathematicians attending in Oslo, such as Franz Alt and Eugene Lukács from Austria, and Juliusz Schauder from Poland, who would be persecuted by the Germans in the years to come.[88]

Lietzmann reported on the various political contacts he had made in Oslo. In accordance with the DKZ guidelines for leaders of German delegations (DKZ 2, p. 14), Lietzmann had immediately visited the German Embassy in Oslo, and, indeed, he could proudly report over-fulfilment of the requirements, for he had also contacted the "district leader of the NSDAP, comrade Kern" and went on an excursion with him one day after the congress.[89] Lietzmann recommended in his report that in the future the DKZ should expressly mention such contact with the local Nazi Party as advisable in its guidelines.[90]

Two symbolic political acts that were undertaken by the German delegation in addition to the programme of the congress stood out in importance: the laying of wreaths at Lie's grave and at Abel's monument (see Sections 2.9.2 and 2.9.3). Both events were discussed with the German Embassy in advance.[91]

The Abel monument (Figure 2.5 on p. 39) stands in the centre of town, near the Royal Castle, and was therefore, as we have noted, very visible — even more so than Lie's tombstone at the *Vår Frelsers Gravlund* cemetery not far away. Laying a wreath with a swastika at Abel's monument was therefore bound to attract public attention, and also possible protest. Indeed, the newspaper *Aftenposten* reported on it on 15 July, with a picture of the wreath (Figure 3.4).

Lietzmann had secured permission from the German Embassy for both ceremonies. Some concerns, particularly about the Abel ceremony, were dismissed by

---

[87]WLR 1936, fol. 391: see Appendix D, p. 274. This remark is related to Weyl's talk on Riemannian matrices and factorial systems, an abstract of which is in the congress proceedings (**ICM, 1936**, vol. 2, p. 3). The remark was originally in Hasse's letter to Lietzmann, 14 August 1936; copy in Hasse Papers (1: 996, attachment).

[88]See, respectively, (**Sigmund *et al.*, 2011**), (**Rohatgi, 1988**), and (**Forster, 2008**), as well as (**Duda, 2014**).

[89]WLR 1936, fol. 385: see Appendix D, p. 272. Kern used the information that he had apparently received from Lietzmann, to write a report (containing many misunderstandings) to his superiors in Berlin; see Appendix E. On Kern, see also Appendix F.

[90]WLR 1936, fol. 385: see Appendix D, p. 272.

[91]*Ibid.*

Tyskerne bekranser Abel-støtten.

Ved en enkel høitidelighet igår morges, før matematikerkongressens åpningsmøte fant sted, nedla lederen for den tyske matematikerdelegasjon, professor Lietzmann, en krans av hvite liljer og med et trefarvet bånd i de tyske farver ved Abelstøtten.

FIGURE 3.4. German wreath on the Abel monument in Oslo (Source: *Aftenposten*, 15 July 1936, p. 3)

the ambassador, the former mayor of Berlin, Heinrich Sahm, who said it was up to the Norwegian police to avoid possible trouble. This is how Lietzmann described one of the ceremonies in his report to the ministry:

> The wreath at the Abel monument was laid by the leader of the delegation, together with the Chairman of the Deutsche Mathematikervereinigung [Erhard Schmidt]. Police were present and a policeman was stationed near the monument throughout the congress. The wreath was reproduced in a daily.[92]

While the unveiling of the bust of Lie had been part of the official programme of the congress, involving a speech by Cartan (p. 73), the ceremony at Lie's tombstone was of a more internal, but no less political, nature. As we noted in Section 2.9.3,

[92] *Ibid.*

Lietzmann was able to report to his superiors that members of the German delegation had been joined at the graveside by Lie's daughters, and that all present had raised their arms in the Nazi salute.

At the non-mathematical events of the congress itself, German mathematicians also had occasionally to represent and thus to defend the cause of the Nazi government: we have already noted Erhard Schmidt's somewhat naive remark about the impending Berlin Olympics (Section 2.9.4). Open criticism of the methods of Nazi rule in Germany was, of course, out of the question, as the nervous remark by Toeplitz in his letter to Courant reveals (p. 68).

But one can safely assume that several mathematicians even in this politically pre-selected German delegation did not approve of Nazi policies at least so far as they had led to the expulsions of so many able mathematicians whom they now met again for the first time at the congress.[93] Indeed, Behnke, who was always very cautious in his political behaviour in Nazi Germany, not least out of concern for his half-Jewish son, later recounted, with hindsight, encounters with Jewish mathematicians such as Georg Pólya of Zurich:

> We both felt: This is going to be the last gathering for a long time of such a big group of mathematicians who have come together only out of scientific interests. And when the last word from the lectern had faded away the anxious question was buzzing inaudibly through the lecture hall: which catastrophes will the coming years bring?[94]

Lietzmann was of course aware that the members of his delegation would have various contacts with Jewish mathematicians, among them many émigrés from Germany, and that most foreign mathematicians were opposed to Nazi doctrines. Even his published report conveys some of the nervousness that he had had about this before the congress: recall his remarks on p. 56 concerning the "excitement" and "scepticism" surrounding the appearance of the Germans. Lietzmann went on to note in his internal report to the ministry a remark that could even be understood as a slight criticism of ideological interference in mathematics in Germany:

> There is one thing that many foreigners have real problems in understanding: our [*sic*] notion of the national peculiarity of science, regardless of the universal validity of its results.[95]

In the published report, however, which is essentially a shortened version of the internal one, Lietzmann was both more cautious and more optimistic. He rephrased

---

[93]This is true even of such conservative mathematicians with German nationalist sentiments as Schmidt. In an internal report on Schmidt to the Berlin NS organisation of docents, written by the Nazi and mathematician Werner Weber, one reads: "I think that Schmidt shows little or no understanding of the Jewish question. It was striking that a few days after the death of the Jew Landau (February 1938), Schmidt was the only one in the entire mathematical faculty who knew about it at all. He was also informed in detail about the circumstances: which disease Landau had had, when he was buried, where the various relatives stayed .... Where he had gathered all that information from (contact with I. Schur??), I do not know." (**Siegmund-Schultze, 2009**, p. 86)

[94]Translated from the German in (**Behnke, 1978**, p. 249).

[95]WLR 1936, fol. 387: see Appendix D, p. 272. After the congress (around 15 August 1936), the office of the Nazi Party in Oslo reported to their leadership in Berlin about a Dutch mathematician (J. A. Schouten?) who had stood up openly against Nazi doctrine — see Appendix E.

the wording "national peculiarity of science" ("nationale Eigenart einer Wissenschaft"), with its clear connotation of the racist theories then *en vogue* with Bieberbach and other Nazis, as "science as a national, not just international, property" ("Wissenschaft als ein nationales, nicht nur internationales Gut"). He alleged that this opinion was not yet shared by all foreigners and expressed his hope that at the next congress "understanding for our position will hopefully be a general one".[96]

It is very interesting to see how far Lietzmann believed he could go with remarks made in public. In both the published and unpublished reports, he stated in connection with Behnke's talk on multidimensional complex function theory (a talk that had almost failed to take place at all: see p. 74):

> This is a topic that has been cultivated for a decade in an exemplary manner in collaboration between German and French mathematicians.[97]

The version of this remark in the internal report is almost identical except for the additional embellishment "friendly" ("freundschaftlich").[98]

Passages that contained unrestricted pride in German mathematical accomplishments could, of course, find a place in both the published and unpublished reports. Thus the following passage about German number-theoretic research, alluding to the three invited lectures by Hasse, Hecke, and Siegel, and not altogether unjustified for the period of time, was contained in both reports:

> The leading position of Germany in number-theoretic research has been kept from Gauss to the present day.[99]

But as a rule, Lietzmann reserved both his more critical political remarks (slight though they were), and his acknowledgement of the accomplishments of foreign mathematicians, for the internal report, thus revealing a certain trust in a residual objectivity on the part of the Nazi-dominated ministry. Lietzmann could certainly not expect such objectivity from the German public, and there was the additional danger of a distortion of published opinions. For instance, Lietzmann admitted internally that "in Norway English is increasingly becoming dominant against German".[100] Lietzmann also reserved acknowledgement for Russian and Polish mathematicians for the internal report. Referring to the geometry section (III) of the congress, Lietzmann observed:

> As to content, topology was clearly dominant. ... the absence of the Russians caused significant gaps.[101]

When he discussed the talks given in section I (Algebra and Number Theory), Lietzmann included a rather unrelated remark about Polish mathematicians, who

[96]Translated from the German in (**Lietzmann, 1936**, p. 304).

[97]Translated from the German in (**Lietzmann, 1936**, p. 302).

[98]The internal version "friendly" is in WLR 1936, fol. 392: see Appendix D, p. 275. The friendly relations between Behnke and Henri Cartan were in fact continued even during the German occupation of France. This helped German mathematicians to regain international contacts after the Second World War: see (**Behnke, 1978**).

[99]WLR 1936, fol. 389: see Appendix D, p. 273. Also (**Lietzmann, 1936**, p. 301). Four years later, in 1940, Weyl spoke about the 'dormant number theory' in the USA that had to be awakened by immigrants from Europe, in particular from Germany: see (**Siegmund-Schultze, 2009**, p. 285).

[100]WLR 1936, fol. 394: see Appendix D, p. 276.

[101]WLR 1936, fols. 393–394: see Appendix D, p. 276.

were more strongly represented in the analysis section. Lietzmann quoted this remark from Hasse's contribution to the report, without mentioning Hasse's name:

> At a social gathering it was stressed [Hasse had not made it clear by whom] with regard to the attending Polish mathematician S. Banach–Lemberg, how admirable it was that Poland had developed such a broad and strong mathematical school of peculiar orientation in a very short period of time.[102]

Earlier in his report, Lietzmann had stressed that Banach — who, incidentally, was deemed "acceptable" by Nazi racial standards — had given his talk on linear operators in German; this was noteworthy, as his influential book on this subject (**Banach, 1932**) had appeared in French, and the congress talk had previously been announced in that language in the printed programme (**Anon, 1936**, p. 378).[103] Also, the following remark by Lietzmann had some political overtones, when he implicitly criticised the state of official contacts with the Poles, and could not avoid referring to the dogma of anti-Semitism that determined many German policies:

> In a similar way — it seems to me — we should improve our contacts with Poland, which is still largely oriented towards France in the scientific respect. However, certain problems may arise, since many Polish mathematicians seem to be non-Aryans.[104]

The internal report contains further "strategic" remarks that showed Lietzmann to be in line with the general cultural aims of the Nazi ministry of maintaining and expanding international influence and which, of course, were not in his published report. Probably partly in an effort to balance his previous remark about "friendly" German-French relations, Lietzmann said in connection with the talk by the analyst Jovan Karamata from Belgrade:

> It would be very desirable indeed to cultivate close scientific relations with the southeast-European countries. It is remarkable that the French — obviously quite deliberately — are making "scientific cultural propaganda" towards the Balkans. The French give regular lectures in that region and visit congresses in the Balkans. On the German side, we have neglected this so far (except for a trip by Herr Blaschke–Hamburg). Perhaps they consider us too proud to go there. [In Oslo] we have tried hard to come into contact with mathematicians who are no enemies of Germany, particularly with younger ones. In this regard Herr Karamata seems important. ... Above all, he is by far the best among the younger "South-East mathematicians". Perhaps it would be possible on occasion to invite him for a lecture tour of German universities.[105]

What is particularly interesting, in its implicit claim of social and political similarities (in addition to the usual claim of kindred blood) between Germany and

[102]WLR 1936, fol. 391: see Appendix D, p. 274. Originally in Hasse's letter to Lietzmann, 14 August 1936; copy in Hasse Papers (1: 996, attachment). On Banach and his place in the development of mathematics in Lwów (Lemberg), see (**Duda, 2014**).

[103]Of course Banach's motive could just have been to amplify the effect of his book by presenting parts of it in a language other than Polish or French.

[104]WLR 1936, fol. 393: see Appendix D, p. 275.

[105]WLR 1936, fol. 392–393: see Appendix D, p. 275.

Norway, is Lietzmann's discussion of the work of section IV: Calculus of Probabilities, Mathematical Statistics, Insurance Mathematics and Econometrics. In both reports, Lietzmann says with respect to the last two (more applied) topics:

> The currently increased scientific activity in Germany in the field of mathematical economy and business cycle analysis is observed, as it turned out, with much sympathy by our Norwegian colleagues.[106]

But only in the internal report does one find information about social contacts between the Norwegians and the Germans who shared an interest in mathematical economics, and there is also some criticism of a comparative German backwardness in the theoretical aspects of the field:

> Together with a small circle of section participants, the members of the German delegation [probably primarily the German specialists presenting in section IV, namely Riebesell, Boehm, and Timpe] were one evening the guests of the Oslo colleagues Ragnar Frisch and Meidell (who, incidentally had not invited the emigrant Gumbel, out of consideration for us). In addition we had the opportunity to visit their university institutes. Mathematical economics has a much larger space in these institutes than in German economic institutes. In the Institute for Economics, which is headed jointly by the representative of mathematical economics and a non-mathematical political economist, 16 assistants are employed, predominantly busy with the analysis of economical statistical data.[107]

While Lietzmann had received help in writing the mathematical parts of the report, he probably wrote the two pages on the historical and pedagogical sections of the congress by himself: in particular a short report on the meeting of ICMI. As a general judgement, we find in the latter the following remark on secondary schools in several countries:

> The situation of mathematics in secondary schools was repeatedly described as not very auspicious.[108]

It was decided to continue the work of ICMI, and the situation of school exams was described as one of the next topics to be discussed. Lietzmann's own report to ICMI in Oslo was published as (**Lietzmann, 1937**) and contains many examples of the changes in school mathematics due to Nazi rule.[109] Among these are the foundation of special National Political Schools (Napola), the priority of the "people over the individual" ("Bedürfnisse des Volksganzen") and a stronger emphasis on biology with its implications for mathematics teaching. (**Lietzmann, 1937**) also mentions several recent German books (by A. Dorner and B. Kerst) on school mathematics,

---

[106] WLR 1936, fol. 395: see Appendix D, p. 276. See also (**Lietzmann, 1936**, p. 302). The unpublished version has the additional phrase: "Collaboration with German journals in the field has been promised."

[107] WLR 1936, fol. 395: see Appendix D, p. 276.

[108] WLR 1936, fol. 397: see Appendix D, p. 278. We refer, for instance, to K. Piene's report on the situation of school mathematics in Norway, which speaks about a strong sentiment against the teaching of mathematics (**Piene, 1937**). Lietzmann's remarks on ICMI, and this quotation in particular, will be revisited in Chapter 10.

[109] See the remarks in note 35 on p. 16.

without confronting the reader with some of the blatant examples of racism and militaristic thinking contained in them.

Finally, it was a must for the leader of a German delegation participating in an International Congress of Mathematicians to report on the state of affairs with regard to the IMU. The boycott policies of the IRC (with its president, the French mathematician Émile Picard) had already developed into a major political focus of German science policy in the Weimar Republic, and was not restricted to mathematics. While international relations in mathematics had been much normalised since the 1928 Bologna ICM (**Siegmund-Schultze, 2016a**), in the 1930s, residual resentment among German scientists against the policies of scientific boycott was used by nationalist politicians to produce a certain identity of interest between them and the regime, and to contribute to the general policies of political revenge. Even during the Second World War, that old conflict was used demagogically in culturally expansionist policies against occupied countries and their science systems.[110] So it does not come as a surprise that in his internal report (but not in the published one), Lietzmann devoted half a page to this problem — in particular to the relation between the IMU and the congresses. He referred to the closing meeting of the congress on 18 July 1936 and declared categorically:

> ... at the Oslo congress, the mathematical "Union", that remnant of the after-effects for science of the Versailles Treaty, was finally laid to rest.[111]

This judgement was not totally unjustified, although the protocol in the proceedings reveals a divided opinion among the delegates, who postponed the decision about a future reactivation of the IMU.[112]

3.3. Further consequences of the congress for the German delegates

Just three days after the congress ended, the German Embassy in Oslo sent a report to Berlin (probably to the Foreign Office), of which a copy also reached REM. Here one finds remarks on the German expatriates and on their political "behaviour":

> A number of "emigrant"-professors participated in the congress, e.g., Prof. Neugebauer of Copenhagen and Prof. Nöther who lives in Omsk [*sic*: should in fact be Tomsk]. These two emigrants gave interviews to a journalist from the *Arbeiderbladct*, but without attacks against Germany. However, the infamous Prof. Gumbel has allegedly talked to Norwegians about Germany and German science repeatedly in an outrageous manner.

[110]For the case of mathematics, see (**Siegmund-Schultze, 2002**); see also the remarks in Chapter 7 about the involvement of German mathematicians in strategies for "restructuring European science". In these plans, contacts with Norwegian mathematicians such as Poul Heegaard and Birger Meidell, who were considered to be "friendly towards the German cause" ("deutschfreundlich"), and who had been among the organisers of the Oslo congress, played a certain role. On Meidell, who was in the Norwegian government under the German occupation, see: *Norsk Biografisk Leksikon*, `https://nbl.snl.no/Birger_Meidell` (accessed 28 August 2018).

[111]WLR 1936, fol. 387: see Appendix D, p. 273.

[112](**ICM, 1936**, vol. I, pp. 46–47). The remaining doubt about the continued existence of the IMU was nevertheless used by Nazi politicians during the war against France, which was allegedly still backing the Union and continuing its policies at the Institut Henri Poincaré in Paris: see (**Siegmund-Schultze, 1993**) and the remarks in Chapter 7.

> The Danish mathematician, Prof. Harald Bohr, claimed in an interview that America is happy to have received quite a few outstanding German emigrants in its universities; just as Göttingen was once the mathematical centre, this is now Princeton University [*sic*: Bohr talked about Princeton as a whole, including the important Institute for Advanced Study], to which the younger generation of mathematicians makes its pilgrimage. The many registered Russian mathematicians cancelled their appearance at the last hour.[113]

Gumbel was one of the few liberal and leftist mathematicians who had already protested against German nationalism and secret rearmament during the Weimar Republic. He had therefore been ousted from Heidelberg in 1932 even before Hitler came to power, and had since been stripped of his German citizenship. Support for him by mathematicians in the Weimar Republic was held against the same mathematicians under the Nazis. In Oslo he tried — apparently without much success — to muster coauthors for his planned volume *Free Science: a collection from the German emigration*, which finally appeared in German in Strasbourg (**Gumbel, 1938**). Gumbel had tried in vain to win Einstein and Weyl as collaborators, and Courant also refused.[114] There was not much enthusiasm among emigrants to speak out publicly against Nazi Germany, and even Einstein, who was generally less hesitant in public, declined to participate in the book.

Another political report was sent from the office of the Nazi Party in Oslo[115] to the headquarters of the NSDAP's "Foreign Organisation" ("Der Auslands-Organisation der NSDAP") in Berlin. The headquarters referred to that report indirectly and approvingly in a letter to the DKZ of 15 August 1936. In addition to some erroneous information, the report also contains criticism of the restricted supply of foreign currency for scientists, and compares this with the relatively lavish sums spent on German youth organisations touring Norway. The report favours support for scientific delegations "whose importance for our cause—provided a right selection is taken in Germany—cannot be doubted".[116] The DKZ, which was responsible for the allocation of foreign currency, sent this information on to several

[113]"Am Kongress nahmen auch eine Anzahl von 'Emigranten'-Professoren teil, so u.a. der in Kopenhagen tätige Prof. Neugebauer und der in Omsk lebende Prof. Nöther. Diese beiden Emigranten haben einem Mitarbeiter von 'Arbeiderbladet' Interviews gewährt, ohne jedoch in diesem Ausfälle gegen Deutschland zu richten. Dagegen soll der berüchtigte Prof. Gumbel in Gesprächen mit Norwegern wiederholt sich in unerhörter Weise über Deutschland und die deutsche Wissenschaft ausgelassen haben. Der dänische Mathematiker, Prof. Harald Bohr, hat in einem Interview behauptet, dass Amerika das Glück gehabt habe, eine Reihe hervorragender deutscher Emigranten in seinen Lehranstalten aufzunehmen; wie Göttingen früher das mathematische Zentrum gewesen sei, so sei dieses nun die Princeton Universität, zu der die jüngere Generation der Mathematiker wallfahre. Die zahlreichen angemeldeten russischen Mathematiker haben in letzter Stunde ihr Erscheinen abgesagt." (REM 2905, fol. 364–365, German Embassy to Foreign Office Berlin, 21 July 1936, copy) Gumbel's interview may be found in *Arbeiderbladet*, 16 July 1936, while Bohr's is in *Aftenposten*, 13 July 1936. See Chapter 6 and Appendices A.1, A.7, and A.9.

[114]Gumbel wrote to Courant after the congress on 19 September 1936, asking for a decision, and Courant finally declined collaboration on 29 September (RCP, Series I, box 4, folder 10). On the fate of Gumbel's collection, see (**Siegmund-Schultze, 2009**, pp. 176–180).

[115]By the "Ortsgruppenführer", i.e., by the Kern who is mentioned in Lietzmann's report. We reproduce this with English translation in Appendix E.

[116]See Appendix E.

political and ministerial offices in Germany.[117] It seems that the German Embassy in Norway, which had been involved in the organisation of the youth travel, felt embarrassed by the report, and the Foreign Office in Berlin reacted with a letter to the DKZ in December. The Embassy supported the youth excursions in principle but admitted that

> balancing objectively ... the reputation of German science abroad is so much more important that the first-mentioned goal has to step behind.[118]

As early as 31 August 1936, the Nazi headquarters in Munich ("Stellvertreter des Führers") had replied in a letter to the DKZ and had supported the view of their "Foreign Organisation". In this letter "highly important congresses" are clearly favoured over support for "youthful wanderers and ... ridiculous doctoral dissertations to be written on the basis of some old manuscript in some obscure library".[119] The Nazi headquarters thus confirmed that scientific communication with potential propagandistic and political value—such as the International Congress of Mathematicians in Oslo in 1936—was closer to their heart than actual scientific research.

[117] DKZ 3: Knapp (DKZ), 25 August 1936, to Foreign Office, Ministry of Propaganda, REM (the letter is available at REM 2905, folio 378), Hitler Youth Organisation (HJ), NSDAP Party Headquarters.

[118] "bei einer objektiven Abwägung ... überwiegt das Ansehen der deutschen Wissenschaft im Ausland ... derartig, dass die erstgenannte Aufgabe dahinter zurücktreten muss." (Hoover Institution, DKZ, box 203, Foreign Office to DKZ, Berlin, 3 December 1936)

[119] "höchst wichtige Kongresse" as opposed to "jugendliche Wanderer [und] ... lächerliche Doktorarbeit ... an Hand einer alten Schrift in irgend einer Bibliothek". (DKZ 3)

CHAPTER 4

The Russian Withdrawal: Isolationism out of Fear and Ideology

In Chapters 2 and 3, we have presented some quite detailed accounts of the preparation for the congress by both the Norwegian organisers and the German participants, as well as discussing their activities once the congress was underway. We were able to do this because of the wealth of sources available to us (see the comments on pp. 15–17). The present chapter, however, is necessarily much shorter: as mentioned in Chapter 1, the sources available to us with regard to the absence of the Soviet participants are considerably fewer. Moreover, there is a limit to what one can say about delegations that did not attend the congress. Nevertheless, we attempt to explain here the Soviet absence, and to place it into the wider context of mathematics in the USSR during the 1930s.

As we will discuss below, the engagement of Russian mathematicians with the international mathematical community had already been hampered by language-related obstacles for some time, which were then compounded by the political issues that followed the October Revolution. However, the political developments of the early 1930s (the takeover of power by Stalin in the Soviet Union and by Hitler in Germany) further increased the isolation of Russian mathematics in the decades to come, leading to an instance of the "isolationist" strategy described in Chapter 1. A crucial event on this path was the non-participation of Russians in the Oslo congress of 1936.

4.1. A conspicuous absence

In early spring 1936, the Norwegian organisers printed an 8-page programme for the impending congress,[1] which named Aleksandr O. Gel'fond (1906–1968; see Figure 4.1) and Aleksandr Ya. Khinchin (1894–1959; see Figure 4.2) as plenary speakers with the following topics, Gel'fond proposing a lecture in French, and Khinchin in German:

- Gel'fond: "Théorie des nombres transcendants" ("Theory of transcendental numbers");
- Khinchin: "Hauptzüge der modernen Wahrscheinlichkeitstheorie" ("Principal features of modern probability theory").

Both speakers were, of course, highly qualified for their tasks, given Gel'fond's recent solution of Hilbert's seventh problem on transcendental numbers, and Khinchin's report *Asymptotische Gesetze der Wahrscheinlichkeitsrechnung* (**Khinchin,**

[1](**Anon, 1936**, p. 378). The programme appeared in *L'enseignement mathématique*, after the death on 15 February 1936 of the original organiser of the congress, Alf Guldberg — see also footnote 3 on p. 302.

FIGURE 4.1. A. O. Gel'fond (1906–1968) (Source: *Uspekhi matematicheskikh nauk*, vol. 11 (1956))

1933) for Springer's *Ergebnisse der Mathematik und ihrer Grenzgebiete* series, together with his work on stochastic processes in general.[2]

When the congress convened in July 1936, however, E. B. Schieldrop, the congress' Norwegian Secretary General, had to report that neither Gel'fond nor Khinchin had appeared, and had expressed their apologies (**ICM, 1936**, p. 45). In the proceedings, no Russian mathematicians are mentioned in the list of official delegates (**ICM, 1936**, vol. 1, pp. 18–28).

However, in the "Liste des membres du Congrès" within the same proceedings (pp. 29–38), we find 11 mathematicians with addresses in the Soviet Union; these are (using the spelling of the proceedings):

- Akimoff, Michel (Leningrad);
- Aleksandroff, Paul (Moscou);
- Bary, Nina (Moscou);
- Bernstein, Serge (Leningrad);
- Jouravsky, André (Leningrad);
- Khintchine, A. (Saratow);
- Kolmogoroff, André (Moscou);
- Kryloff, M. Nicolas [*sic*][3] (Kieff);

[2]For a few further details of the work of Gel'fond and Khinchin, and speculations as to what their plenary lectures would have contained, see Sections 9.21.1 and 9.21.4, respectively.

[3]Krylov's first names were Nikolay Mitrofanovich, so the listing in the congress proceedings is either a mistake, or the "M." is supposed to stand for "Monsieur" — though in the latter case it would be curious that his is the only name to which such a title has been added.

FIGURE 4.2. A. Ya. Khinchin (1894–1959) (Source: *Uspekhi matematicheskikh nauk*, vol. 1 (1946))

- Noether, Fritz (Tomsk);
- Orloff, Chr. (Kieff);
- Pfeiffer, Georg (Kieff).

The enumeration of the members of the congress by country on page 39 of the proceedings speaks of 11 "participants" from the USSR (and five from Italy, who, as we will see in Chapter 5, were also absent), but the word "participant" is clearly only symbolic. Contemporary reports of the congress all record that only one of the above 11 Soviet "members of the congress" actually took part, namely the ill-fated German refugee Fritz Noether.[4]

What had happened?

4.2. Political obstacles

After the congress had ended, the Jewish German émigré Richard Courant received a letter (dated 8 August) from his close Russian friend, the topologist Pavel S. Aleksandrov. The letter was written in German, in which Aleksandrov was fluent, and began with the very intimate salutation "Lieber Simplizialkater" ("Dear simplicial cat"), evidently using a nickname which Courant, Aleksandrov,

[4]For example, Lietzmann mentions Noether explicitly in his reports of the congress, and records also that "[t]he Russians, who had announced a number of lectures had, at the last moment, with one exception, cancelled" (see Appendix D, pp. 272, 279). See also Appendices A.3 and A.7, together with the accompanying commentary in Chapter 6.

and Heinz Hopf (1894–1971) had created during their discussions of topological questions in Göttingen of 1926.[5] Aleksandrov continued his letter with the following remark:

> Our trip to Oslo was prevented quite unexpectedly by the fact that just at that moment very important and, for our scientific life, crucial problems, in particular regarding our Academy of Sciences, had occurred. The presence of all mathematicians of the Soviet Union in the moment of a restructuring of the highest mathematical organism of the country — the mathematical department of the Academy — was basically [so gut wie] necessary.[6]

As if he wanted to balance this less-than-convincing, quasi-political statement with something personal,[7] and maybe trying to evade criticism for not having tried harder to come to Oslo in spite of all the odds, Aleksandrov added:

> As far as I am personally concerned I could not have come to Oslo anyway because I have been suffering for about a month from a heavy and unpleasant carbuncle (on my stomach) ...[8]

Aleksandrov had apparently written in the same vein to Heegaard, who expressed to others his lack of understanding.[9]

And yet Aleksandrov's statement was basically correct, and the connection between certain political events in Russia with the withdrawal of Soviet delegates from the congress in Oslo was valid. To be sure, there are no remarks in the literature,[10] or indeed in Aleksandrov's own "Pages from an autobiography" (**Aleksandrov, 1979**) concerning the restructuring (in a real sense) of the Mathematical Institute of the Soviet Academy of Sciences at that particular point in time. In fact the Academy's headquarters had already been moved from Leningrad to Moscow in

[5](**Reid, 1976**, p. 107). Indeed, three months before the congress, Aleksandrov was still announcing a large Russian delegation in a very personal and joyful letter to Courant of 16 April, containing the following allusion to their old friendship, and signed "Ocean-arse" ("Meerarsch"): "This coming summer, the 'simplex' will have existed for 10 years. A celebration of that event together with you in Oslo is a matter of course." ("In diesem Sommer sind es 10 Jahre, dass das 'Simplex' existiert. Eine diesbezügliche Feier mit Deiner Teilnahme darf in Oslo nicht fehlen." RCP, Series I, box 2, folder 1: P. S. Aleksandrov to Courant, 16 April 1936)

[6]"Lieber Simplizialkater ... Unsere Reise nach Oslo wurde in einer ganz unerwarteten Weise dadurch verhindert, dass gerade in diesem Augenblick gewisse sehr wichtige und für unser wissenschaftliches Gesamtleben bedeutungsvolle Probleme, betreffend hauptsächlich unsere Akademie der Wissenschaften entstanden sind. Die Anwesenheit aller Mathematiker der Sowjet-Union in diesem Augenblick der Neugestaltung des höchsten mathematischen Organismus des Landes — der mathematischen Abteilung der Akademie — war so gut wie notwendig." (RCP, Series I, box 2, folder 1: P. S. Aleksandrov to Courant, 8 August 1936) Since "so gut wie" means "about" or "basically", it expresses some doubt about this alleged necessity, although one may allow here for insufficiency of German.

[7]Although it appears to have been true in Aleksandrov's case (see below), such "I-am-needed-urgently-at-home" excuses came to be used frequently in later decades by Soviet scientists who were denied the right to travel: see the comments in (**Hollings, 2016**, p. 35).

[8]"Was mich persönlich betrifft, so könnte ich sowieso nicht hin, seit fast einem Monate leide ich an einem ausserordentlich schweren und unangenehmen Karbunkel (am Bauch) ..." (RCP, Series I, box 2, folder 1: P. S. Aleksandrov to Courant, 8 August 1936)

[9]Størmer Papers, box 10: Heegaard to Størmer, 8 August 1936.

[10]For instance in (**Vucinich, 2000**).

1934. However, in all likelihood, the infamous Luzin affair of 1936,[11] which had been fuelled by eight long articles with attacks on Luzin and other mathematicians in the leading party newspaper *Pravda* between 2 July and 16 July (recall that the Oslo congress started on 13 July), was the real reason for cancelling participation in the congress. Aleksandrov, who in his letter to Courant and even in his later autobiographical notes, managed to avoid mentioning the Luzin affair altogether, was, in line with other prospective Russian participants in Oslo, heavily involved. And in a sense, this affair could indeed be considered as one of the "important and, for our scientific life, crucial problems", which Aleksandrov mentioned in his letter. Aleksandrov himself — in spite of his "carbuncle" — and his close friend Kolmogorov, had to appear several times in July 1936 at political meetings which discussed the case of their teacher Nikolay N. Luzin (1883–1950), who was accused of "submissiveness to foreigners and wrong treatment of young scientific cadres" (as we note below; see also the comments in (**Lorentz, 2002**, p. 205)).[12]

In fact, the publication in Russian (by (**Esakov, 2000**)) of the protocol of the political decision of the Politburo of the Communist Party of the Soviet Union (CPSU) against the participation of Russian mathematicians in Oslo shows an immediate connection. Before we present and discuss this document, however, some remarks about the development of Soviet mathematics and its international connections are necessary.

4.3. Russian and Soviet mathematics to 1936

There is no doubt that Russian mathematics and science, even before the revolution of 1917, had never been fully integrated with Western European science, not least due to the language barrier, which exists even today.[13] It is also clear that after the revolution of 1917, and due to political pressure in the early 1920s, several Russian mathematicians, such as J. Tamarkin (1888–1945), G. Yu. Rainich (1886–1968), and A. S. Besicovich (1891–1970), left the country. However, before Stalin strengthened his grip on power in the late 1920s, there was still an atmosphere of free and liberal discussion in Soviet Russia, even about the philosophical interpretation of mathematics (such as intuitionism, formalism, von Mises' theory of probability, etc.) which paralleled experimental and modernist tendencies in the

[11]For a short introduction to the Luzin affair, see, for instance, (**Vucinich, 2000**, pp. 65–67) or (**Lorentz, 2002**, pp. 199–208); a much longer and more detailed account, featuring valuable primary sources, may be found in (**Demidov and Lëvshin, 1999/2016**), which has recently been translated into English. We will not try here to give another detailed description of this decisive event in the political conditions for mathematics in Soviet Russia — the reader is directed instead to the resources found in (**Demidov and Lëvshin, 1999/2016**).

[12]The testimony of both Aleksandrov and Kolmogorov was not altogether positive for their teacher, from whom they had been estranged for quite some time for personal, political, and mathematical reasons. S. Bernshteyn, however, another mathematician who had planned to go to Oslo, gave unconditional support to Luzin, which was probably not without danger for himself (for more details, see (**Lorentz, 2002**)). He was increasingly silenced politically by the regime in the following years (**Vucinich, 2000**, pp. 64–65).

[13]Work by Chebyshëv, Lyapunov, and Markov — although partly published in French — had not been fully recognised in its importance, although Russians at different times have tended to exaggerate the accomplishments of their national ancestors and heroes. For a detailed discussion of the Russian-language barrier in science, see (**Gordin, 2015**); see also (**Hollings, 2016**, Chapter 4).

arts (Eisenstein, Mayakovskiy).[14] The sharp political turn between 1928 and 1931, the "Great Turn" ("Великий перелом"), had consequences for Russian mathematics and for its international communication. While several Russian mathematicians (among them Aleksandrov and Luzin) had received stipends from the IEB in the 1920s, the many more Russian mathematicians proposed for stipends around 1930 (Gel'fond, Shnirel'man, Kolmogorov) were put on a waiting list by the Rockefeller Foundation, not least because the philanthropists were afraid that the Russians would defect to the West, as others had before, and thus would become a burden for the Foundation.[15] From the Russian side, increasing isolationism set in, which paralleled the political tendencies on the Soviet side to abandon previous internationalism and to focus on Stalin's policy of building "Socialism in One Country" ("Социализм в одной стране").[16] An anonymous editorial in the leading Russian mathematical journal *Matematicheskiy sbornik* (**Anon, 1931**) appealed to its Russian readership not to publish abroad in "bourgeois" countries. However, publication abroad was not yet strictly forbidden, as shown by the appearance of the influential Springer *Ergebnisse* reports on probability by Kolmogorov and Khinchin in Berlin in 1933.[17]

Currency problems also seem to have played a role in Soviet self-isolation, at least as one cause of the low participation of just three Soviet mathematicians in the ICM in Zurich in 1932. In fact, when, on 25 August 1932, Luzin was belatedly nominated by a decision of the Politburo of the CPSU to attend the Zurich congress, which began on 4 September, the protocol said explicitly:

> Foreign currency is only required for the travel of Khinchin and Kol'man.[18]

This means that Luzin, as well as Aleksandrov, Bernshteyn, and N. G. Chebotarëv (1894–1947), who had been nominated a month before,[19] were supposed to pay for their trip to the congress from their own resources of Western currency, thus relying on their Western connections, officially despised. This may have been among the reasons that Luzin and Bernshteyn did not ultimately go to Zurich, thus reducing the number of Russian participants to three: Aleksandrov, Chebotarëv, and Ernst Kol'man (1892–1979).[20] Four years earlier, in Bologna in 1928, there

[14]Some of this atmosphere is conveyed by (**Yushkevich, 2007**), the historian of mathematics, who himself took part in these discussions.

[15]See (**Siegmund-Schultze, 2001**, pp. 125–133). This was a time of widespread academic unemployment in the West, including Germany and the USA.

[16]These new Stalinist policies were, incidentally, designed by intellectuals such as N. I. Bukharin (1888–1938), who led the delegation of Soviet historians of science at the International Congress of the History of Science and Technology in London in 1931. The delegation and their Marxist presentations made a big impression on the leftist intellectuals there: see (**Cross Roads, 1931**). Bukharin himself later became a victim of the purges.

[17]But this fact was occasionally used against them later (see below).

[18]Translated from the Russian in (**Esakov, 2000**, p. 125).

[19]This was confirmed in a protocol of 23 July 1932 (**Demidov and Lëvshin, 1999/2016**, English translation, p. 14).

[20]The proceedings of the Zurich congress (**ICM, 1932**) include papers by many Russians who do not appear to have participated in person: their names are not mentioned in the list of participants. Volume I mentions explicitly that the plenary talk by Bernshteyn (p. 11) and the talk in the analysis section by Luzin (p. 45) were presented by B. Hostinský (Brno) and W. Sierpiński (Warsaw), respectively. Khinchin is not mentioned at all in the proceedings, and thus was probably (for unknown reason) not there. The one mathematician from Leningrad who was apparently in Zurich and who is mentioned as an official delegate of the Soviet government (p. 24), namely

had apparently (based upon examination of the proceedings) been about 15 participants from Russia, including Aleksandrov and Luzin.[21] As to the absence of Luzin in 1932, one might also suspect that he was not interested in meeting or even in subordinating himself to Kol'man, the political watchdog among the Russian participants. Campaigns against Luzin, organised by party functionaries and dilettante mathematicians such as Kol'man, who delivered two talks in Zurich (a rather elementary mathematical one, and another with strong political overtones), had already been underway since about 1930, not least because of Luzin's and his teacher D. F. Yegorov's (1869–1931) conservative (by Soviet standards) philosophical and religious positions.[22]

The fact that Luzin had only belatedly been nominated for the congress in Zurich was explained in the protocol by the following words in the application to the Politburo:

> Kul'tprop requests the approval of an additional participant at that congress ... Academician Luzin is a very great mathematician, a specialist in the theory of functions of real variables, who has created the Moscow mathematical school in this area. Among his students are several outstanding Moscow mathematicians. One has also to consider that it would provoke bewilderment and possibly unfavourable criticism of Soviet organisations if academician Luzin did not go to the congress.[23]

Thus, in 1932, Soviet politicians were still concerned about the possible adverse propagandist effect of isolationist actions.

As it turned out, 1932 was almost Luzin's last chance to go abroad, given the difficulties surrounding international relations and the increasing campaign against him from at least 1934 onwards. In fact, when he applied to go for medical treatment in Paris together with his wife in 1934, this was turned down, by decision of the Politburo on 9 June 1934.[24]

The seizure of power by the Nazis in Germany in 1933 undoubtedly had consequences for Soviet involvement in international mathematical communications, not least because the Soviets considered themselves as diametrically opposed to the new German regime. A stipend that the Rockefeller Foundation had given to Kolmogorov for study in Göttingen could not be taken up: in January 1934, he was still waiting for an exit visa from the Russian side (**Siegmund-Schultze, 2001**, p. 132). Some mathematicians who were expelled from Nazi Germany, especially those with leftist political leanings, explored the possibility of finding a

H. Müntz (1884–1956), had not been nominated by the Politburo and was expelled from the Soviet Union in 1937 because of his German citizenship (**Siegmund-Schultze, 2009**, p. 134).

[21]For Luzin's own comments on the 1928 congress, see (**Demidov and Lëvshin, 1999/2016**, English translation, p. 139).

[22]On which, see (**Graham and Kantor, 2009**). In a presentation at the 1931 London congress mentioned above in footnote 16, Kol'man had attacked Luzin's book *Leçons sur les ensembles analytiques et leurs applications*, published in Paris in 1930. He said that Luzin tried to solve the "crisis in the mathematical sciences" by going in the wrong direction, namely discontinuity and real functions, and he added that "absolute discontinuity reflects with remarkable fidelity the extreme individualism of idealist philosophy" (**Kol'man, 1931**, pp. 217–218). (**Demidov and Lëvshin, 1999/2016**, English translation, p. 13) suggest that Kol'man had also manoeuvred to exclude Luzin from the delegation to the 1932 ICM.

[23]Translated from the Russian in (**Esakov, 2000**, p. 125).

[24]Document 106 in (**Esakov, 2000**, pp. 142–143).

job in Russia, but these hopes were fulfilled only partially and temporarily. From the outset there were bureaucratic and political hurdles in the Soviet Union with respect to the appointment of foreigners. Aleksandrov describes this in his obituary of Emmy Noether, who had originally tried to go to the USSR after her dismissal (**Aleksandrov, 1935**). Other mathematicians who had hoped to find a position in Soviet Russia but who did not make it primarily due to political obstacles included Kurt Mahler (1903–1988), Felix Pollaczek (1892–1981), and Wolfgang Sternberg (1887–1953).[25] Sternberg even made the very expensive trip from Palestine (where he lived temporarily but without a job) to the congress in Oslo primarily in order to talk to the Russians about his job prospects — in vain.[26] Those among the persecuted in Germany who did make it to the Soviet Union had to leave that country in 1937 at the latest, if they had not received Russian citizenship by then. As early as 1935, Sternberg in Jerusalem had received information about

> German scholars, who had settled in Russia, but who all of a sudden and without being given a reason were forced to give up their positions and leave Russia.[27]

At least one of the refugees to Russia, the Oslo participant Fritz Noether, was later murdered by Stalin's secret police under false accusations.[28]

4.4. Attempts to participate in Oslo

In his memoirs (**Aleksandrov, 1979**), Aleksandrov mentioned as international highlights after 1933 only two conferences, both of which took place in the Soviet Union. In 1934, the still relatively young A. N. Kolmogorov (1903–1987) became the director of the Mathematical Institute of Moscow University, concerning which event Aleksandrov reported:

> One of his first administrative steps as Director of the Institute in the area of international mathematical contacts was a plan to set up a whole series of international conferences on various branches of mathematics. Only the first two links of this broadly conceived chain of conferences came into being: a conference (in 1934) on differential geometry and tensor analysis, with V. F. Kagan as president of the organizing committee, and in 1935 a topology conference under my leadership. (**Aleksandrov, 1979**, p. 331)

The first conference was attended by, among others, É. Cartan, W. Blaschke (from Nazi Germany), J. A. Schouten, and V. Hlavatý. The topology conference was

---

[25]See (**Siegmund-Schultze, 2009**) for details.

[26]See RCP, Series I, box 2, folder 1, Courant to Aleksandrov ("Liebes Väterchen"), 16 September 1936: "Sternberg is now in Jerusalem and is pretty desperate. Bohr and I both ask whether you could, either directly or indirectly, determine how the matter really stands. Sternberg had an extra reason to make the trip from Palestine to Oslo, which was for him a huge sacrifice: to meet with Russian colleagues there." ("Sternberg sitzt jetzt in Jerusalem und ist ziemlich verzweifelt. Bohr und zugleich ich fragen hiermit an, ob Du nicht direkt oder indirekt feststellen könntest, wie die Sache eigentlich steht. Sternberg hatte extra deswegen das für ihn riesige Opfer gebracht, von Palästina nach Oslo zu kommen, um die russischen Kollegen dort zu treffen.")

[27]Sternberg to Courant, 1 July 1935: "deutsche Gelehrte, die in Rußland untergekommen waren, plötzlich ohne Angaben von Gründen ihre Stellung aufgeben und Rußland verlassen müssen" (RCP, Series I, box 8, folder 5).

[28]See (**Segal, 2003**, p. 60) and (**Schlote, 1991**).

attended by, for example, G. D. Birkhoff's student H. Whitney (see (**Whitney, 1988**)), G. Nöbeling (again from Germany), W. Sierpiński, and S. Lefschetz.[29]

Aleksandrov went on to write about his close collaboration with Kolmogorov: "[t]he second half of the 30's went by peacefully" (**Aleksandrov, 1979**, p. 333). But before it became "peaceful" — if one can trust such a description of a time of increased Stalinist terror and the Moscow trials[30] — there was the Luzin affair, as mentioned above.[31] Indeed, it was not just Luzin who was in danger: his friend, the mathematician, priest and philosopher P. A. Florenskiy (1882–1937), was shot in 1937. Others, including Aleksandrov, had to be wary about their political behaviour as well. In the midst of the political trials against Luzin, the Communist newspaper *Pravda* wrote on 9 July 1936, under the title "Traditions of Servility":

> While Luzin published his papers abroad as a form of sabotage, also Aleksandrov and Kolmogorov, Bernshtein and Khinchin did this, forgetting their national pride and Soviet patriotism. New scientific centers are established in countries like Poland and Holland, putting the USSR at a disadvantage. (**Lorentz, 2002**, p. 205)

One day after the first publication against Luzin appeared in *Pravda*, on 3 July 1936, there was a meeting of the Politburo of the Communist Party, which discussed the participation of Russian mathematicians in the congress in Oslo. The protocol of that meeting, which was first published in 2000 by V. D. Esakov from the archives of the Politburo,[32] reads as follows in English translation:[33]

[29]The topology conference in September 1935 was mentioned during the Oslo congress by S. Lefschetz, who reported that W. Sierpiński had, in the name of the Polish government, extended an invitation to a follow-up conference in Warsaw in 1939 (**ICM, 1936**, vol. I, p. 48). This conference seems to have fallen victim to the political conditions of the times, just like the larger international congress planned for 1940. On the 1935 meeting, see also Whitney's comments on p. 145.

[30]See the letter by Neugebauer on p. 100 concerning increasing Russian nationalism in 1937.

[31]Moving beyond mathematics, we mention also in passing that by the mid-1930s a different section of the Soviet scientific community was under assault, namely geneticists, following the rise of Lysenkoism. The latter had no direct effect on Soviet mathematics, however, perhaps because Lysenko held that mathematics had no relevance for biology: displaying little knowledge of the nature of mathematics, he noted "we biologists do not take the slightest interest in mathematical calculations that confirm the useless statistical formulas of the Mendelists" (quoted in (**Joravsky, 1970**, p. 211)).

[32]In general, Esakov used the Russian State Archives, which also contain the protocols of the Politburo (**Esakov, 2000**, p. 21). However in the introduction to his edition (**Esakov, 2000**, p. 15), he quoted from the Archives of the Academy of Sciences when he said that only one out of twelve proposals to the Academy for conferences, namely the one for Oslo, was supported and sent on to the Politburo, but then rejected.

[33]All footnotes are our commentary.

3 July 1936

On the International Mathematical Congress in Oslo

[Decision] to reject the proposal of comrade Bauman to send a delegation of Soviet scholars to the International Mathematical Congress in Oslo.

Protocol № 41, p. 55. *D. 979. L. 14.*

Adopted in connection with the following note by K. Ya. Bauman:

"TO THE SECRETARY OF THE CENTRAL COMMITTEE OF THE CPSU comrade I. V. STALIN

ON A DELEGATION OF SOVIET SCIENTISTS AT THE INTERNATIONAL MATHEMATICAL CONGRESS IN OSLO (NORWAY)

The next International Mathematical Congress takes place in Oslo (Norway) from the 13th to the 18th of July of the present year. Scholars from all the major countries of the world (France, England, Italy, Germany and others) will take part in the work of the congress.[34]

All countries sent large delegations to the previous congress in 1932 (Zurich): France — 69 persons, Germany — 118, USA — 66 etc. The Soviet Union was represented by only 3 mathematicians.[35] Meanwhile, the work of Soviet mathematicians can claim one of the leading positions in world mathematics.

The Academy of Sciences of the USSR, Narkompros[36] and the Academy of Sciences of the Byelorussian SSR have approached the Central Committee of the CPSU with a proposal to send a delegation of 29 persons to the XI International Mathematical Congress.[37]

The Department of Science, Scientific-Technical Investigations and Discoveries [of the Central Committee of the CPSU] proposes to send a delegation of 10 persons, composed of the following: Bernshteyn S.N., full member of the Academy of Sciences of the USSR, I.M. Vinogradov — full member of the Academy of Sciences of the USSR, Burstin, Ts.L. — full member of the Academy of Sciences of the Byelorussian SSR, Prof. Aleksandrov P.S., Prof. Gel'fond A.O., Prof. Khinchin A.N.,[38] Prof. Kovner S.S., Prof. Kol'man E., Prof. Kolmogorov A.N., Prof. Muskhilishvili N.I.

[34] Note that the USA is not mentioned — but see below.

[35] As mentioned before: Aleksandrov, Kol'man, and Chebotarëv.

[36] People's Commissariat for Education: Народный комиссариат просвещения.

[37] One can only reach the number XI for the ICM in Oslo by including the "unofficial" congress in Chicago (1893) and both "boycott congresses": Strasbourg 1920 and Toronto 1924.

[38] There is an error here: the initials should be A.Ya.

> In addition, the Academy of Sciences asks permission for the Soviet delegation to propose that the future XII International Mathematical Congress convene in the USSR (Moscow – Leningrad) in 1940.
>
> The Department of Science of the Central Committee of the CPSU supports the proposal of the Academy of Sciences to convene the next mathematical congress in the USSR.
>
> *Attachment:* Draft resolution and list of scholars.
>
> Head of the Department of Science, Scientific-Technical Investigations and Discoveries of the Central Committee of the CPSU, Bauman".[39]

Esakov, as editor, adds the following commentary:

> On the note of the resolution (with simple pencil): "Against St[alin]", below (in blue) — V. Molotov with note that "comrade Mikoyan is in favour of the proposal of comrade Stalin, comrade Kalinin — in favour."[40]

The protocol clearly shows the downgrading of the wishes of the Russian mathematicians from the original 29 delegates proposed by the Academy, through the 10 delegates supported by Bauman, to the flat rejection by Stalin. It nevertheless requires some explanation.[41]

Karl Yanovich Bauman (1892–1937) was an old Communist Party functionary. Since 1929, he had been a member of the secretariat of the CPSU. When, on 4 June 1934, the Politburo had decided on a new division of responsibility among the secretaries of the Central Committee, Stalin had become personally responsible for three sectors:[42]

(1) Kul'tprop (Культпроп, concerned with culture and propaganda);
(2) the "Special Sector" (the secret department, charged, amongst other things, with distributing Party documentation);
(3) the Politburo as a whole.

Within Kul'tprop, a Department of Science, Scientific-Technical Investigations and Discoveries was created under Bauman. Bauman had already been removed as a secretary of the Central Committee in 1932. In 1937, with the arrest and subsequent execution of Bauman, the Department of Science was dissolved.

4.5. After Oslo

The years in Russia following the Luzin affair and the Oslo congress of 1936 were certainly not "peaceful" politically, nor, indeed, were the years after the German invasion of Russia in 1941. Even though, as mentioned above, Aleksandrov wanted the readers of his autobiography to believe that "the second half of the 30's went by peacefully", there was an increasing self-isolation of the Soviet Union, partly born out of a fear of foreigners and of the foreign ideologies of both Nazi Germany and the Western democracies. For their part, Russian mathematicians

[39] Translated from the Russian in (**Esakov, 2000**, p. 240).

[40] Translated from the Russian in (**Esakov, 2000**, p. 241).

[41] Following (**Esakov, 2000**, pp. 15–16).

[42] See Document 6.8 in (**Boobbyer, 2012**, p. 92).

were often forced to philosophise about their or their colleagues' theories in order to prevent an unfavourable interpretation of mathematics by self-proclaimed "Marxist" philosophers such as Kol'man.[43]

Otto Neugebauer wrote to Courant from Copenhagen in March 1937 that Kolmogorov and Khinchin had experienced problems because of their Springer publications of 1933, and that it would not be possible at that time for Courant to engage them to write comprehensive books on probability in Springer's *Grundlehren* series, because it was too dangerous for them:

> In Russia there is now flourishing the same idiotic nationalism as in the Third Reich.[44]

Finally, there is an interesting episode after the Second World War which sheds indirect light on the influence of the Luzin affair on the non-participation of Soviet mathematicians in the Oslo congress; (**Esakov, 2000**, pp. 385–386) includes the following as his document 359:

> *18 September 1948*
>
> **On the trip to Warsaw of Soviet mathematicians for the VI. Congress of Polish Mathematicians** [45]

Here the decision of the Politburo was affirmative:

> To allow the delegation proposed by the Academy of Sciences of the USSR to attend the VI. Congress of Polish Mathematicians in Warsaw: comrades Kolmogorov A.N. (head of the delegation), Aleksandrov P.S. and Mardzhanishvili K.K.[46]

The material edited by Esakov contains a letter from the Department of Propaganda and Agitation (Agitprop) of the Central Committee of the CPSU to G. Malenkov, secretary to the Central Committee of the CPSU, which says, among other things:[47]

> The President of the Academy of Sciences of the USSR, Academician Vavilov, requests permission to send to Warsaw a delegation of Soviet scientists consisting of Academician Kolmogorov, A.N., corresponding member of the Academy of Sciences of the USSR Aleksandrov, P.S. and candidate of physico-mathematical sciences Mardzhanivshili K.K., in order to participate in the VI.

---

[43] See, for example, (**Siegmund-Schultze, 2004**) on Khinchin's criticism of the "Machist" theories of von Mises.

[44] "Lieber Courant, es wird Dich sicher interessieren zu erfahren, dass Kolmogoroff und Khintchine grosse Skandale in Russland hatten wegen Ihrer in Deutschland veröffentlichten Ergebnis-Berichte. In Russland blüht nämlich jetzt auch derselbe idiotische Nationalismus wie im dritten Reich. Du sollst natürlich über solche Sachen nicht nach Russland schreiben, aber es wird Dir vielleicht wegen Gelber Bücher [the *Grundlagen* series had yellow covers] wichtig sein, dies zu wissen. Ich glaube z.B. nicht, dass einer der beiden jetzt ohne Gefahr ein Gelbes Buch schreiben könnte." (RCP, Series I, box 7, folder 1: Neugebauer to Courant, Copenhagen, 14 March 1937) Indeed, Courant had problems in finding an author for such a monograph on probability; his negotiations in Oslo with the Swedish probabilist Harald Cramér failed. This follows from a letter which Courant wrote to Harald Bohr concerning Cramér (RCP, Series I, box 2, folder 4: 13 December 1936). Springer would not come up with such a book for decades.

[45] Translated from the Russian in (**Esakov, 2000**, p. 385).

[46] *Ibid.*

[47] This letter also appears in (**Demidov and Lëvshin, 1999/2016**, English translation, p. 41).

> Congress of Polish Mathematicians and in the jubilee celebrations of V. K. Sierpiński.[48]
>
> The Polish professor Sierpiński is known as one of the most reactionary Polish mathematicians and bourgeois nationalists. In 1936, in connection with the article in the newspaper "Pravda" which criticised academician N. N. Luzin for submissiveness before foreigners and wrong treatment of young scientific cadres, Sierpiński stood up in print in defence of Luzin, attacking the Soviet press. Sierpiński hindered the actions of the progressive part of Polish students.[49]
>
> The secretary of the party organisation of the V. A. Steklov Mathematical Institute of the Academy of Sciences of the USSR, Mardzhanishili, who has been recommended for travel to Poland, rightly deems the sending of Soviet scholars to the VI. Congress of Polish Mathematicians inexpedient.
>
> Considering that the VI. Congress of Polish Mathematicians is associated with the celebration of the reactionary Polish mathematician Sierpiński, Agitprop does not support the request by Academician Vavilov to send a delegation of Soviet mathematicians to the congress.[50]

Nevertheless, we learn from (**Anon, 1949a**, pp. 307–308) that Aleksandrov, Kolmogorov, and Mardzhanishvili did go to Warsaw, as the Politburo had decided. K. K. Mardzhanishvili (1903–1981) was a number theorist, later remembered as a friend by Aleksandrov in his memoirs (**Aleksandrov, 1979**, pp. 340–341). At this time (1948), he was the party secretary of the Steklov Mathematical Institute, but only had the relatively low scientific grade of a candidate. What is interesting is that the party functionary Mardzhanishvili, who had obviously already been selected by the Academy as a compromise delegate or watchdog to supervise the other participants, acted against his own delegation for political reasons.

In any case, the decision was taken the other way round this time: first refusal by the party department of the Central Committee, but finally approval by the secretaries of the Politburo. The comparison of Oslo and Warsaw shows the arbitrariness of the decisions taken by the Politburo, which did not have any superiors. It also shows the Luzin affair still hovering but no longer effective in the end against Soviet participation in international conferences.

And yet, Stalin was still alive, and the Stalinist period of Soviet science policy lingered on even beyond his death in 1953. Indeed, when the proposed 1940 Harvard ICM was finally held in 1950, the Soviets sent a curious telegram to the gathered mathematicians. It is strongly reminiscent of the formulation which Aleksandrov had used in his private letter to Courant in 1936:[51]

> USSR Academy of Sciences appreciates receiving kind invitation for Soviet scientists take part in International Congress to be

---

[48] This should be V. F. or W. F.

[49] In 1936, Luzin had been criticised specifically for his links to Sierpiński: see, for example, (**Demidov and Lëvshin, 1999/2016**, English translation, pp. 103–105).

[50] Translated from the Russian in (**Esakov, 2000**, p. 385).

[51] Cf. the comment in footnote 7 above.

> held at Cambridge. Soviet mathematicians being very much occupied with their regular[52] work unable attend congress. Hope that impending congress will be significant event in mathematical science. Wish success in congress activities.
>
> S. Vavilov, President, USSR Academy of Sciences. (**ICM, 1950**, vol. 1, p. 122)

Here also a decision of the Politburo stands in the background.[53]

This telegram from the president of the Academy of Sciences, S. I. Vavilov, had been approved, maybe even drafted, by the Politburo, as a protocol of 15 August 1950 shows.[54] What we learn from this protocol is that it was the Academy itself that had deemed Soviet participation in the congress "inexpedient" ("нецелесообразно" — note the same choice of word as that of Mardzhanishvili in connection with the 1948 Warsaw congress), and had merely asked the Politburo to approve the telegram. What was really behind this decision might be uncovered in future historical work.

[52]The word "regular" does not appear in the (Russian) version of the protocol edited by (**Esakov, 2000**, p. 474). The word seems to reveal and underline the absurdity of the apology which could not refer to any extraordinary circumstances.

[53]It is interesting to note that Soviet scientists were similarly absent from another conference (on cell biology) held in the USA that year, although they were represented at a meeting on physiology held in Copenhagen: see (**Hollings, 2016**, p. 35).

[54]Document 426 in (**Esakov, 2000**, p. 474).

CHAPTER 5

The Italian Case: Mathematics as a Victim of World Politics

5.1. Introduction

In the preceding chapter, we were able to go some way towards explaining the absence of all Soviet mathematicians from the Oslo congress. We must now turn to the other absence that we have already noted several times during this book (and a further example of the "isolationist" strategy described in Chapter 1): that of the Italians. Just like the expected Soviet delegates, several Italians are in fact listed in the "List des Membres du Congrès" (see Table 5.1), but again this is not a reflection of reality. In contrast to the Soviet situation, however, in which there is a passing acknowledgement of the absence (at least for the cases of the would-be plenary speakers Gel'fond and Khinchin (**ICM, 1936**, vol. I, p. 45)), there is no reference in the proceedings to the Italian absence; one of the few hints that we get is a note that indicates that the congress sent a telegram of greeting to Vito Volterra in Rome (**ICM, 1936**, vol. I, p. 46), despite the fact that, if the list of members were to be believed, Volterra should have been present in Oslo.[1]

In order to find acknowledgements of the Italian absence from the congress, we must turn instead to the published reports, most notably those of Lietzmann, Fehr, and Dunnington. The latter, for instance, after a note about the Soviet absence, remarked further:

> The writer was told that Italy was not represented since Norway had belonged to the sanctions [*sic*]. (**Dunnington, 1936a**, p. 88)

The report of (**Fehr, 1936**) makes a similar comment, indicating further that the Italian absence may have been due to an official move by the Italian government or scientific authority:

> Italy officially abstained because of the sanctions.[2]

The report by Lietzmann, of which we made so much use in Chapter 3, adds little to the picture, merely making another vague reference to absence "for reasons of sanctions".[3] However, in his later reminiscences, Lietzmann was a little more forthcoming, using slightly inflammatory language:

---

[1]See also the remark regarding Severi's absence in note 44 below.

[2]Translated from the French in (**Fehr, 1936**, p. 373).

[3]See pp. 272, 279 of Appendix D. Later on in his report, Lietzmann remarked in passing that Italian was not used as a working language of the congress (pp. 273, 280). He also noted the impact of the Italian absence on the representation of algebraic geometry at the congress (pp. 276, 282), a point that we will pick up again in Section 8.4.

TABLE 5.1. Italian delegates listed in the proceedings of the Oslo congress (**ICM, 1936**, vol. I, pp. 29–39)

Fasciotti, Ernestina	(Milan)
Palazzo, Elena	(Rome)
Scire, Pietro	(Palermo)
Tortorici, Pietro	(Palermo)
Volterra, Vito	(Rome)

> The Italians had not appeared, since Norway had joined the repressive measures of the League of Nations in connection with the Abyssinian conflict.[4]

Thus, at least on the surface, the reason for the Italian absence is rather clearer than that for the Soviets: the Italians appear to have boycotted, or otherwise been prevented from attending, the congress because of the economic and cultural sanctions imposed on Italy by the League of Nations (of which Norway was a member) in response to Italy's actions in Ethiopia, leading to the Second Italo-Ethiopian War (1935–1939). The precise nature of the internal deliberations within Italy in regard to the congress remains largely to be determined, although some broad indications may be found in Section 5.3. It is difficult to know what to make of the list of names in Table 5.1: none of these, with the obvious exception of Volterra,[5] were major Italian mathematical figures (cf. the Italian names mentioned in Section 5.2 in connection with earlier congresses).[6] On the other hand, it is curious that Volterra should be listed as a potential delegate despite his refusal to swear an oath of loyalty to the fascist regime in 1931 (on which, more in Section 5.3). It may indeed be that what we see here is a list of delegates who were not officially approved, but then why were they ultimately prevented from attending? Strictly speaking, we cannot rule out the possibility that some of these mathematicians *did* in fact attend the congress — Volterra is the only person in Table 5.1 whom we know for certain did *not*. Moreover, we know by name other mathematicians from Italy who *were* prevented from attending (see p. 114). The absence of Italians from the proceedings (beyond the list of members) could simply be a reflection of the sanctions. There is much here that remains to be explained.

---

[4]Translated from the German in (**Lietzmann, 1960**, p. 61).

[5]For a survey of Volterra's mathematical contributions to the early ICMs, see (**Guerraggio *et al.*, 2016**).

[6]Our basis for declaring the other people in Table 5.1 to be lesser-known figures is the fact that they receive only fleeting mention (only in connection with the Oslo ICM) in (**Guerraggio and Nastasi, 2005a**, p. 248). On Fasciotti, we have no further information, nor have we any personal details concerning Scirè or Tortorici, though we have identified several textbooks authored by the former (for example, (**Scirè, 1942**), (**Scirè, 1946**)), and zbMATH (accessed 8 February 2018) lists 40 publications for the latter, mostly on broadly geometrical topics. Geometry and its history appear also to have been the focus of interest for Palazzo (see, for example, (**Palazzo, 1961**)), who delivered a lecture entitled "Una proposizione di geometria elementare e le sue applicazioni" at the First Italian Mathematical Congress in Florence in 1937 (**Palazzo, 1938**). We note that both Tortorici and Palazzo had previously attended the 1932 ICM, the former as a representative of the Circolo matematico di Palermo (**ICM, 1932**, vol. I, pp. 21, 33, 35).

In contrast to the Soviet case, with the document unearthed by Esakov and explored in our Chapter 4, we have as yet found no source that is as substantive in the Italian case, and that *fully* explains their absence from the Oslo ICM — some relevant archival sources will be discussed in Section 5.3, but while these do add detail, they do not provide a complete picture. Nevertheless, by way of providing background for these sources, we devote the first part of this chapter to a discussion of the Italian mathematics of the first half of the twentieth century, particularly earlier Italian engagement with the ICMs (notably Rome 1908 and Bologna 1928) and the experiences of Italian mathematicians under Mussolini. We could certainly have spent equally as much space in Chapter 4 discussing the venerable traditions of Russian and Soviet mathematics before Oslo. We have chosen, however, to give more space to the Italian case, because the much broader involvement of the Italians in previous congresses gives us a better opportunity to discuss the history of the Oslo ICM in this context. We offer this chapter as a starting point for further investigation, and hope that some future scholar will discover archival material that adds to the picture of the Italian withdrawal.

5.2. Italian mathematicians and the early ICMs

Although it had had its well-known mathematical figures in the past (we might name, for example, Cardano or Ruffini),[7] it was only at the end of the nineteenth century, following the unification of the 1860s, that Italian mathematics gained prominence at an international level,[8] a prominence that grew further during the early decades of the twentieth century. These were of course the years of the formation of the ICM movement, and some authors have viewed Italian mathematics and the ICMs as having matured together.[9] Indeed, a strong international outlook appears to have accompanied the development of mathematics in late-nineteenth-century Italy.[10] Italian mathematicians were certainly a major presence at the early ICMs, with their membership of organising committees often perhaps serving to balance French and German interests. The Italian link to the ICMs was cemented by the hosting of two congresses in Italy prior to 1936: Rome 1908 and Bologna 1928, meetings that took place under rather different circumstances. There are parallels also between the internationalism of the ICMs and that of Italian mathematics: both grew during the early years of the twentieth century, only to go into decline around the time of the 1936 congress.

[7]For a survey of Italian mathematics of an earlier period, see (**Grattan-Guinness, 1986**).

[8](**Struik, 1962**, p. 281) commented, for example, that "[t]he Risorgimento, the national rebirth of Italy, also meant the rebirth of Italian mathematics". For more detailed accounts, see (**Bottazzini, 1989**), (**Bottazzini, 1994**), (**Brigaglia *et al.*, 1998**), (**Guerraggio and Nastasi, 2005a**, Chapter 1), (**Guerraggio and Nastasi, 2008**, Chapter 2), (**Tazzioli, 2011**), (**Bottazzini and Nastasi, 2013**), and (**Rogora, 2016**). For the views from France and Germany, see, respectively, (**Brechenmacher *et al.* 2016**) and (**Neuenschwander, 1986**). On the place of mathematicians within late-nineteenth-century Italian politics, see (**Durand, 2017**); on Luigi Cremona specifically, see (**Brigaglia and Di Sieno, 2009**).

[9]See, for example, (**Guerraggio and Paoloni, 2008**, English transl., p. ix).

[10]See, for example, the remarks in (**Brigaglia, 2002**, p. 180) — in particular a quotation from Volterra which supports this view: Volterra dated the development of modern Italian mathematics from the 1858 tour by Enrico Betti (1823–1892), Francesco Brioschi (1824–1897), and Felice Casorati (1835–1890) of foreign universities. See also the remarks in (**Goodstein, 2007**, p. 118).

Within our survey of Italian involvement in the ICMs, special mention must of course be made of the Rome congress of 1908.[11] The hosting of this fourth ICM was viewed as an informal measure of which country was to be considered the world's third mathematical nation, after France and Germany.[12] Much to the disappointment of the British delegate George Greenhill,[13] the Heidelberg congress elected to accept the invitation brought by Volterra from the Accademia dei Lincei to host the next meeting in Rome (**ICM, 1904**, p. 53). In Section 1.2.1, we outlined the three respects in which the Rome congress was an important one for internationalism in mathematics: namely, the formation of ICMI, and early suggestions regarding both an international mathematical organisation and a mathematical prize.

Overall, the impression of Italian mathematics that appears to have been instilled in foreign visitors by the Rome congress was one of great strength and prosperity: for instance, after lamenting the relative under-representation of his own countrymen,[14] the British applied mathematician G. H. Bryan, writing in *Nature*, noted of the Rome congress that

> [t]here were many proofs that the Italian as well as the German mathematicians present constituted a powerful and influential body, and the presence of the King at the inaugural meeting—an honour rarely conferred on such occasions—together with the brilliancy of the receptions, afforded evidence of the esteem in which Italy holds the mathematical professors of her universities.[15]

Following the Rome congress, the participation of Italian mathematicians in the future ICMs appears to have been assured: the proceedings of the 1912 Cambridge congress give a figure of 35 for Italian attendance (plus six accompanying spouses and other relatives (**ICM, 1912**, vol. I, p. 28)). Italians were represented in official positions at the congress, including one plenary speaker: Federigo Enriques (1871–1946) (**ICM, 1912**, vol. I, pp. 67–79). Italy was also well represented in the sectional lectures. Had the proposed 1916 Stockholm congress gone ahead, it is reasonable to speculate that this trend would have continued, and that there would have been many Italian mathematicians in attendance.

At the start of the First World War, Italy remained neutral, although opinions were divided as to whether the nation should enter the war. As to which side Italy would join, this was not a foregone conclusion: political tensions existed between those who believed that Italy should remain within the *Triple Alliance* (*Dreibund*), with which it had been associated since 1882, and those who felt that it would be more appropriate to align with the *Triple Entente*, particularly France, to which

[11]For a detailed account of the congress, see (**Guerraggio and Nastasi, 2008**). See also the recent remarks in (**Cioli, 2016**, §III.1), as well as the survey in (**Guerraggio and Paoloni, 2008**, §4.5), whose focus is the activities of the principal organiser of the congress: Volterra.

[12]See, for example, the comments in (**Bolondi *et al.*, 2011**, p. 11) or (**Guerraggio and Nastasi, 2008**, p. 76).

[13]See his remarks in (**ICM, 1904**, pp. 53–54), including: "Disappointed in our expectation we must congratulate Italy and Rome on its good fortune, and we must content ourselves with the next best in our wish, and hope that England may be selected at this Assembly as the meeting place in 1911 or 12."

[14]Twenty-two, according to (**ICM, 1908**, vol. I, p. 20).

[15](**Bryan, 1908**, p. 584). According to (**Pancaldi, 1993**, p. 29), the "'Italian school of mathematics' [is] often depicted by historians as one of the few fields of Italian science which achieved eminence in the nineteenth century."

Italy had been growing closer since the beginning of the century. Opinions were no less mixed among mathematicians.[16] Most prominent of those who advocated Italian intervention on the side of the *Entente* was Volterra (see Figure 5.1); indeed, upon Italy's revoking the Triple Alliance and entering the war in May 1915, he immediately enlisted and devoted his mathematical abilities to the war effort, particularly to aeronautical developments. Moreover, his efforts were so firmly set in this direction that when Mittag-Leffler attempted to organise an international mathematical meeting in Stockholm in 1916 in spite of the war (see Section 2.1.1), and duly invited Volterra, the latter expressed his reservations about continuing mathematical work of an academic nature at such a time.[17]

Volterra's position was certainly not shared by everyone: the historical links to Germany that had given some Italians pause for thought over involvement in the war made it more difficult for Italy than for the other Allies to distance itself in the aftermath from its wartime enemy — many prominent Italian mathematicians, such as Pincherle, Bianchi, and Fano,[18] had studied in Germany. Even during the war, anti-German sentiment had not been universal among Italian mathematicians: for instance, Michele de Franchis, Guccia's successor as president of the Circolo matematico di Palermo, had refused to expel German members, despite pressure from French colleagues to do so.[19] After the war, Tullio Levi-Civita (1873–1941) was a strong advocate of the readmission of German scientists to international meetings (**Nastasi and Tazzioli, 2005**, pp. 224–225); together with Theodore von Kármán in Aachen, he organised an international congress on fluid mechanics in Innsbruck in Austria in 1922, at which the mathematicians of all nations were welcome.[20] Volterra was invited, but politely declined (**Mazliak and Tazzioli, 2015a**, p. 13). This meeting gave rise to a further one in Delft in 1924 (**Battimelli, 2016**, §C), thereby launching the sequence of International Congresses for Applied Mechanics (ICAM) — later the International Congresses for Theoretical and Applied Mechanics (ICTAM).[21] German mathematicians were active participants in the Delft congress, although this apparently led to the withdrawal of the French.[22]

[16] See, for example, (**Mazliak and Tazzioli, 2015a**). Indeed, there is a quite extensive literature on Italian mathematicians during the First World War: see also (**Guerraggio and Nastasi, 2005a**, Chapter 2) for a general overview, (**Mazliak and Tazzioli, 2009**) on Volterra's advocacy of Italian intervention and his relationship with French allies, (**Tazzioli, 2013**) for a concise summary of the divergent views with regard to the war, (**Nastasi and Tazzioli, 2013**, §2) on the so-called "national polemic" ("polemica nazionale") regarding intervention, (**Nastasi and Tazzioli, 2014**) for a broad survey of the associated debates and institutional consequences for Italian science, (**Mazliak and Tazzioli, 2015b**) for a selection of wartime letters written by Italian mathematicians, (**Linguerri, 2015**) for a discussion of Volterra's war work and its post-war consequences, and (**Comoglio, 2015**) for the individual story of one mathematician who was killed in action.

[17] See (**Stubhaug, 2010**, p. 588) and (**Mazliak and Tazzioli, 2015a**, p. 9); the latter observe that Volterra invited Hadamard to lecture in Rome that year, although his reasons for doing so were at least partly propagandistic (**Mazliak and Tazzioli, 2015a**, p. 11).

[18] See, respectively, (**Tricomi, 2008**), (**Carruccio, 2008**) and (**Struik, 2008**).

[19] See (**Nastasi and Tazzioli, 2013**, §4.2.2) or (**Nastasi and Tazzioli, 2014**, pp. 210–211).

[20] See (**Battimelli, 2016**, §B) or (**Nastasi and Tazzioli, 2013**, §4.3.2).

[21] On Levi-Civita's role in the creation of these congresses, and their connection to Italian science, see (**Battimelli, 1996**).

[22] See the remarks in (**Battimelli, 2016**, p. 26), as well as the very many German names in the list of participants found in the congress proceedings (**ICAM, 1924**, pp. XIX–XXII). There does not appear to be any explicit acknowledgement of the French absence in the proceedings, though no French names appear in the list just mentioned. Only one participant is indicated as

FIGURE 5.1. Vito Volterra (1860–1940) (By unknown photographer, Public Domain, via Wikimedia Commons)

TABLE 5.2. Italian delegates listed in the proceedings of the 1920 Strasbourg congress (**ICM, 1920**, pp. ix–xiv).

Boccardi, Jean (Giovanni)	(Turin)
Boffa Tarlatta, Mlle. Ortenza	(Turin)
Verde	(La Spezzia)
Volterra, V.	(Rome)

---

having come from France: Prof. E. Hahn of Nancy, whose name of course suggests a German origin, although his contribution to the congress was in French (**Hahn, 1925**). This is probably the Eberhard Hahn for whom zbMATH (accessed 27 February 2018) lists seven publications (three in French, four in German) in applied mathematics between 1913 and 1958.

As we know from Chapter 1, an "international" congress of mathematicians was organised in Strasbourg in 1920, at which the first IMU was founded. Volterra's support for Strasbourg as a venue despite Mittag-Leffler's continued hopes for Stockholm, together with his involvement in the IRC and its exclusion of German scientists, led to a cooling of relations between the two men (**Stubhaug, 2010**, p. 612). Volterra delivered a plenary lecture at the 1920 ICM,[23] as well as serving as a vice president (under Picard's presidency) and official representative of Italy (**ICM, 1920**, pp. vii, viii).[24] Attendance at the Strasbourg congress was low compared to that of the ICMs before the First World War,[25] having been by invitation only (**Lehto, 1998a**, p. 33), and the Italian delegation was certainly much thinner than it had been (see Table 5.2).[26]

5.3. Italian mathematics under Mussolini

We do not deal here in detail with the experiences of Italian mathematicians under fascism — detailed analyses of this subject can be found elsewhere.[27] Nevertheless, it would be remiss of us not to include some comments on this highly relevant topic, upon which there has been much research in recent years, and which will not only add colour to the foregoing, but also help to integrate it into the story told in the rest of the book.

In the years following the First World War, the organisation of Italian science underwent several changes, most notably the establishment of the Unione Matematica Italiana (UMI) in 1922 and of the Consiglio Nazionale delle Ricerche (CNR) in 1923. These new foundations took place against a background of profound political change: Benito Mussolini had been appointed Prime Minister in October 1922, assuming dictatorial powers in 1925, thus inaugurating the fascist era. Both the UMI and the CNR would play an important role in bringing Italian science under state control; it has been argued, moreover, that the connections between science and fascism as an ideology were even stronger in Mussolini's Italy than links in Hitler's Germany between science and national socialism (**Epple *et al.*, 2005**, p. 145).

A number of mathematicians were at the time prominent figures on the Italian political scene: Volterra, for example, was a senator, just like Betti, Beltrami, Cremona, and Dini before him (**Durand, 2017**). It was in this capacity that he participated in a vote of no confidence against Mussolini in the Senate in 1924, following

---

[23]With the title "Sur l'enseignement de la physique mathématique et de quelques points d'analyse" (**ICM, 1920**, pp. 81–97).

[24]This was also the year in which Volterra issued his 1900 and 1908 plenary lectures as a booklet entitled *Saggi scientifici* (**Volterra, 1920**) — but with all references to German mathematicians removed: see (**Guerraggio and Paoloni, 2008**, English transl., p. 102) or (**Guerraggio and Nastasi, 2008**, p. 143).

[25]The proceedings give a round figure of 200 (**ICM, 1920**, p. xv); cf. the 574 for Cambridge (**ICM, 1912**, vol. I, p. 28), or the 535 for Rome (**ICM, 1908**, vol. 1, p. 20); see also Figure 1.3 on p. 7.

[26]We cannot fully reconcile the figures: a table on p. xv of the Strasbourg congress proceedings states that there were five delegates from Italy, but we can only find four in the list. Of these, we have no information on either Boffa Tarlatta or Verde, other than what is given in the list of members, where Boffa Tarlatta is listed as "docteur ès sciences" and Verde (with no indication of first name or initial) as "commandant" (**ICM, 1920**, pp. ix, xiv).

[27]For example, (**Nastasi, 1998**), (**Guerraggio and Nastasi, 2005a**) or (**Guerraggio and Nastasi, 2005b**).

the murder of the socialist leader Giacomo Matteotti (**Capristo, 2015**, p. 641). The following year, when the fascist Minister of Education, Giovanni Gentile (1875–1944), issued his "Manifesto degli intellettuali del Fascismo", laying down the ideological principles of Italian fascism, Volterra, Levi-Civita, Castelnuovo, Severi, and several other mathematicians signed their names to a counter-manifesto written by the philosopher Benedetto Croce (1866–1952) (see (**Guerraggio and Nastasi, 2005a**, §4.3)). Around this same time, Castelnuovo, as a member of a commission on education at the Accademia dei Lincei, warned of the dangers to Italian scientific instruction posed by Gentile's proposed reforms.[28] Levi-Civita was another prominent anti-fascist, motivated by a strong internationalism (see p. 107).

This is not to say that all Italian mathematicians were opposed to fascism, however: Pincherle and Gini, for example, both supported the Gentile manifesto (**Guerraggio and Nastasi, 2005a**, p. 93). The analyst Mauro Picone (1885–1977), who, like Volterra, had been an active figure in Italy's scientific war effort,[29] was closely aligned with the regime, and as a result enjoyed great success as head of the Istituto Nazionale per le Applicazioni del Calcolo[30] (**Nastasi and Tazzioli, 2014**, pp. 219–221), and also advancement at the University of Rome following Volterra's eventual removal (**Nastasi, 2015**, p. 25). In light of his wartime experiences, Picone was a strong advocate of the use of applied mathematics in the service of the state (see, for example, (**Epple *et al.*, 2005**)), and was very active within the CNR (**Nastasi, 2001**). Alongside Picone, Francesco Severi (see Figure 5.2), who had originally signed the anti-Gentile manifesto, was a leading mathematical supporter of the regime: initially a socialist, who, indeed, had resigned from his post as rector of the University of Rome in protest at Matteotti's assassination (**Babbitt and Goodstein, 2009**, p. 802), Severi eventually moved into the fascist camp. With direct access to Mussolini, Severi became an extremely influential figure in Italian mathematical life. In 1939, for example, he successfully sought state funds for the establishment of the Istituto Nazionale di Alta Matematica (INdAM) in Rome,[31] and served as its first president.[32] Severi was further able to use his position to his advantage in, for instance, his long-running (initially mathematical) feud with Enriques over the proof of the "completeness theorem" in algebraic geometry.[33]

---

[28]See (**Guerraggio and Nastasi, 2005a**, §4.2). On the report of the commission, see (**Guerraggio and Paoloni, 2008**, English transl., pp. 119–120). On the opposition to the reforms more generally, see (**Capristo, 2015**, §2.1).

[29]See the references in footnote 16.

[30]This institute was founded in Naples in 1927 as the Istituto di Calcolo per l'Analisi Numerica, later transferring to Rome, and was one of the first research centres devoted to computational mathematics. From 1932 to 1975, it was known by the name given above; it is now the Istituto per le Applicazioni del Calcolo "Mauro Picone": `http://www.iac.rm.cnr.it/` (accessed 6 March 2018). On the history of the institute, and Picone's role in its foundation, see (**Ghizzetti, 1986**).

[31]Now the Istituto Nazionale di Alta Matematica "Francesco Severi": `http://www.altamatematica.it/` (accessed 6 March 2018). On the history of INdAM, see (**Roghi, 2005**); for the wider context of Severi's work at this time, see (**Zappa, 1984**).

[32]On Severi's changing attitude towards fascism, and the consequent effects on his career, see (**Guerraggio and Nastasi, 2005a**, §§5.2–5.3); see also (**Goodstein and Babbitt, 2012**).

[33]See, for example, (**Goodstein and Babbitt, 2012**, pp. 1066–1067). On the mathematical context of the feud, see (**Babbitt and Goodstein, 2011**). Indeed, this feud, both political and mathematical, may have been the reason for the "absence from Italy" ("l'assenza dall'Italia") that prevented Enriques from delivering his planned plenary lecture at the Bologna ICM (**ICM,**

FIGURE 5.2. Francesco Severi (1879–1961) (By unknown photographer, http://www.dm.unito.it/sism/m_italiani/immagini/ritratti/Severi1.JPG, Public Domain, https://commons.wikimedia.org/w/index.php?curid=16108910)

It was against the backdrop of such developments that the 1928 Bologna ICM took place, with its clear indications of fascist involvement, as noted in Section 1.2.3. Further state influence over the Italian scientific establishment was enabled by the abolition in 1929 of the Accademia dei Lincei and its replacement by the Accademia d'Italia, through offers of membership of which Mussolini hoped to win the allegiance of scientists who had not yet pledged themselves to the regime (**Guerraggio and Nastasi, 2005a**, §5.1). One of the first, and particularly visible, members of

1928, p. 68) — although the reason given by Enriques himself was that a conference tour of Latin America had been extended (**Capristo, 2016**, p. 298).

the Accademia d'Italia was Severi.[34] The establishment of the Accademia d'Italia was followed in 1931 by the imposition of the oath of loyalty on Italian academics: this had been a requirement for state employees for some time, though academics had been exempt (**Guerraggio and Nastasi, 2005a**, §6.5). In the event, only twelve academics out of a total of around 1,200 refused to swear the oath,[35] Volterra being one of the most prominent to decline, this resulting in his removal from most of his official positions.[36] Others, such as Levi-Civita, took the pragmatic approach and swore the oath to protect their careers and families (**Nastasi and Tazzioli, 2005**, p. 223).

In Section 1.2.3, we have brought the story of Italian involvement in the ICMs as far as the Bologna congress of 1928, a congress at which, according to (**Guerraggio and Nastasi, 2005a**, p. 160), "Italy's international prestige reache[d] its peak". Thereafter, fascist interference in the academic sphere led to political influence and the growth of suspicion of foreign contacts (see Chapter 7). Nevertheless, Italian mathematicians were still a significant presence at the 1932 ICM in Zurich, with 64 delegates out of a total of 667 (**ICM, 1932**, vol. I, p. 36). Among the 21 plenary speakers, however, there was only one Italian: Severi.[37] Indeed, although Severi was not the only Italian speaker at the congress, he was certainly the most conspicuous:[38] in addition to delivering his plenary lecture, he also attended the congress as an official representative of the Italian government, of the Circolo matematico di Palermo, of the Accademia d'Italia, and of the Società Italiana delle Scienze (**ICM, 1932**, vol. I, pp. 21–22), as well as being explicitly indicated as the "president" of the Italian delegation (**ICM, 1932**, vol. I, p. 62), in which capacity he conveyed Italy's greetings to the congress (**ICM, 1932**, vol. I, p. 75). One is tempted therefore to see Severi in Zurich in 1932 as occupying a somewhat similar

[34]See, for example, (**Goodstein and Babbitt, 2012**, pp. 1066–1067), and also (**Guerraggio and Nastasi, 2005a**, §5.3). For a brief survey of the establishment of the Accademia d'Italia and an account of the available archival materials, see (**Cagiano de Azevedo and Gerardi, 2005**).

[35]The names of the twelve are listed in (**Guerraggio and Paoloni, 2008**, English transl., p. 131).

[36]See (**Guerraggio and Paoloni, 2008**, English transl., §7.3). In fact, Volterra had been progressively marginalised for some time, owing to his visible anti-fascist stance, although his strong international connections had made it difficult for him to be side-lined completely (**Capristo, 2015**). Indeed, while his refusal to swear the loyalty oath signalled Volterra's withdrawal from Italian academic life, he nevertheless remained influential in international mathematics through his involvement with the Rockefeller philanthropy: see (**Siegmund-Schultze, 2001**).

[37]His lecture was entitled "La théorie générale des fonctions analytiques de plusieurs variables et la géométrie algébrique" (**ICM, 1932**, vol. I, p. 10), although his text in the congress proceedings is headed "Le rôle de la géométrie algébrique dans les mathématiques" (**ICM, 1932**, vol. I, pp. 209–220). Severi also spoke in the section on geometry with the lecture "Nuovi orizzonti della geometria sopra una superficie algebriche" (**ICM, 1932**, vol. II, pp. 156–158).

[38]We may contrast Severi's involvement in the congress with that of Pincherle, for example, who was a vice president and representative of several Italian organisations, but did not deliver a lecture (**ICM, 1932**, pp. 21–22, 42). Levi-Civita was one of eight representatives of the Circolo matematico di Palermo and chaired one sectional session, but similarly did not speak (**ICM, 1932**, vol. I, pp. 21, 37). Picone represented the Italian government, the Accademia Pontaniana, and the University of Naples (**ICM, 1932**, vol. I, p. 21), as well as delivering two lectures in the section on analysis (**ICM, 1932**, vol. II, pp. 89–93). We note that while the proceedings list Vito Volterra as having been present at the congress (**ICM, 1932**, vol. I, p. 35), he did not deliver a lecture, although his son Enrico did (**ICM, 1932**, vol. II, pp. 247–248).

role to that played by Lietzmann within the German delegation in 1936.[39] Indeed, a document quoted by (**Guerraggio and Nastasi, 2005a**, pp. 188–189) seems to hint in this direction.[40] On 14 April 1932, the CNR and Italian Foreign Office approved an official delegation for the Zurich congress consisting of Gaetano Scorza (1876–1939), Enrico Bompiani (1889–1975), and Picone, but this decision appears to have been challenged by Severi. The quoted document is a report by Bompiani, dated 11 May, outlining Severi's concerns: that his exclusion from the appointed delegation would not create a good impression abroad, given that he was to deliver the only Italian plenary lecture at the congress, and that it was he, Severi, who should be the leader of the delegation, in place of the chosen Scorza — Severi even seems to have threatened to withdraw his plenary lecture from the congress if he were not appointed lead delegate. In the event, as we have seen, Severi's demands were met.[41]

Given his stature within Italian mathematics by this time, we must question the reasons for Severi's absence from the list of would-be Oslo delegates four years later; indeed, recall from Section 2.8.1 (see in particular Table 2.5 on p. 41) that Severi had been considered as a speaker for the Oslo ICM, and had even made it as far as the preliminary programme with a talk entitled "Théorie et questions nouvelles dans la géométrie algébrique", to have been delivered in parallel with[42] that of Oseen.[43] Moreover, Severi's appointment in Zurich in 1932 as the chairman of a committee tasked with considering the future of the IMU (**Lehto, 1998a**, p. 58) gave him particular reason to go to Oslo (Section 1.2.3), as did his position as a member of the Fields Medal awarding committee (on which, see Section 9.22). Some archival materials shed a little light on the matter: on 21 April 1936, the UMI considered the case for sending an Italian delegation to the upcoming ICM, and resolved to place the matter in the hands of the government, which evidently decided that participation would not be appropriate in light of the sanctions. Severi appealed directly to the Minister of Education for permission to attend the congress, but this was denied on 30 May 1936, in a letter to the rector of the University of Rome

[39]It must be noted, however, that in contrast to Severi, Lietzmann was no first-class mathematician, and therefore needed the support of Erhard Schmidt when it came to the official representation of German mathematics at the Oslo congress — see the quotation on p. 57.

[40]The quotation appears also in (**Nastasi, 2001**, pp. 250–251).

[41]Although Scorza apparently communicated to Severi his intention *not* to go to Zurich (**Guerraggio and Nastasi, 2005a**, p. 189), the congress proceedings give every indication that he did in fact attend the ICM (for example, he is listed as having chaired a session: (**ICM, 1932**, vol. I, p. 37)). We note that Scorza was a prominent figure within ICMI (**Giacardi, 2009**); he was elected a vice president in Zurich. Bompiani also attended the congress — like Scorza, he represented Italian organisations and acted as a session chair (**ICM, 1932**, pp. 21, 37), but did not deliver a lecture, although he had previously spoken in both Cambridge (**ICM, 1912**, vol. II, pp. 22–27) and Bologna (**ICM, 1928**, vol. IV, pp. 277–284).

[42]See the remark on the notion of "plenary" in footnote 44 on p. 40.

[43]Recall that draft lists of names in the Størmer Archive were the main source of the information in Section 2.8.1. Similar documents in the Bessel-Hagen Papers in Bonn allow us to be a little more precise about the timeline of Severi's proposed involvement in the ICM. His name first appears in a draft programme that must post-date February 1936, since Guldberg (d. 15 February 1936) no longer appears on the list of organisers. An earlier list of proposed invitees, dated Autumn 1935, does not feature Severi's name at all — he must still have been on the "reserve" list at this stage.

(**Guerraggio and Nastasi, 2005a**, p. 249).[44] Two days before, on 28 May 1936 the Italian Ministry of Foreign Affairs had dispatched a letter to the Norwegian Foreign Ministry saying, in reply to a letter sent to Mussolini (who was both Prime Minister and Foreign Minister) on 16 March, that "the Royal government is not willing to send [non à in grado di inviare] an official delegation" (KUD). Some further information about the background on the Italian side is contained in files kept in the Central State Archives in Rome concerning the Council of Ministers, which was led by Mussolini as Prime Minister.[45] The file contains a request from the Minister of Education, dated 7 April 1936, for a ruling on an inquiry from the president of the Circolo matematico di Palermo about the possibility of taking part in the Oslo congress. Several reminders of this request follow in the file, ending in a circular letter from the Foreign Minister (i.e., Mussolini) to the Minister of Education and the Interior Ministry, among others, dated 19 May 1936:

> ... on the basis of decisions taken by the chief of the government and foreign minister ... the Royal government is not willing to send an official delegation to the International Congress of Mathematics ...
>
> Because the decisions taken pertain also to unofficial [apparently non-governmental] participation, please note that ... clearance cannot be given for the participation of the Circolo Matematico di Palermo, the University of Naples and the professors Angelo Tonolo, Allessandro Terracini, Berzolari and Gino Fano, and that it will be necessary to cancel the clearance for Prof. Gaetano Scorza ...
>
> It would also be convenient that the Royal Minister of National Education should kindly inform of the above His Excellency, the Member of the Academy of Italy, Prof. Severi, who had the intention of taking part in the international meeting under discussion.[46]

In the file, there follow several appeals by Severi against this decision, with negative telegraphic responses from the government. The last refusal in the file, dated 2 June, concludes with the sentence:

---

[44]As a result, we find Gaston Julia, vice chair of the commission of which Severi was chair (see above), explicitly lamenting the latter's inability to attend the congress in his closing remarks in a session concerning the future of the IMU (**ICM, 1936**, vol. I, p. 47).

[45]Archivio Centrale dellos Stato, Roma, Presidenza del Consiglio dei Ministri, Gabinetto (1934–1936), b. (folder) 2022, f. (file) 14.3.6375 "Oslo. Congresso Internazionale dei Matematici". Source kindly provided by Annalisa Capristo.

[46]"... in base a disposizioni impartite da S.E. il Capo des Governo, Ministro degle Affari Esteri ... non è i grado di inviare una delegazione ufficiale al Congresso Internazionale di Matematica ...

"Poichè le disposizioni impartite concernono anche gli interventi non ufficiali, si fa presente, ... che non potrà accordarsi il nulla osta per la partecipazione al Congresso del Circolo matematico di Palermo, del l'Università di Napoli e dei Proff. Angelo Tonolo, Allessandro Terracini, Berzolari e Gino Fano, e che converrà ritirare il nulla osta relativo al Prof. Gaetano Scorza ...

"Sembrerebbe inoltre conveniente che il R. Ministero dell Educazione Nazionale si compiacesse informare di quanto precede S.E. l'Accademico d'Italia, Prof. Severi, il quali avrebbe manifestato l'intenzione di assistere alle riunione internazionale die cui si tratta." (*Ibid.*)

> Of course, if in the period mentioned the international situation changes, the question of participation might be raised again.[47]

At the time, the sanctions against Italy hadn't yet been lifted. This happened officially on 15 July 1936, i.e., during the congress (**Baer, 1976**, pp. 298–299), by which time it was probably too late for Italian participation. It seems also that Mussolini wanted to demonstrate to the Norwegians the consequences of the sanctions in cultural-scientific domains. As Halvdan Koht, who saw these implications directly at the ICM, later noted, Norway was one of the first countries to reach a new trade agreement with Italy (**Bakke, 2016**, p. 84). This happened on 25 August 1936, after the ICM, and had been pressed for by Norway, among several reasons being a desire to be able to sell their stockfish.

By way of bringing this chapter to a close, we observe that even in their absence from the congress, Italian mathematicians were still called upon to participate in the activities of the international mathematical community: for example, Levi-Civita was appointed a member of the commission that would consider the award of the next Fields Medals (**Nastasi and Tazzioli, 2005**, p. 226) — but of course no further award would be made until 1950, nine years after Levi-Civita's death. Although a number of foreign applied mathematicians were hosted at the Fifth Volta Congress in Rome in 1935,[48] the overall trend post-1936 was the withdrawal of Italian mathematicians from the international scene. It seems likely that even if the ICM proposed for the USA in 1940 had taken place, no Italian delegates would have been present.[49] An "international congress" on a lower level took place in November 1942 in Rome (DKZ 3); it was attended by a German delegation headed by Hasse.[50] Other mathematical congresses were taking place in Italy around this time, namely the congresses of the UMI, but these were staged as largely national affairs, with little international interest (**Guerraggio and Nastasi, 2005a**, p. 267). It would not be until the revival of the ICMs in 1950 that Italian mathematicians once again resumed their place in the international community.[51]

[47] "Naturalmente, se per l'epoca predetta la situazione internationazionale si fosse modificata, potrebbe essere ripresa in esame la questione della partecipazione italiana al Congresso di cui si tratta." (*Ibid.*)

[48] On the place of the congress in the development of supersonic flight, see (**Anderson, 2006**, pp. 78–79) and (**Ferrari, 1996**). According to (**Eckert, 2017**, p. 235), who quotes the participant Theodore von Kármán, the Volta congress in 1935 was a highly propagandistic event for Italy. Mussolini used it to proclaim the invasion of Ethiopia.

[49] Indeed, the Italians actively dissuaded the Germans from attending: see p. 133.

[50] Around the same time, discussions of plans for the post-war restructuring of European science that were being held at the German Ministry of Education were taking into account the possible positions of Italy within these arrangements (**Siegmund-Schultze, 2002**, p. 346).

[51] For some brief comments on Italian attendance of the post-war ICMs, see (**Bottazzini and Nastasi, 2001**, §5).

CHAPTER 6

The Congress in the Norwegian Dailies

6.1. Mathematical refugees and other political and social issues

The Oslo ICM received a great deal of attention in the Norwegian dailies. The newspapers reported on the general programme of the congress, and covered its social events, including the honouring of Abel and Lie, as discussed earlier (see, for example, Section 2.9.1). Indeed, we have already made use of these Norwegian newspaper reports several times in this book. They were also mentioned in a report of the German embassy to Berlin that we quoted in Section 3.3.

We believe that it is interesting to look at those newspaper reports that go beyond the record of the congress that appeared in the two volumes of the proceedings in 1937. Among these are reports based on interviews with participants in the congress, often containing photographs of them.[1] These interview-based articles are at the same time most relevant to one of the main aims of our book, namely to describe the peculiar political circumstances under which the Oslo ICM was held. We have therefore decided to include here, as Appendix A, full English translations of all Norwegian newspaper articles that we could find that concerned the participants in the congress. With much more help from H. K. Sørensen, we have found a total of nine articles in Norwegian dailies — all published during the days of the congress (from 13 to 18 July) — which give biographical information about (for the most part) foreign participants in the congress. The articles appear as Appendices A.1 through A.9, following the chronological order of their appearance. Two of the articles (the very short A.5 and A.6) address summarily and speculatively the absence of Russian mathematicians from the congress, another one (A.8) is based on an interview with the Norwegian mathematical educator Kay Piene. The Norwegian dailies where these articles appeared include the conservative *Aftenposten*, the liberal *Tidens Tegn* ("Sign of the Times"), and the social democratic *Arbeiderbladet*. While emotions ran high in these outlets otherwise in connection with Ossietzky and Trotskiy, we find that the congress did not provoke much controversial opinion, at least in Norway.

The overwhelming political background to the reports in the dailies was of course the massive expulsion of mathematicians from Nazi Germany in those years.[2] Along with the increasing importance of mathematics for industry and war, emigration was among the most consequential social factors for change in mathematics in the first part of the twentieth century. The Nazi expulsions started with the infamous anti-Semitic "Law for the Restoration of the Civil Service" of 7 April 1933. They were extended as a result of the anti-Semitic Nuremberg Laws (proclaimed

[1] The photos are often of too poor a quality for reproduction here, although some images are included in Appendix A in order to give an impression of these articles.

[2] For details, see (**Siegmund-Schultze, 2009**).

at a Nazi Party conference in September 1935), reached another peak with the annexation of Austria in March 1938, and continued during the war from 1939 with Germany's occupation of almost all of Europe.[3]

During the early period of expulsions, the Nazi bureaucracy paid pensions to the dismissed professors; "private docents" such as Emmy Noether did not have this social status. The Nazis often used a stipulation in the 1933 law to reduce pensions because of alleged "political unreliability", which affected mathematicians such as Fraenkel and Emmy Noether's brother Fritz (see below). Since pensions became void in case of employment abroad, the Nazis supported emigration in the first years after 1933. Refugees had usually to leave behind all other belongings as well. Later, in particular during the war, many refugees lost their German citizenship, although the Jews among them had after the 1935 laws become second-class citizens anyway.

In one of the articles appended below (A.7), the journalist remarked:

> Among the German mathematicians participating in the congress of mathematicians, no fewer than ten are emigrants.

One can, however, safely say that the figures for refugees were much higher and that already during the congress the following 19 participants (according to the proceedings) were refugees from Germany: F. Behrend, H. Busemann, R. Courant, W. Feller, W. Fenchel, A. Fraenkel, E. J. Gumbel, H. Heilbronn, K. Hirsch, B. Kaufmann, P. Nemenyi, O. Neugebauer, B. H. Neumann, R. Rado, W. Sternberg, O. Szász, F. Noether, A. Weinstein, and H. Weyl.

Five of the nine articles use interviews with non-Norwegian participants in the congress who worked at the time in five different countries, namely with Harald Bohr and — briefly — Johan Frederik Steffensen (A.1, Denmark), Philipp Frank (A.2, Czechoslovakia), Abraham Adolf Fraenkel (A.4, Palestine), Emil Julius Gumbel (A.7, France), and Fritz Noether (A.9, Soviet Union). It is revealing that — apart from Steffensen — all the mathematicians (or physicists, in Frank's case) interviewed in these five articles had either at the time already been expelled from Nazi Germany, or would be forced to flee abroad after the occupation of their countries (Bohr, Frank). It does not seem accidental that no scientist living in Germany was interviewed, not even A. Rosenthal or M. Dehn, who had already been dismissed from their positions. Such an interview would probably have been burdened by embarrassment on both sides, the journalist and the interviewee.

In the interviews that were performed in the five articles mentioned above, Fraenkel and Noether claimed to take an apolitical stance, while Bohr only passingly revealed his grudge against the growing nationalism in international relations. But in the end, political themes could not be avoided in any of these interviews with the five scholars who all had Jewish backgrounds and also shared clear political dissent from the Nazi regime. Gumbel, who had already left Germany before 1933, was a special case. During the 1920s, he had been the most outspoken critic of the rising Nazi movement[4] and was dismissed by German-nationalist and anti-Semitic colleagues from his position in Heidelberg even before the Nazis came to power. Gumbel's fate is in a particular way intertwined with the dismissal of other German

[3] Among the few countries spared occupation were the United Kingdom, Switzerland, and Sweden. The latter received many refugees during the war from occupied Norway and Denmark.

[4] We have seen him mentioned (p. 78) in Lietzmann's congress report as "the infamous Herr Gumbel" (see also Appendix D, pp. 271, 278).

mathematicians after 1933. Those who had publicly opposed Gumbel's expulsion were then punished by the Nazis who used the occasion to reduce their pensions after dismissal. Fritz Noether was one of those against whom the Nazis took revenge in this manner. Noether was also the one refugee who suffered the most: he lost his life under the Soviet regime, an indirect consequence of his previous flight from Nazi Germany (see below).

Political issues were raised also in the other four articles in Appendix A. The article in A.3 covered short "lightning interviews" with the historian of ancient mathematics and astronomy Otto Neugebauer (Denmark), the president of the congress Størmer (Norway), the mathematicians Øystein Ore (USA), Fritz Noether (then Tomsk, Soviet Union), and Schur (Japan?), together with the physicists Georges Lemaître (Belgium) and Manuel Sandoval Vallarta (Mexico). Again, it was the emigrants among them (namely Neugebauer and Noether) who raised some political issues. Neugebauer had been a decided political dissenter from the Nazi regime when he left Göttingen in 1934.

Several interviewees (Bohr, Ore, Gumbel) stressed that the USA had meanwhile become the centre of mathematics (in place of a European centre such as Göttingen); this development was of course to a considerable degree influenced by the mass expulsions and thus inherently political as well.

Apart from the expulsions and Nazi rule in Germany, further political background to the 1936 ICM were provided by fascism in Italy and Stalin's dictatorship in the Soviet Union, which, as we have seen, led in different ways to the non-participation of Italian and Russian mathematicians in the congress. As mentioned above, the conditions in Russia are conjured up in the two interviews with Fritz Noether (A.3 and A.9), and in A.5 and A.6. A very short allusion to the sanctions against Italy due to the occupation of Ethiopia may be found in the lightning interview with the Norwegian-American Øystein Ore (A.3), who expressed indirectly his admiration for the leader of the Ethiopian resistance, Haile Selassie.

Even the mathematician from the host nation, Kay Piene — who would be deported to a northern province of Norway (Finnmark) after occupation of the country in 1940 — did not refrain from raising political issues in his newspaper interview (A.8). Piene commented on the peculiarity of educational reform in Nazi Germany and the growing emphasis there on applications (see also Chapter 10).

Allusion to mathematical and physical content in the newspaper reports was of course very limited. Some journalistic wording (related to the squaring of the circle in the contribution on Bohr, and on non-Euclidean geometry in the article on Frank) lent itself to possible misunderstanding.

We will add in Appendix A occasional very short footnotes to the nine articles from Norwegian dailies. All but two (A.1 and F) of the ten appendices drawn from Norwegian newspapers appeared anonymously; the two exceptions were signed by initials only. The remaining part of this chapter gives — in the chronological order of the articles in Appendix A — some additional biographical information and some selected commentary on the interviewees.

6.2. Commentary on Appendix A

A.1: Harald Bohr (1887–1951). Bohr was a noted authority on quasi-periodic functions (**Bohr, 1932**). In Oslo he was elected as one of twelve vice presidents of the congress (**ICM, 1936**, vol. I, p. 44), but, according to the proceedings, he did not give a talk. He had been involved for decades in international organisational work for mathematics (**Ramskov, 1995**). Together with his famous brother, the theoretical physicist Niels Bohr, he was instrumental in helping refugees from Nazi Germany to find positions in Denmark and elsewhere (**Sørensen, 2014**). After the German occupation of Denmark in 1940, and being "half-Jewish" according to Nazi terminology, he and his brother had to flee in 1943 (to Sweden and the USA, respectively).

What is scientifically interesting in the newspaper article is Harald Bohr's denial of mathematical collaboration with his brother Niels (because their fields differed too much), and his criticism of non-mathematical Austrian economics.

Bohr's anger about the growing nationalism is evident from one particular remark: "how should I express myself to avoid becoming too aggressive" ("hvad skal jeg si for ikke å bli for aggressiv"); for the full quotation, see p. 51.

For the Norwegian readers of *Aftenposten*, the most interesting parts of the interview were probably Bohr's reflections on the architectural style in Oslo and his admiration for Abel.

A.1: Johan Frederik Steffensen (1873–1961). Steffensen — listed erroneously in the proceedings as J. E. Steffensen (**ICM, 1936**, vol. I, p. 36) — is mentioned in passing in the same article. He was a Danish mathematician, statistician, and actuary who did research in the fields of calculus of finite differences and interpolation. He was professor of actuarial science at the University of Copenhagen from 1923 to 1943. Steffensen's inequality and Steffensen's method (an iterative numerical method) are named after him. Steffensen chaired one of the sessions at the congress (**ICM, 1936**, vol. I, p. 17), but apparently gave no talk.

A.2: Philipp Frank (1884–1966). Frank was a mathematical physicist from Vienna, a friend of Einstein and successor to him as a professor in Prague (1912). He later focused mainly on the philosophy of physics. He became a Czechoslovak citizen with the emergence of the formerly Habsburgian provinces as a republic after the First World War. Together with his life-long friend, the applied mathematician Richard von Mises, he contributed to a positivistic philosophy of mathematics and physics (**Siegmund-Schultze, 2007**). In 1932, Frank published the influential book (**Frank, 1932**), which is mentioned in the article in *Aftenposten*. As also revealed in the article, Frank took part in the Second International Congress for the Unity of Science, which was held in Copenhagen in June 1936 (see (**Werkmeister, 1936**) and also our p. 157). This gave him the opportunity to visit the ICM in Oslo, while his friend von Mises, a refugee in Istanbul at the time, found the distance to Norway too great.[5] Following the German occupation in 1939, Frank, who was of Jewish origin, had to leave Czechoslovakia for the USA where he met von Mises at Harvard.

The journalistic description of non-Euclidean geometry in this short article is misleading. It could create the impression that the truth of Euclidean geometry

[5]See footnote 12 on p. 148.

(where the sum of the angles in a triangle is 180°) can be proved by mathematics ("on the blackboard"), whereas Frank certainly meant the opposite: non-Euclidean geometries, which can be in good agreement with measurements and thus suitable as mathematical models, do not contradict mathematics and logic.

A.3: Otto Neugebauer, Carl Størmer, and Øystein Ore. These three scholars — the historian, and the applied and pure mathematicians — gave plenary talks in Oslo which are, together with some scientific-biographical information, discussed in Sections 9.9, 9.2, and 9.20, respectively.

The interviews in *Arbeiderbladet* with Neugebauer and Størmer contain some allusions to the contemporary political discussion about "racial differences" in mathematics. Neugebauer seems to refer directly to the infamous Nazi theories which served as a "justification" for the expulsions (**Siegmund-Schultze, 2009**). Generally speaking, the ease with which the notion of "races" was connected to "nations" ("nasjoner og raser") in this newspaper article (by the journalist or by the interviewees) shows the ubiquity and rather innocent meaning of wording such as "race" in those years before Auschwitz.

The short interview with the Norwegian-American mathematician Øystein Ore, which is contained in the same article, is historically very interesting. Ore, apparently an old acquaintance of the journalist from school years, had emigrated to the USA in 1927 due to insecure job prospects in Norway. Ore confirmed the newly acquired leading position of American mathematics, and related it to emigration from Germany. His jocular remark about a costume party which he missed due to the congress seems to indicate sympathies for the Ethiopian Emperor Haile Selassie I and his resistance against the Italian aggression. Selassie had become a popular figure that year following a speech to the League of Nations in May 1936 condemning Italy's use of chemical weapons in his country. The League of Nations voted to lift its sanctions against Italy on 15 July 1936 (**Baer, 1976**).

A.3: Georges Henri Joseph Edouard Lemaître (1894–1966). Lemaître was a Belgian cosmologist and priest, and precursor of Hubble with respect to the theory of the expanding universe. He was also a researcher on cosmic rays, a phenomenon that came under increased scrutiny in physics in the 1930s. Størmer found that cosmic rays lent themselves even better to his own numerical method than the Northern Lights (on Størmer's work on these, see Section 9.2).

A.3: Manuel Sandoval Vallarta (1899–1977). Vallarta was a Mexican physicist who collaborated closely with Lemaître on cosmic rays but was at the time at MIT in Boston/Cambridge, Massachusetts. Their joint work was quoted by Størmer in his plenary talk. The two physicists apparently had a joint presentation in section V (Mathematical Physics, Astronomy and Geophysics) of the ICM under the title "Results of Calculations of Asymptotic Trajectories in the Field of a Magnetic Dipole with Applications to Cosmic Radiation" (**ICM, 1936**, vol. II, pp. x, 237). This was closely related to Størmer's lecture. But Lemaître and Vallarta refrained from publishing an abstract of their talk in the proceedings, and it appeared instead in *Physical Review* shortly afterwards (**Lemaitre and Vallarta, 1936**).

As indicated in the article in *Arbeiderbladet*, Vallarta worked on Størmer's theory using Vannevar Bush's analogue computer, the "Differential Analyzer" in

Boston. A copy of this machine would be installed in the Astrophysical Institute in Oslo the next year, mainly organised by the Norwegian physicist Svein Rosseland (**Holst, 1996**). This machine was obviously very relevant for Størmer too, who worked partly at that institute (see also p. 71).

A.3: Fritz Noether (1884–1941). On the applied mathematician Fritz Noether we say more below in our commentary on the more detailed interview in *Tidens Tegn* (A.9). But the short interview with him in *Arbeiderbladet* provides interesting additional political points. The reference to the "small pension" offered by the Nazis seems to confirm the point made above about political persecution against him in addition to the anti-Semitic discrimination to which he was exposed. His claim of being "leftist" but no communist can be confirmed from his biography in Germany, but was also meant to reassure readers of the social democratic paper and to reduce some of the prevalent fear of Soviet Russia.

A.3: The Japanese mathematician Schur (?) and Matsusaburô Fujiwara (1881-1946). The list of participants in the congress proceedings, which we have shown to be rather unreliable (see, for example, Section 4.1), lists four mathematicians from Japan, and just one more Japanese who was apparently at that time in Germany.[6] There is no Schur (or any alternative transliteration thereof) among these, nor does the name appear elsewhere in the proceedings. The article refers briefly also to Fujiwara. He was a professor at the Tohoku Imperial University in Sendai. He is listed on zbMATH (accessed 15 August 2017) with 109 publications between 1906 and 1942, covering a range of topics including geometry, numerical analysis, and number theory; a printed list of his publications appears in (**Anon, 1949b**). Fujiwara appears also to have had an interest in the traditional Japanese mathematics Wazan (see, for example, (**Fujiwara, 1941**)). For a brief biography, see (**Kubota, 1949**).

At the ICM in Oslo, Fujiwara gave a sectional talk in German with the title "Ein Problem aus der Theorie der Diophantischen Approximationen", in which he referred repeatedly to the work of the German mathematicians A. Hurwitz and O. Perron (**ICM, 1936**, vol. II, pp. 13–14).

A.4: Adolf Abraham Fraenkel (1891–1965). As a mathematician, Fraenkel is most famous for his work in set theory, in particular the so-called Zermelo–Fraenkel axioms (1922). He was a refugee to Jerusalem where in 1925 the Hebrew University had been founded. Because the academic tradition in Palestine was not old or established, Fraenkel and other refugees in this region had to face particular problems connected to scientific communication, availability of academic jobs, etc.[7] The interview is revealing in this respect. Fraenkel told the interviewer that he had already been employed in Jerusalem from 1929 and thus that his departure from Germany had been "entirely voluntary".[8] But he or the journalist did not tell the reader that Fraenkel had returned to Germany in between and that in 1933 he was

[6] See footnote 48 on p. 42.

[7] We have already noted the problems of Sternberg in finding employment as a mathematician, which caused him to go to the Oslo ICM in search of alternatives: see p. 96.

[8] Likewise other refugees stressed repeatedly the voluntariness of their departure: for instance Hermann Weyl (**Siegmund-Schultze, 2009**, p. 177). This seems to have been psychologically connected to the stigma (however unjustified) of failure which is often connected to emigration. Fraenkel's memoirs (**Fraenkel, 1967/2016**) contain several slips of memory which might have had psychological reasons and thus have been signs of the trauma of emigration.

given no choice but to leave again. Fraenkel's declared intention in the interview not to "touch upon anything that has to do with politics" is clearly illusory.

A.5 and A.6: Russians who did not come. These two very short notes in the morning and regular issues of the same newspaper on 15 July 1936 reveal the almost total lack of information about the reasons for the non-appearance of the Russian mathematicians (on which, see Chapter 4).

The subtitle in A.6, "Were they afraid they would not return to the paradise?", suggests that the Soviet authorities were afraid about possible defection of mathematicians to the West. The one "Russian scientist",[9] who according to Poul Heegaard had defected a year before, may well have been the theoretical physicist George Gamow (1904–1968), although he had in fact already escaped in 1933.

A.7: Emil Julius Gumbel (1891–1966). The newspaper article on Gumbel deserves a somewhat more detailed commentary, because of its size and because of the particular political role that Gumbel played, as partly outlined above in the commentary in Section 3.2 on Lietzmann's report and in Section 6.1. Gumbel had been a leftist political activist, dating from the time after the First World War and the rise of nationalism in Germany. For his courage he was admired by Einstein.

Gumbel was an international authority on the statistics of rare events (**Gumbel, 1958**). The aim of a statistical theory of extreme values is to analyse observed extremes and to forecast further extremes. In the text of the newspaper article, Gumbel referred to the observed values of the maximum weight of whales and maximum age of people in a given population (for example, 100), taking the latter data from Sweden. Gumbel explained in particular that his theory allowed him to forecast the age of the oldest man in Sweden in one hundred years but did not enable conclusions for the case of an individual.

Gumbel presented two talks from his statistical theories in section IV (Probability) of the congress:

- "Der größte Wert einer statistischen Veränderlichen" (**ICM, 1936**, vol. II, pp. 200–203);
- "Das Grenzalter" (**ICM, 1936**, vol. II, pp. 203–205).

The journalist apparently misunderstood Gumbel on various points, particularly with respect to the alleged dismissal of Gumbel by the Nazis.

In the interview in *Arbeiderbladet*, Gumbel alluded to the mathematical journal *Deutsche Mathematik*, run from 1936 by the Nazi mathematicians Ludwig Bieberbach and Theodor Vahlen. This journal carried valuable mathematical contributions like those by Oswald Teichmüller. But it also published ideological articles in the spirit of Bieberbach's racist theories of the same name, "Deutsche Mathematik", according to which there was an incompatibility between the German and Jewish minds in mathematics. These theories were of course used to separate Jewish professors from German students and to justify expulsions. Gumbel alluded to the following motto taken from a speech by Hitler at the 1935 Reich Party Conference of the NSDAP in Nuremberg, which was published as p. 1 in the first issue of

[9]Note that Heegaard did not allude to a mathematician in particular, but to a "scientist" ("videnskapsmann").

Deutsche Mathematik:

> Who wants to instill pride in the people has to create visible items of pride for the people.
>
> The Führer at the Reich Party Conference of Freedom.[10]

"Freedom" was here a code-word for rearmament and "items of pride" were apparently mathematical works, as arcane as they may have appeared to the majority of "the people".

Gumbel also referred to the keynote article in the first issue of *Deutsche Mathematik*, which was written by the former mathematics student Fritz Kubach, now responsible for mathematics within the NS students association (Reichsfachabteilungsleiter Mathematik der Deutschen Studentenschaft): "Students to the front!" There one finds for example the following passage:

> Decisive ... are not the formulas and otherwise rationally understandable results but only and exclusively the question of the creative act, which leads to these results. ... In barely any other field the situation is such as in mathematics where the overwhelming number of the present lecturers has no understanding at all of these questions which are today for us the most elementary and the most decisive. ... It suffices to say that the influence of Jews in mathematics which in previous times had been growing immensely — both with respect to the selection, development and forming of scientific questions and with respect to the assigning of the professorships — is still palpable today.[11]

In his interview Gumbel pointed to the absurdity of the race theories, and to the paradox that among the quintessential Nordic people, the Norwegians, the theories were not very popular. Interviewed for *Arbeiderbladet*, the paper of the Norwegian social democratic *Arbeiderparti* (founded 1887), Gumbel connected this fact to the existence of the social democratic government in Norway (see our p. 49).

Apart from possible misunderstandings by the journalist, some remarks by Gumbel on the dismissal figures in Heidelberg (for instance that less than fifty per cent of the dismissed were Jewish) or that all new professors appointed by Nazis were "nobodies instead of scientific researchers" have to be taken with a pinch of salt. They do not fully agree with later historical work.[12] At the very least they are of a more peculiar local character and are not typical of developments in other places in Germany at that time.

A.8: Kay Waldemar Piene (1904-1968). Piene came from an old Norwegian family of artists and scientists.[13] Although originally interested in mathematical research, the poor job prospects in the late 1920s induced Piene to take the state exam for teachers in 1929. In 1931–1932 he studied mathematics in Göttingen. At the time of the congress he was a teacher at a famous private school in Oslo (*Vestheim skole*). Piene became a member of the "Norwegian Carnegie Committee for Exams", which was partly funded by the Americans. This work was interrupted

[10] Translated from the German in *Deutsche Mathematik* **1**(1) (1936), p. 1.

[11] Translated from the German in (**Kubach, 1936**, p. 6).

[12] See (**Weckbecker, 1985**).

[13] See also Chapter 10, as well as R. Siegmund-Schultze: Kay Waldemar Kielland Piene, `http://www.icmihistory.unito.it/portrait/piene.php` (2012, last accessed 11 June 2018).

by the war and by the German occupation of Norway in 1940. Together with other so-called "Kirkenes teachers", Piene was arrested by the Norwegian Nazis in 1942 and deported to the far north, to Kirkenes at the Russian border, for eight months. After the war, Piene became a professor of mathematics education in Oslo.

At the 1936 ICM, Piene presented the report on the state of Norwegian school mathematics to the International Commission for Mathematical Instruction (ICMI) (**ICM, 1936**, vol. II, p. 288). Like other such reports, this one, written in German, was not published in the congress proceedings but in *L'enseignement mathématique* (**Piene, 1937**). The report contains rather critical remarks about the endangered place of mathematics in Norwegian schools (see Chapter 10). Some of these remarks also appear in the newspaper interview in *Arbeiderbladet*, although when comparing Norway with other countries, Piene came to the conclusion that "we are not behind any of these countries". Probably the most politically interesting aspect of this article is Piene's remark about Germany when he alluded to the German report to ICMI:

> It becomes evident that even such a neutral subject as mathematics is influenced by politics, and you can indirectly trace Nazism also in the teaching of mathematics these days. Schools are now no longer supposed to be for the individual pupil but primarily serve the country. Germany now deliberately aims to strengthen applied mathematics in schools whereas the rest of us emphasise theory more. The fatherland must be furthered through education, that is the purpose of the entire German teaching plan now.

A.9: Fritz Noether (1884–1941). Fritz Noether, the younger brother of the much more famous Emmy Noether, was an accomplished applied mathematician and mathematical physicist. At the Oslo ICM he presented a talk in section V (Mathematical Physics) with the title "Über elektrische Drahtwellen" (**ICM, 1936**, vol. II, pp. 234–235). In the published abstract, he referred to the classical handbook on integral and differential equations edited by P. Frank and R. von Mises (**Frank and von Mises, 1925–1927**).

Fritz Noether had been one of the few refugees from Nazi Germany (Breslau) who ended up in the Soviet Union (Tomsk). He apparently kept his German nationality, a fact that enabled him to visit Oslo from Russia in 1936, while other mathematicians from the Soviet Union were not allowed to go. But his stay in Oslo later probably counted against Noether when he was treated as a suspicious foreigner under Stalin's terror. (**Schlote, 1991**, p. 40) intimates that the stay of Stalin's main critic, Trotskiy in Norway, at exactly the same time as the congress may have contributed to suspicions against Noether.[14] He was finally sentenced under wrongful accusations and executed by the Russian secret police in 1941.

The article in *Tidens Tegn* from 18 July appeared under the title "A rail journey of six days to speak for 15 minutes in Oslo". Somewhat contrary to the journalist's claim, there were other participants, such as the Japanese, who had to take even

[14] The logician Jan van Heijenoort, who was Trotskiy's secretary at the time, did not mention the congress or any mathematical event in his book *With Trotsky in exile* (**van Heijenoort, 1978**). Nor did Trotskiy himself in his memoirs about his stay in Norway from 1935 to shortly after the congress in 1936, which are on the following website: `https://www.marxists.org/archive/trotsky/1936/12/nor.htm` (accessed 11 June 2018).

longer to reach Oslo. With transatlantic flights not being the regular means of transport, even the Americans from the East Coast of the USA would still have needed about five days for the ship transfer to Europe.

The interview with *Tidens Tegn*, which complements other information already given by Noether four days before (A.3) in *Arbeiderbladet*, provides interesting, if personally coloured, information about the mentality of Russian students and professors, and about the material conditions of teaching and research in Russia at the time. Direct commentary on the political conditions in the Soviet Union, in particular the reasons for the absence of Russian mathematicians, is avoided in the paper.

CHAPTER 7

International Mathematics Shortly Before and After the Second World War: A Glimpse Ahead and Back Again

We finish our investigation of the political side of the Oslo congress by giving a few hints about how mathematical communication on the eve of the Second World War became ever more stifled. We consider how mathematical communities worldwide became more and more segregated and isolated before the war, but then how mathematicians after the Second World War tried to reestablish the international community and find a way to handle the new situation of the Cold War. We look finally at how they looked retrospectively at developments before the war.

On p. 51, we mentioned a newspaper interview with Harald Bohr, dated 13 July 1936, which carried the subtitle "Princeton is the centre of the mathematical world, but for the moment it is Oslo" ("Princeton den matematiske verdens centrum, men i øieblikket er det Oslo"). In the interview, Bohr was asked which continent currently held the leading position in the mathematical world. He answered that at present, the USA had been fortunate enough to attract a number of German emigrants, whereby the mathematical epicentre had shifted from Göttingen to Princeton (on this theme, see also Section 2.10). However, Bohr also acknowledged leading mathematical schools in Russia and Japan, and he argued that the presence of role models (such as Abel or Lie in Norway) was important for the promotion of mathematical research.

Nevertheless, the Oslo congress was the first ICM in Europe at which US mathematicians were present in such large numbers. The major part of the delegation came from prominent universities on the East Coast, in particular from the large mathematical centres in Cambridge, Massachusetts (12), Princeton (8), and New York (8). Each of these centres sent contingents of delegates that would outnumber many smaller European nations. However, mathematicians from the West Coast were also represented, and one of the official representatives of the US government and the American Mathematical Society was Hans Blichfeldt (1873–1945), who had Scandinavian ancestors.[1] When he arrived in Copenhagen en route to Oslo, he was interviewed for a leading Danish newspaper; the article said nothing of the ICM in Oslo — instead, Blichfeldt described his journey to Copenhagen as a pilgrimage

[1]Blichfeldt had been born in Denmark but had emigrated to the USA at the age of 15 and worked his way westwards before being able to pay for his education at Stanford. He had then spent some time with Sophus Lie in Leipzig where he earned his doctorate in 1898, before returning to a career at Stanford, which made him a full professor in 1913, and chairman of the mathematical department in 1927 — see (**Dickson, 1947**) and (**Royden, 1989**).

to visit Harald Bohr.[2] This self-made mathematician must therefore have seen the ICM in Oslo as a return to his Scandinavian roots, exemplified by the unveiling of the bust of his former teacher, Lie (Section 2.9.1).

When the Oslo congress approached its conclusion, the issue of where to organise the next congress became a delicate matter. As noted in Section 2.3, Oslo had been preferred over competing invitations from Greece and the USA in the past, and these two bids were again clashing behind the scenes in Oslo. Two rather intimate letters from Heegaard to his close German friend Engel shed interesting light on the matter. We have already seen the first on p. 33, concerning the peculiarities of the organising committee. After the conclusion of the congress, Heegaard divulged, in private, some of those quirks to Engel; they concerned the placing of the next congress:

> The Greeks had sent an invitation for the next congress to be held in Athens, but Zervos and Hatzidakis had not obtained funding to travel [to Oslo]. A couple of young Greek mathematicians hid away nervously, but on the last day Størmer — who had gotten an appetite for going to Athens and did not want to promote the offer of the Americans — called upon Papaioannou. When asked whether the invitation was meant seriously, he [Papaioannou] struck his chest and spoke well and at length. Then I whispered a few words in the ear of Lefschetz and within moments, P. was surrounded by Americans who aggressively and indelicately asked whether Greece had the money: "We already have $15,000 to start with". Then P. bent his head and declared that of course Greece would — although with great regret — bend to the will of the rich America. Thus, the 1940 congress will be in America.[3]

As one might expect, the securing of the next congress for the USA engendered a positive outlook among American mathematicians. For instance, G. D. Birkhoff sounded optimistic in his talk at the Harvard Tercentenary later the same year:

> Let us hope that by the year 1940 conditions will have so much improved that we will be able to think again of the questions, such as that of a reasonable organization, without fear of such [political] entanglements. Without wishing to argue for the desirability of some kind of organization, I would merely point out that lacking it, it will be very difficult for any delegates at all to come from certain countries like Japan where the expenses have to be provided with government subsidies.[4]

Even two years later in 1938, in connection with the celebration of the 50th anniversary of the AMS, the historian of the latter, R. C. Archibald, was still up-beat:

> No political taint marred the delights of the Congress at Oslo. The IMU swansong was faintly heard at the last session. ... Plans are well advanced for holding the Congress 4–12 Sept.

[2] *Politiken*, 23 June 1936.

[3] EN: NE150305, Heegaard to Engel, 24 July 1936; see Appendix B for the full letter.

[4] Birkhoff Papers, 4213.4.5, box 1, personal (1936), p. 3.

> 1940 in Cambridge, Mass., at Harvard U. and Mass. Institute of Technology. (**Archibald, 1938a**, p. 20)

On 15 April 1938 a "Subcommittee on Conferences" referred to an earlier decision by the "Organizing Committee" for the 1940 ICM, led by Harvard mathematician W. C. Graustein (29 December 1937, Indianapolis). There it had been decided that the congress

> should consist of six sections, as follows (I) Algebra and Theory of Numbers; (II) Analysis; (III) Geometry and Topology; (IV) Probability, Statistics, Actuarial Science, Economics; (V) Mathematical Physics, Mechanics, Astronomy, Geophysics; (VI) Logic, Philosophy, History, Didactics.[5]

The subcommittee, headed by M. H. Stone, now made detailed proposals for more specialised so-called "conferences", to be organised within the forthcoming congress, basically restricting them to the more theoretical topics from the first four of the six "sections". The explanations given are interesting and deserve more commentary in detail, which cannot be given here. We refer to the excerpts in Appendix G. There was scepticism in the subcommittee *vis-à-vis* organising a conference on number theory "where a conference without the participation of Russian mathematicians would be disappointing"[6] while the success of the other "suggested topics is to a considerable degree independent of eventual difficulties, however grave, in securing the attendance of foreign speakers".[7] There was also scepticism with regard to the more applied fields, apparently including mathematics education, hoping that "appropriate scientific associations are sufficiently interested to cooperate financially and otherwise in organizing them".[8]

As a kind of commentary on the subcommittee's proposal, R. G. D. Richardson, the secretary of the AMS, wrote on 2 June 1938 to the mathematics educator W. D. Cairns (1871–1955) in Oberlin, Ohio:

> As you perhaps know, there is indifference and to some extent, hostility, on the part of some of the members of the Organizing Committee to having anything to do with pedagogy so the very best thought must be given to the program of the international commission to justify its inclusion in the Congress.[9]

One year later, on 17 May 1939, the Organizing Committee could report on many invitations which had meanwhile been sent to mathematicians both abroad and at home, but restricted to plenary sessions and the four "conferences" envisaged before. In a round letter with a special letter head "INTERNATIONAL CONGRESS OF MATHEMATICIANS. LOW MEMORIAL LIBRARY, 531 WEST 118TH STREET NEW YORK CITY, U.S.A." one reads about the foreign mathematicians invited:

> The Committee has invited the following mathematicians from other countries to deliver scientific addresses before the Congress: Alexandroff, Cramér, Denjoy, Dirac, Hodge, H. Hopf,

[5]MAA Headquarters, Austin, round letter, 15 April 1938, 5pp., M. H. Stone, chairman, p. 3.

[6]We may contrast this view with that earlier expressed by both Lietzmann and Weyl on our p. 82 concerning the (former) leading position of German number theory.

[7]*Ibid.*, p. 1.

[8]*Ibid.*, p. 3.

[9]Richardson to Cairns, 2 June 1938, MAA Headquarters, Austin, TX.

> Khintchine, E. S. Pearson, M. Riesz, Tarski, G. I. Taylor, Tinbergen, Tonelli, Vinogradoff, van der Waerden.
>
> In addition, it has invited the following mathematicians from abroad to speak at Conferences on special topics: Algebra -- Aitken, Chevalley, Deuring, P. Hall, Krull; Theory of Measure and Integration, Probabilities, and Allied Topics -- Banach, Cantelli, Hostinsky, Kolmogoroff, P. Lévy, von Mises, Ridder, F. Riesz, Steinhaus, L. C. Young, Zygmund; Topology -- Borsuk, Čech, Kuratowski, Pontrjagin, Stoilow; Mathematical Logic -- Bernays, Chwistek, Gentzen, Heyting, Kalmár, Scholz, Skolem.
>
> A full list of invited speakers is attached.[10]

Again, like in the letter of the Subcommittee for Conferences quoted above, the very pure orientation and the hugely extended format compared to Oslo are clearly visible. The only one of the invited mathematicians who had taken a prominent role in Oslo was Banach.

There were also plans for awarding the Fields Medals at the ICM in 1940, as transpires from a letter written by G. D. Birkhoff in March 1939. In full awareness of the responsibility of the Americans for the forthcoming congress of 1940, of his personal standing in the international community, and of the tediousness of the job, Birkhoff proposed G. C. Evans of Berkeley as chairman of the Fields Medal committee for 1940 (see also Section 9.22). Birkhoff explained to Evans the problems that had occurred after Hardy had been nominated chairman by the Oslo congress:

> At Oslo I was asked to make up a new Committee and I suggested a Committee headed by Professor G. H. Hardy with Professors Alexandroff, Hecke, Julia, and Levi-Civita. Hardy promptly resigned since he dislikes such things intensely and so Lefschetz, who was one of my alternates, was made Chairman in his place. The other alternate as a possible member of the Committee is Rolf Nevanlinna. Recently Lefschetz has become so disgusted with the state of affairs that he too refuses to serve on the Committee. ... A few days ago, when I saw Lefschetz ... we hit upon you as the best possible successor to him.
>
> It seems to me obvious that there should be an American on the Committee and this is another reason why you in your capacity as President of the Society would be the best choice.[11]

While there was optimism in America with respect to the forthcoming congress, in Germany and Italy, however, things only got worse when it came to international mathematical communication, both with respect to publications and personal meetings. To pick up on the discussion of Section 5.3, and to draw a parallel with what we saw in Chapter 4 for the Soviet case, the foreign contacts of Italian mathematicians came increasingly under suspicion during the years of fascism — particularly those contacts via the Circolo matematico di Palermo, still one of the most

[10]MAA Headquarters, Austin, 17 May 1939, signed Graustein and Stone, the copy from which we quote was addressed to W. D. Cairns. See Appendix H for a list of all mathematicians invited, including the ones in the USA.

[11]Birkhoff to Evans, 2 March 1939, Birkhoff Papers, HUG 4213.4.5. box 1, file personal (1938/39). Evans had been nominated and would act as president of the AMS for 1939/40. See also Section 9.22: in particular, footnote 170 on p. 226.

outward-looking of Italian scientific organisations, whose foreign membership was capped (**Brigaglia, 2002**, p. 197). Membership was reduced still further in 1938 with the introduction of Italy's anti-Semitic race laws, resulting more generally in the expulsion from any remaining official positions of a large number of Jewish mathematicians, including Volterra, Pincherle, B. Segre, Castelnuovo, Enriques, and Levi-Civita.[12] The situation was compounded on 10 December 1938, when the Italian Mathematical Union (UMI) issued a statement in which it asserted that the ousting of Jewish mathematicians would have no effect on Italian science — nevertheless there was concern, which indeed proved justified, that the university chairs left vacant in this way would be diverted to other disciplines (**Guerraggio and Nastasi, 2005a**, p. 265).

As one would expect, the image of Italian mathematics abroad suffered greatly from the effects of the expulsion of Jewish mathematicians. The race laws were enacted around the time (1938) that the Fifth International Congress for Applied Mechanics was due to take place in the USA, thereby preventing Levi-Civita from attending, and provoking the criticism of mathematicians elsewhere.[13] The dismissal of Levi Civita from the editorial staff of *Zentralblatt für Mathematik und ihre Grenzgebiete* similarly provoked international outcry.[14] In response to this image problem, the fascist government made 50,000 lire available for the publication of the collected works of the great Italian mathematicians of the nineteenth century, and funded the creation of the above-mentioned National Institute for Higher Mathematics, INdAM (**Nastasi and Tazzioli, 2005**, p. 229). The result of the race laws was, however, much the same as it was in Germany: the severe depletion of the Italian mathematical community.[15]

In Germany, international congresses for pure and applied mathematics were increasingly hampered by the Nazi dogma of anti-Semitism. As early as 1934, there had been plans to invite the International Congress for Applied Mechanics to Germany. In preparation for the 1938 congress, which ultimately took place in the United States at Harvard and MIT (see above), Ludwig Prandtl of Göttingen had explored the question of whether a German invitation for 1942 had a chance. He had the full support of Göring's Ministry of Aviation in Germany, and mechanics

[12]See, for example, (**Israel, 2000**) or (**Guerraggio and Nastasi, 2005a**, §8.2); (**Nastasi and Tazzioli, 2005**, p. 228) provide a list of Italian mathematicians who were prevented from teaching because of the race laws. On Volterra's exclusion, see (**Guerraggio and Paoloni, 2008**, §9.2) and (**Capristo, 2015**, §4). On the impact of the race laws on Italian science more generally, see (**Maiocchi, 1999**).

[13]See (**Nastasi and Tazzioli, 2005**, p. 227). Levi-Civita appears in the congress proceedings as a member of the congress' international committee; C. Guido of Turin is listed as another, although it is noted that he did not attend the congress either (**ICAM, 1938**, p. XXIII). In the membership list (**ICAM, 1938**, pp. XXIV–XXXI), we find three people with Italian addresses, namely F. Burzio (Turin), P. Ferretti (Naples), and M. G. Salvadori (Rome), although the latter two are explicitly indicated as not having been present at the congress. Nevertheless, the proceedings feature an article (**Salvadori, 1939**) by Salvadori concerning recent work at his home institution, the Istituto Nazionale per le Applicazioni del Calcolo (on which, see footnote 30 on p. 110).

[14]See (**Nastasi and Tazzioli, 2005**, p. 229) or (**Siegmund-Schultze, 2016b**, pp. 85–86). Levi-Civita was replaced by Severi (**Goodstein and Babbitt, 2012**, p. 1070).

[15]We note, however, that the relative timings of the enactments of the German and Italian race laws meant that some mathematicians fleeing Nazi Germany had gone initially to Italy: for example, the function theorist Peter Thullen (1907–1996) — see (**Siegmund-Schultze, 2009**, Appendix 6).

scientists internationally did not appear averse to coming to Germany, provided no distinction was made between Jews and non-Jews among the delegates. Despite intensive efforts, however, Prandtl could not guarantee the latter condition to his colleagues abroad. The Reich Ministry of Education, which was responsible for such matters, communicated to Prandtl on 28 August 1938 that "Jews of foreign citizenship" who took part in the congress would not be regarded "as Jews here", but that there would be no place for "non-Aryans of German citizenship".[16] The decision destroyed the hope for a congress in Germany in 1942, which, due to the war, would not have taken place anyway.[17]

Things were no better with regard to the preparations for the ICM in 1940. In February 1939 the Nazi Ministry of Education tasked the chairman of the DMV, Wilhelm Süss (1895–1958), with preparing a German delegation for the ICM in the USA. However, in a letter dated 20 February, the ministry asked first for information as to whether "the chairman of the organizing committee, Prof. W. C. Graustein, was a non-Aryan".[18] In his reply, Süss informed the ministry about the recent conflicts with the Americans around *Zentralblatt* and the American plans to found an independent reviewing journal.[19]

The 1938 dismissal of Levi-Civita as an editor of *Zentralblatt* (noted above), and the ensuing foundation of *Mathematical Reviews* in the United States have been much discussed in the literature.[20] Suffice it to say here that even mathematicians like Helmut Hasse, who were no old comrades of the Nazis, supported the Nazi measure to discontinue reviewing the work of "German authors" by Jewish mathematicians. Like Bieberbach in an attack on Harald Bohr in 1934,[21] Hasse promoted an apartheid-notion of internationalism. Fearing a German boycott of the planned 1940 ICM, Graustein argued against the proposed American reviewing journal (**Reingold, 1981**, p. 332). *Mathematical Reviews* finally appeared in 1940 under Otto Neugebauer's lead.

But this happened after the war began; in early 1939 the situation was different and — as we have seen above from the American activities — the plans for the ICM in 1940 had not yet been cancelled. In a letter to Süss, dated 7 June 1939, the Nazi ministry announced plans to send 15 mathematicians to the congress, a limited number due to problems with foreign currency (fol. 418). In another letter to Süss, dated 4 July, the ministry reported on the results of its enquiry concerning Graustein:

> As Prof. [Friedrich Karl] Schmidt, Jena reports, Graustein is German friendly, as are Stone and Bliss; Veblen, although <u>no</u> Jew, organizes the American fight against German mathematicians.[22]

[16]Quoted in (**Mehrtens, 1989**, p. 57).

[17]See, for example, (**Doel *et al.*, 2005**) or (**Eckert, 2017**).

[18]REM 2905, fol. 406.

[19]*Ibid.*, fols. 411–416.

[20]See, for example, (**Reingold, 1981**) or (**Siegmund-Schultze, 1993**; **2016a**).

[21]Bieberbach described Bohr as a "damager of all international cooperative work" (quoted in (**Mehrtens, 1989**, p. 52)).

[22]"Wie mir Prof. Schmidt, Jena, mitteilt ist Graustein deutschfreundlich, ebenso Stone und Bliss; Veblen, obwohl <u>kein</u> Jude, organisiert den amerikanischen Kampf gegen die deutschen Mathematiker." (REM 2905, fol. 423, emphasis in original)

One week later, on 11 July, the ministry was informed by the German Foreign Office that the Italian embassy had asked Germany not to participate in the congress:

> At this congress, which took place last in Oslo 1936 with German but without Italian participation ... Jews and anti-Fascists played a big role.[23]

In the ministerial file there follow some distancing remarks concerning the Italian request, but also the following commentary: "In principle, agreement with Italy in congress matters is desirable".[24]

Four months later, on 10 November 1939, the Italian request was forwarded to Süss for comment (fol. 435). This was of course after the outbreak of the war, following the German invasion of Poland on 1 September 1939, and the entire political situation had changed. One finds now inner-ministerial commentary, such as the following from 22 December 1939:

> I am in favour of not participating. It would be reasonable to bring Italy, Russia and Japan to the same opinion.[25]

At the same time, however, the German embassy in the USA still recommended the participation of German mathematicians in the planned ICM (fol. 443) — after all, the two countries were not yet at war with each other. It was only one year later, on 14 December 1940, that the Nazi Ministry of Education, based on information from the Foreign Office, informed Süss that the congress had been postponed indefinitely (fol. 450). In fact, Süss had known about this for almost a year, so one gets the impression that the Nazi ministry lacked international contacts.

Indeed, already on 23 November 1938, after the German Kristallnacht pogrom and the *Zentralblatt* affair, Archibald had become less optimistic with respect to the congress in America. In contrast to a passage that he had published in the context of the AMS celebrations earlier in 1938 (see above), he now wrote to colleagues who were involved in the preparation of the didactical section of the congress:

> I doubt very much if any German will be permitted to attend an International Congress in the United States ... One of the chief reasons is that the congress is to be held at Harvard University and the Massachusetts Institute of Technology. When Professor Hecke of the University of Hamburg visited the United States last winter he had the strictest orders not to accept any invitation to lecture at Harvard University. The reason for this was that Harvard University had not seen fit to accept from a pro-Nazi a sum of money[26] which would assist an American student to carry on study at German universities as administered by the present regime.[27]

[23] "Auf diesem Kongress, der zuletzt 1936 in Oslo mit deutscher, aber ohne italienische Beteiligung stattfand, spielten, nach Ansicht des Herrn Dr. Ridomi, Juden und Antifaschisten eine grosse Rolle." (fol. 426)

[24] "An sich ist ein Zusammengehen mit Italien in Kongressangelegenheiten grundsätzlich erwünscht." (*Ibid.*, handwritten note)

[25] "Ich bin dafür, dass wir uns nicht beteiligen. Zweckmässig wäre es, Italien, Russland und Japan zu der gleichen Ansicht zu bewegen." (fol. 438)

[26] This is an allusion to the so-called Hanfstaengl affair when a German Harvard alumnus who was close to Hitler had offered money, which Harvard did not accept.

[27] MAA Headquarters (Austin, TX), Box 80, folder 4.

The declaration of war on Germany by Britain and France altered plans abruptly. At the AMS council meeting of 6 September, Graustein, speaking for the ICM organising committee,

> proposed suspension of activity, postponement of the congress to a more favorable time, and the appointment of an Emergency Executive Committee to announce postponement of the congress at a time that it should determine and to take the initiative for resumption of activity. Moreover, invited speakers were to be notified immediately both of the possibility of postponement and of the termination of invitations to speak in the event of postponement. These proposals were adopted, as were proposals aimed at conserving the contributions and pledges of funds for the Congress when it was rescheduled.[28]

Following this decision, the AMS council passed the following resolution on 28 December 1939, which was endorsed at the February 1940 meeting of the society and published in its *Bulletin* (**Hollcroft, 1940**, p. 369):

> On recommendation of the Council, the Society also approved sending the letter quoted below to sister scientific organizations in other parts of the world:
>
> The American Mathematical Society, gathered at New York City for its February Meeting, marks this occasion by giving collective expression to the deep sympathy felt by its members for their many scientific colleagues who today are suffering the deprivations and the agonies brought upon them by the spreading cancer of war; and by giving collective expression also to their grave, even desperate, concern for the preservation of the cultural values and the effective organs of scientific research throughout the world during these days of destruction. In the cause of simple humanity and in the interest of the unborn generations for whom science can prepare benefits as yet but dimly descried, the American Mathematical Society now appeals to its sister societies in every land—and most particularly to those in lands which are at war with one another—to exert all possible effort toward the conservation of the scientific resources of the world against the day when peace shall reign once more. The Society calls upon its sister societies to concern themselves especially for those scientists who, by the fortunes of war, may fall prisoner or may come under new flags, to the end that their individual sufferings may be mitigated and their scientific activity continued to the benefit of all men; and it declares its readiness to lend its good offices and its assistance, both moral and material, to those of its sister societies which may respond to this urgent appeal.

[28]Quoted in (**Pitcher, 1988**, pp. 147–148). Although the German invasion of Poland led to the cancellation of the 1940 congress, (**Barany, 2016**, pp. 78–79) asserts that "the Polish refugee mathematician Alfred Tarski capitalized on the work that had already been done in preparation for the congress to advance his case for a permanent visa to reside in the USA. ... William Graustein wrote on International Congress of Mathematicians letterhead to certify that Tarski was among the sixteen foreign mathematicians already invited for plenary addresses."

On 4 March 1940, the AMS secretary R. G. D. Richardson sent a copy of the resolution to Conrad Müller, "Schriftführer" of the DMV, in Hanover, adding:

> Your American colleagues fervently hope that the time will speedily come when they can welcome mathematicians from all parts of the world to an International Congress in this country.[29]

On 24 May 1940, Müller forwarded this to Süss, Sperner, and Hasse, the leading figures in the DMV, and proposed to publish the appeal in its next newsletter.[30]

On 24 October 1940, Richardson informed Evans (AMS president 1939–1940) and M. H. Stone about a reply he had received from Süss, dated 30 September 1940, which was apparently the only response that he had received to date after he had sent the resolution to 39 mathematical societies. As chairman of the DMV, Süss had written the following in his German letter, clearly siding with the German cause, stressing the necessity of revising the Treaty of Versailles, putting all the blame on the English and French, claiming victory (apart from over the "last war enemy, England") and having a slightly paternalistic tone *vis-à-vis* the Americans:

> The board of the German Mathematical Association has noted with great interest the content of this call and feels with grateful joy that the American Mathematical Society has become aware of the unfortunate and destructive effect that the declarations of war of France and England have upon the scientific work of the peoples suffering from the war. Our Association confidently hopes that it has been possible to peacefully carry out the necessary revision of the Versailles Treaty, which brought so much misfortune on the people. I express the hope that despite the explicit war-mindedness of our last war enemy, England, war will soon be ended without too much damage to scientific values and scholars in the war zone.[31]

The internal reaction within the AMS to Süss' letter is interesting. A letter from Stone to Evans and Richardson of 31 October 1940 is even amazing in its prescience. At a time when neither the Russians nor the Americans had yet joined the war against Nazi Germany, Stone recommended continuing the contacts with the Germans and to "keep in mind the probability that the war situation will continue for a long period of time in such a way as to make our efforts less and less unwelcome". Stone was even slightly critical of the British who had not replied to the AMS resolution:

> I think we should elicit some response from the London Mathematical Society. From personal information I know that the

[29] Freiburg University Archives, E4 37, DMV (Schriftführer Müller) 1936–1942.

[30] *Ibid.*

[31] "Der Vorstand der Deutschen Mathematikervereinigung hat mit grossem Interesse von dem Inhalt dieses Aufrufs Kenntnis genommen und empfindet es mit dankbarer Freude, dass die American Mathematical Society sich der unglücklichen und zerstörenden Wirkung bewusst geworden ist, welche die Kriegserklärungen Frankreichs und Englands auf die wissenschaftliche Arbeit der unter dem Kriege leidenden Völker besitzt. Unsere Vereinigung würde sich glücklich schätzen, wenn es gelungen wäre, die notwendige Revision des Versailler Vertrages, der so viel Unglück über die Menschen gebracht hat, auf friedlichem Wege durchzuführen. Ich gebe der Hoffnung Ausdruck, dass trotz des betonten Kriegswillens unseres letzten Kriegsgegners, England, der Krieg ohne allzu starke Beeinträchtigung der wissenschaftlichen Werte und der Gelehrten im Kriegsgebiet bald beendigt werden." (Stone Papers at AAMS, box 41, folder 9A)

> British government has shown scant consideration for refugees whom they felt obliged to intern at the beginning of the summer. I believe that our English colleagues have a definite obligation in this matter and that we should expect an acknowledgment at least.[32]

In Germany at this time, mathematicians were becoming more and more involved in Nazi plans for "reordering European science".[33] In December 1940, Harald Geppert, who had attended the Oslo congress, was sent on a trip to German-occupied Paris to inquire about the Institut Henri Poincaré, which, it was thought, would continue the job of the IMU under French dominance.[34] This was of course mostly propagandistic, since the IMU was long dead.[35] The former head of the IRC, Émile Picard, was very old (he died one year later in 1941) and had little influence at the Institut Henri Poincaré, which was directed by Émile Borel. In his report on his Paris trip, Geppert stated that there was no political danger coming from the Institut, but he proposed to neutralise certain French mathematical journals by founding parallel German publications.[36] Geppert envisioned

> a restoration of international mathematical communication in terms of smaller international meetings in special disciplines, which would be most appropriately organized by the DMV, preferably in connection with a larger institute. The financial means should be provided by the Ministry of Education. (**Siegmund-Schultze, 1993**, p. 180)

As this suggestion makes clear, the plan to have an institute in Oberwolfach[37] — today one of the most attractive international meeting places for mathematicians — was already looming large, at least in the minds of mathematicians. On 5 July 1941 a letter from Hitler's main quarters (*Reichskanzlei*) to all ministers of the government referred, under the title "Internationales Kongress- und Verbandswesen", to the "indispensable experiences" of the Deutsche Kongreßzentrale which, as we saw in Chapter 3, had controlled the participation of German delegates in Oslo in 1936. However, according to the letter, there was now a new situation, and Hitler's state secretary Lammers wrote:

> Due to the military and political turns of the past two years the possibility and the necessity have developed to re-regulate the international system of congresses and unions and — should occasion arise — to move the seat of international unions to Germany.[38]

---

[32]Stone Papers at AAMS, box 41, folder 9A.

[33]"Neuordnung" ("reordering" or "restructuring") was the notion often used in the Nazi plans. Occasionally "Neuregelung" was used, as in the letter from Hitler's main quarters (*Reichskanzlei*) from 5 July 1941, as quoted below.

[34]See footnote 112 on p. 85.

[35]See (**Lehto, 1998a**). See also the remarks on the IMU in Lietzmann's report, recorded here on our p. 85.

[36]See (**Siegmund-Schultze, 1993**, p. 183). Geppert particularly stressed the need for an equivalent to the short reports in the *Comptes rendus des séances de l'Académie des Sciences de Paris*, which did not exist in Germany at that time.

[37]On the early history of which, see (**Senechal, 1998**).

[38]"Durch die militärischen und politischen Umwälzungen der letzten 2 Jahre haben sich die Möglichkeit und Notwendigkeit ergeben, dass internationale Kongress- und Verbandswesen neu

In March 1942, both Geppert and Süss took part in a meeting at the Ministry of Education that aimed to prepare the overall strategies of German science and mathematics for the restructuring of European science, since

> it is very likely that after the War, world science will split into one of the European continent, and another of the American continent.[39]

They also mused whether America might take over the lead in the IMU after the war — and they were not far from the truth (see below).

The political after-effects of the alienation of German mathematicians from foreign mathematicians were felt well into the 1960s (**Siegmund-Schultze, 2009**), despite efforts by mathematicians from allied countries immediately after the war to involve the Germans once again in international communication.[40] Developments in mathematics, like the French Bourbaki movement that had been very much inspired by German algebra around 1930, had to be "re-imported" into Germany in the 1950s and 1960s, enriched with newer concepts such as topological ones.

As suspected by the Germans when losing out in international communication during the war, the preparation for a renewed IMU took place under American lead (in particular, that of Stone) in a period of three years before the postponed ICM in Cambridge, Massachusetts in 1950.[41] We will not go in detail into this process or into the 1950 congress itself, which both deserve independent investigations, not least due to the totally changed political conditions of the Cold War. We have indicated above that participation in the 1950 ICM was still politically affected, which even temporarily endangered the participation of one of the Fields Medallists, Laurent Schwartz. While the Russian mathematicians sent a telegram to the congress with a rather embarrassing explanation for their absence (see Section 4.5), a limited number of Germans went to the meeting. Under the conditions of the Cold War, these were of course basically West Germans.[42]

The appearance of the (West) Germans was anticipated with some nervousness by German refugees in the USA. Alfred Brauer, the brother of the more famous Richard Brauer, wrote to Hermann Weyl on 3 April 1950:

> It was my intention to go to the International Congress of Mathematicians. However, I have become hesitant due to the letter which is enclosed. As much as I would like to meet again several German mathematicians, there might occur embarrassing moments. Because the list contains, if I am properly informed, a number of active Nazis, namely Kneser, Schmieden, Süss, and

zu regeln und gegebenenfalls den Sitz internationaler Verbände nach Deutschland zu verlegen." (REM 3190, fol. 231)

[39]Translated from the German quoted in (**Siegmund-Schultze, 1993**, p. 181). See also footnote 50 on p. 115.

[40]For example, John Todd, F. J. Weyl, and others were invited to visit Oberwolfach in the late 1940s; see (**Senechal, 1998**).

[41]For more details on this, see (**Barany, 2016**, pp. 170–178).

[42](**ICM, 1950**, vol. 1, p. 12) lists 13 German participants, among them Ernst Hölder from the East German Saxonian Academy of Sciences (Sächsische Akademie der Wissenschaften) in Leipzig. We have no confirmation and find it doubtful that Hölder was really there.

> Ullrich.[43] In the decisive years Süss was chairman of the DMV and carried out the political coordination [Gleichschaltung] of the DMV.[44]

While we leave things here only with this short impression from 1950, we glimpse further back to the first large international mathematical meeting after the Second World War: a three-day event in Princeton in December 1946, under the lead of Solomon Lefschetz and other mathematicians based there, on the occasion of the celebration of 200 years of Princeton University.[45] Lefschetz set the tone in his foreword to the proceedings:

> Owing to the spiritual and intellectual ravage caused by the war years, it seemed exceedingly desirable to have as many participants from abroad as possible. As the list of members shows, considerable success was attained in this. Our conference became, as it were, the first international gathering of mathematicians in a long and terrible decade. The manifold contacts and friendships renewed on this occasion will, we all hope, in the words of the Bicentennial announcement, "contribute to the advancement of the comity of all nations and to the building of a free and peaceful world." (**Duren, 1988/1989**, vol. 2, p. 310)

At this point of time, no German or Russian was attending, however, there were some Polish and Czech mathematicians from behind the slowly emerging Iron Curtain. Among the 93 participants in Princeton only nine were working in Europe at the time. Twenty-two of the 93 had also been in Oslo in 1936, to name just (in addition to leading Americans such as Birkhoff, Lefschetz, Veblen, and Wiener) Hermann Weyl and Antoni Zygmund. The names of the following 23 mathematicians remind the reader of those important representatives of logic, algebraic geometry, topology, probability, and modern algebra who had *not* been present in Oslo: A. A. Albert, E. Artin, S. Bochner, R. Brauer, A. Church, P. Dirac, J. L. Doob, K. Gödel, H. Hopf, H. Hotelling, M. Kac, S. C. Kleene, S. Mac Lane, J. von Neumann, W. Quine, H. Rademacher, T. Radó, N. E. Steenrod, A. Tarski, A. W. Tucker, H. Whitney, S. Wilks, O. Zariski.

In accordance with the interests of many of these participants, the stated aim of the event was "to help mathematics to swing again for a time toward unification"

---

[43] Of the four Germans mentioned by Brauer, two (Kneser and Süss) appear as participants in the proceedings of the 1950 ICM. As to Kneser, Schmieden, and Ullrich, one has to add that they were certainly not active for the Nazis in a sense comparable to Süss, let alone Bieberbach.

[44] "Es war meine Absicht zum Internationalen Mathematiker Kongress zu fahren. Ich bin auf Grund des einliegenden Briefes wieder schwankend geworden. So sehr ich mich freuen würde, eine Reihe der deutschen Mathematiker wiederzusehen, so mag es doch auch peinliche Augenblicke geben. Denn die Liste enthält, wenn ich recht orientiert bin, eine Reihe strammer Nazis, nämlich Kneser, Schmieden, Süss und Ullrich. Z.B. war Süss in den entscheidenden Jahren Vorsitzender der D.M.V. und hat die Gleichschaltung der D.M.V. durchgeführt." (Archives AIS General folder Brauer, Alfred) In the end, both Alfred Brauer and his brother Richard apparently went to the 1950 ICM, according to vol. I, p. 21 of the proceedings. For the policies of the DMV and of Süss under the Nazi regime, see (**Remmert, 2012**).

[45] Our sources are simply the report on the meeting that was originally published by Lefschetz in 1947 (*The Princeton University Bicentennial Conference on the Problems of Mathematics*), but reprinted in 1989 with some modern mathematical commentary in (**Duren, 1988/1989**, vol. 2, pp. 309–359). A further commentary on the republication can be found in (**Rowe, 2003**).

(**Duren, 1988/1989**, vol. 2, p. 309), and indeed that this tendency towards unification should characterise mathematics for several decades to come. Applied topics were nearly absent, and so even less represented than in Oslo in 1936. Thus the increased importance of mathematics due to war applications, and the developments in computing going on in Princeton, at Courant's institute in New York, and other places, were not really reflected in the Princeton meeting. In a conference session entitled "New Fields", Hermann Weyl even commented with some irony on these new developments, viewed primarily from the perspective of pure mathematics:

> The deeper one drives the spade the harder the digging gets; maybe it has become too hard for us unless we are given some outside help, be it even by such devilish devices as high-speed computing machines. (**Duren, 1988/1989**, vol. 2, p. 326)

Thus, certain diverging tendencies between pure and applied mathematics were foreshadowed in Princeton and would last over the decades until the 1970s and 1980s when new technologies created a new historical and mathematical environment.

It was at this late moment that some Americans, such as Garrett Birkhoff (1911–1996), son of G. D. Birkhoff, looked back into history. In view of the overall scientific dominance of the United States in the Western Hemisphere after the war, which was also economically grounded, the younger Birkhoff wrote:

> After the defeat of Hitler the U.S. (along with Canada) was standing triumphant, virtually unscathed in a world in which most advanced countries were prostrate and in ashes. Its scientific preeminence was taken for granted and most Americans thought of computing machines as almost a national monopoly. They had largely forgotten the tradition of European scientific superiority that had been accepted only a decade earlier. What is now known about ENIGMA and the Zuse machines should help dispel this illusion. (**Birkhoff, 1980**, p. 29)

As this quote suggests, the isolation of German mathematicians during dictatorship and war had consequences not only for Germany. It also prevented foreigners from becoming aware of German scientific developments. This further underscores the complexity of any analysis of international communication and participation.

Birkhoff in this impressive quotation looked back at the past and showed how difficult it was for some mathematicians to get the proper perspective on pre-war mathematics after all the dramatic and often violent change, emigration, and war emotion. Our look at the events surrounding the ICM in Oslo in 1936 has hopefully contributed to broadening this historical perspective as well.

Part 3

The Mathematics of the Congress

CHAPTER 8

Assessing the Mathematics of the Congress

In the earlier parts of this book, we have described the broad political circumstances that surrounded the 1936 ICM, and also the personal interactions between mathematicians with different alignments. In the present chapter, and the one that follows, we turn to the *mathematics* of the congress, and its place within the wider mathematical progress of the twentieth century. Further chapters in this part of the book provide a brief account of the activities of ICMI in Oslo (Chapter 10), and some conclusions as to the representation of mathematics at the congress (Chapter 11).

The period between the two world wars had a clear impact on the development of mathematics and an influence on the decades to come. Writing in 1948, the mathematician Roger Godement identified the general characteristics of mathematics in the interwar period:

> The period from 1919 to 1940 saw the accomplishment of a veritable revolution in mathematics. Principally under the impetus of the German, Polish and Russian schools — of course, there are exceptions, represented in France by Elie Cartan — theories of an essentially new character were born: modern algebra, spaces of an infinite number of dimensions, topology, non-Euclidean spaces. Towards the end of this period the French school — which until then had remained almost entirely apart from these studies — participated itself; and it is, doubtless, to the young and multiple genius of Bourbaki, that the future will owe the systematization of a multitude of results from which the legendary French clarity was until then unfortunately missing. The recent political upheavals have, moreover, not slowed down the mathematicians' labors, they have simply altered their geographical distribution so that nowadays the United States, along with the Soviet Union, are the two principal sources of progress (from the point of view of our present topic, at least).[1]

By analysing the content of the plenary lectures in particular, we assess the impact of the Oslo congress on these advances in mathematics.

This chapter serves as a preparatory discussion for the detailed treatment of the plenary lectures in the following chapter. We begin with a brief study of the developments in mathematical communication that took place at the end of the nineteenth century and the beginning of the twentieth (Section 8.1), before turning

[1](**Godement, 1948/1971**, pp. 337–338). The "present topic" to which Godement referred was the contrast between modern (abstract) methods and "concrete" mathematics. The "mathématiques concrètes" in the title of the original article was unfortunately mistranslated as "applied mathematics" in the English version of 1971.

to the question of how to measure "mathematical impact" (Section 8.2). Following some brief remarks on the languages and geographical distribution of the plenary speakers in Section 8.3, the chapter concludes in Section 8.4 with some comments on the broad areas of mathematics that were and were not covered in Oslo.

8.1. Developments in mathematical communication

In the 120 years since their inauguration, the ICMs have served, and perhaps also driven, an urgent need in the international community of mathematicians. This need, of course, is primarily the promotion of mathematics and, increasingly, of its applications in a broad sense, the latter including teaching. The Oslo congress appears to have occupied a special place within these developments.

In recent years, modern means of communication such as the internet and email have dramatically changed the character of scientific, and in particular mathematical, communication. However, even by the end of the Second World War, the growth of science and of its societal importance (revealed not least during the war), the enlargement of available funding, and changes in the technological basis for travel (in particular air traffic) had deeply changed the character and needs of mathematical communication. Indeed, the options for mathematicians to meet personally (stipends, smaller conferences, etc.) had already begun to expand during the 1920s and 1930s.[2] In addition to the changing technological infrastructure, there have been varying political conditions which influenced the character and relative importance of the ICMs.

The Oslo congress of 1936 is historically located on the eve of the Second World War, in a period of political unrest, marked by persecution and emigration of scientists primarily from Germany, and by growing control of science and mathematics within nation states. This control, however, was not generally matched (and thus made acceptable to the practitioners) by generous or systematic support from the various governments. Mathematics still bore all the marks of a "little science", having, compared with today, a small attendance of just a few hundred mathematicians at the ICMs and a limited number of countries participating in them. The numbers had actually gone down, apparently due to the economic situation around 1932 and the political situation around 1936, from 836 in Bologna, through 667 in Zurich to an official figure of 487 in Oslo.[3] At the same time, the 1930s saw an increase in smaller, specialised international mathematical meetings (on topology, foundations, probability, applied mathematics) which of course had implications for the decisions of mathematicians, in particular their willingness to attend big congresses as well. As Rolin Wavre, a participant in Oslo (see Figure A.1), wrote in 1948:

> In 1933, there was established in Geneva, with the name of the International Conferences of Mathematical Sciences, a series of

[2]In an article of 5 September 1932 in the *Neue Zürcher Zeitung*, entitled "Die Bedeutung des Internationalen Mathematikerkongresses", the president of the 1932 ICM (G. Motta) greeted the participants and explained that the success of any such meeting lies in the personal communication of results, because it is difficult, especially in mathematics, to write down thoughts in an intelligible form (**Burckhardt, 1980**, p. 11).

[3]Of course, as we have seen in Chapters 4 and 5, even the 487 participants recorded in the proceedings were an exaggeration, owing to the absence of the Soviets and the Italians. It was at the 1950 ICM in Cambridge, Massachusetts, that over 1000 mathematicians took part for the first time: in fact, over 2000 (**ICM, 1950**, p. 135).

> conferences and of international colloquia devoted to the study of particular fields; a colloquium on quantum theory, October 1933; on mathematical logic, June 1934; on topology, October 1935; on partial differential equations, June 1935; on theoretical probability, October 1937, on applied probability, July 1939; this last colloquium was organized with the cooperation of the Commission for Intellectual Cooperation which had, in 1938, organized at Zurich the "Conferences on the Fundamentals of Mathematics". In May 1934, a conference on tensor differential geometry met at Moscow. The conference resolved to form groups of scholars for a planned work. The members of each group would maintain a scientific correspondence and would communicate interesting problems with one another. In August 1935, there met at Brussels the first I.C. of mathematical recreations. In September 1935 at Moscow was held the first international conference on topology and at about the same time in Paris the first I.C. on scientific philosophy. The necessity of uniting the specialists of the whole world on particular problems then made itself imperiously felt. We must hope that a fragmentation of mathematics does not result but that, on the contrary, following Hilbert's views, the study of a particular subject will only be able to strengthen the whole. (**Wavre, 1948/1971**, p. 316)

Hassler Whitney, who published his celebrated paper on differentiable manifolds in the year of the Oslo congress (**Whitney, 1936**), said in his recollection of the Moscow conference on topology of September 1935:[4]

> It was the first truly international conference in a specialized part of mathematics, on a broad scale. (**Whitney, 1988**, p. 97)

For the period after 1945, Bruce Chandler and Wilhelm Magnus devoted a chapter to "modes of communication" in their book on the history of combinatorial group theory, in which they commented:

> Like research institutes, specialized conferences are a novelty in mathematics, appearing in rapidly increasing numbers and in many countries only after World War II and replacing, in effectiveness, not only the international congresses of mathematicians but, to some extent, even the meetings of the various mathematical societies as a means of communication. (**Chandler and Magnus, 1982**, p. 189)

In fact, the organising committee for the ICM in Cambridge, Massachusetts, in 1950 decided to host specialised conferences in addition to the usual programme.[5]

As well as specialised conferences, the 1930s saw several international conferences on a level lower than the ICM, for instance the 8th and 9th Scandinavian congresses (1934 in Stockholm, 1938 in Helsinki), and several conferences organised by the Germans and Italians, even after the outbreak of the war but necessarily

[4] Recall the remarks on this conference on p. 96; see also (**Aleksandrov, 1936**) and (**Tucker, 1935**). For further references to this, and to the above-mentioned Moscow congress on tensor analysis, see also (**Hollings, 2016**, p. 20).

[5] These were four conferences covering algebra, applied mathematics, analysis, and topology; see (**Curbera, 2009**, p. 130).

with the participation only of their wartime allies (see Chapter 7). One specialised international conference on applied mathematics not mentioned by Wavre above is the Volta Congress in Italy (1935) on high speeds in aviation.[6]

In looking at the specific mathematical impact of the Oslo congress one has therefore to consider both the general and necessary limitations for mathematical communication that big congresses by their nature imply, and the particular conditions of the 1936 congress with its peculiar place and historical environment.

8.2. Measuring mathematical impact

There are several possibilities for finding historical markers to measure the value of the Oslo congress in its mathematical context of the mid-1930s.[7] A useful list of mathematical achievements in the first part of the twentieth century may be found in (**Dugac *et al.*, 1994**). This chronological list was compiled by leading mathematicians and historians and comprises 34 pages with a total of about one thousand milestones in the history of (for the most part) "pure" mathematics, often referring to monographs that appeared in the relevant years. Looking at the subperiod 1930–1938, which is symmetric as to time with respect to the Zurich and Oslo ICMs, one may check how many of the results listed were produced by participants in either the Zurich or the Oslo congress. One finds that of the 287 major accomplishments (some of them produced by a number of mathematicians collectively, with quite a few mathematicians being involved in several collaborations), 125 were achieved by researchers who attended neither congress. By way of comparison, 122 of the 287 were achieved by Zurich attendees, and 93 by participants in Oslo (with considerable overlap between the two groups). For 1936 in particular, of the 35 results named for this year, nine were delivered by prior participants in Zurich and nine by participants in Oslo, with five mathematicians having been at both congresses. As to Oslo, four of the 19 plenary talks (Hasse, Cartan, Birkhoff, Ore) were immediately connected by topic to the 35 achievements listed for 1936. By way of contrast, of the 32 results listed for 1932 there were 15 produced by participants in Zurich and 10 by future participants in Oslo. Of the 20 plenary talks given in Zurich just two (H. Bohr, E. Noether) were connected to the 32 results for that year. But what do these numbers tell us? They say, in essence, that a great part (only a little less than 50%) of the mathematicians who contributed to the results listed for the nine years 1930–1938 bracketing the Zurich and Oslo congresses attended at least one of them, and that only a slightly smaller proportion of the results are attributed to mathematicians who attended both congresses. The numbers tell us also, however, that the plenary lectures given at the two congresses do not account by themselves for many of the outstanding results of the respective years 1932 and 1936. This seems natural given that much of the research outcome of any given year is produced by younger scholars whose merits are usually rewarded only much later by inviting them to give plenary lectures at future ICMs. In any case, the ICMs of the 1930s were obviously still able — in spite

[6]See the remarks on the Volta Congress on p. 115.

[7]Although we select measures that are appropriate for our current historical study, we note that the notion of "impact" is a pertinent one for all present-day academics, particularly within the context of the research assessment exercises that are run periodically in the UK; for a commentary on the way in which impact is measured within these, see (**Meagher and Martin, 2017**).

of the growing tendency towards specialised conferences — to draw a representative sample of leading researchers.

Another benchmark for judging the extent to which the Oslo congress reflected the top level of mathematical research might be the representation there of authors of the then-newly founded Springer series *Ergebnisse der Mathematik*. The series had been started in 1932 by the editors of *Zentralblatt für Mathematik und ihre Grenzgebiete* around Otto Neugebauer in Göttingen and was continued under the latter even after his emigration to Copenhagen. By 1936, this series of rather short books on cutting-edge areas of mathematical research could claim several very successful publications, although their real influence would only be visible in the decades to come. By the time of the Oslo congress, 19 of these *Ergebnisse* ("result-reports") had appeared in print. One of the most influential of these was Kolmogorov's 1933 report on the foundations of probability (**Kolmogoroff, 1933**). But the reports of (**Reidemeister, 1932**), (**Bohr, 1932**), (**Veblen, 1933a**), (**Radó, 1933**), (**Bonnesen and Fenchel, 1934**), (**Behnke and Thullen, 1934**), (**Heyting, 1934**), (**Zariski, 1935**), (**Deuring, 1935**), (**Krull, 1935**), and (**Koksma, 1936**) were probably no less influential in their respective areas of mathematical research.[8] Many of the reports were referred to and used in Oslo, definitely and explicitly so the ones by Reidemeister, Deuring, Behnke and Thullen, Veblen, and Koksma, the latter for instance being used as the basis for van der Corput's plenary lecture with the same title (see Section 9.16); indeed, it was cited also by Mordell (Section 9.14)

Looking at how many of the 21 authors of the 19 *Ergebnisse* attended the Oslo congress, one finds just four: Bohr, Behnke, Fenchel, and Veblen. Only one of them, namely Veblen, gave a plenary talk, which was in fact related to the content of his *Ergebnisse* report (see Section 9.6). Of the other 18 authors of reports that had appeared by 1936, four (Reidemeister, K. Hohenemser, Deuring, van der Waerden) lived under political suspicion or racial discrimination in Nazi Germany with no realistic chance of, or support for, going to Oslo; one author was a refugee from Nazi Germany (P. Thullen), whose career as a mathematician was destroyed;[9] two were Russians (Kolmogorov and Khinchin), and for this reason were unable to come. Thus, a mixture of political and scientific conditions gave rise to the rather low participation of *Ergebnisse* authors in Oslo. However, taken together with the strong dominance of number theory in the plenary talks (see Section 8.4 below), one might well also recognise a certain conservatism among the Norwegian organisers, who had, from the beginning, only planned to invite two of the 19 *Ergebnisse* authors, namely Khinchin and Veblen, to give plenary sessions (see Section 2.8).

A third marker for the importance of the Oslo congress — which, however, seems more ambiguous in its value — might be the number of political refugees from Germany who participated. For this group of mathematicians, there were several factors at work which may have stimulated participation; others, however, may have militated against them going to Oslo. Appendix 1.1 of (**Siegmund-Schultze, 2009**) lists, with an aim towards completeness, 145 mathematicians of very different ages, expertise, and degree of research orientation, who fled Hitler's

[8] Roquette, for example, sees Deuring's report as the one with the broadest impact, and mentions Krull's as being one of similar importance (**Roquette, 1989**, p. 110).

[9] He went to South America in 1935 and never returned to mathematical research: `http://www.unifr.ch/news/de/2960/` (accessed 16 August 2016).

Germany. Although mathematical achievement beyond the completion of university studies was not a criterion for selection for this list, there were indeed many research-minded mathematicians among the refugees who had been expelled by the Nazis without regard for their achievements. Of the 145 refugee mathematicians listed, 31 took part in Oslo.[10] A small number were former mathematics school teachers; several, however, were applied mathematicians. It seems to us that this number says something about the high level of research-consciousness among the refugees and, possibly, their need to cultivate personal connections in order to find new jobs (explicitly mentioned by some refugees, such as Sternberg — see pp. 96–96). It probably says less, however, about the importance of the Oslo congress within the mathematical communication of its time. This becomes clear if one looks at the names in detail. One realises that several top-notch researchers among the refugees did not take part in Oslo, for instance Reinhold Baer, Richard Brauer,[11] and Richard von Mises.[12] One may assume that refugees who had just settled in new surroundings or had been travelling much in recent years, and who had plenty of foreign contacts anyway, but often only meagre funds to live on, were not too keen on (and maybe financially incapable of) attending another international meeting.

We owe some further revealing commentary on participation in Oslo to Liliane Beaulieu, concerning specific reasons for the absence of any member of the Bourbaki group of young French mathematicians (to be discussed further below):

> The most pragmatic explanation is that those who would have been more likely to attend Oslo were not ready to do so: [Henri] Cartan was ill, Weil had just fallen in love with de Possel's wife and others were mostly at the start of their career. It is also an expense that not everyone could foot at a time when very few stipends were available. Moreover they may have perceived these meetings as a reunion of established mathematicians, and not of people who are at the beginning of their career ... Indeed, attendance to these meetings is something one does when one has (had) both feet solidly into the profession.[13]

As to the general limitations on the impact of international congresses, one has to be aware that most of the talks, even before Oslo, and particularly the plenary ones, were and are based on the previous publications of the speakers. In Oslo

[10]Though not all of these had yet left Germany at the time of the congress: for example, Max Dehn did not leave Germany until the end of the 1930s. Of the 31 delegates mentioned here, only 19 had already left Germany by 1936: these are listed on p. 118.

[11]In the years prior to the Oslo congress, Brauer's work in group theory had, along with that of Hasse and Noether, inspired much pioneering research on the structure of algebras.

[12]The Swiss mathematician H. Fehr wrote to von Mises on 30 June 1936 in French thanking him for having asked the Turkish government to send a Turkish representative for the ICMI meeting in Oslo (von Mises was at that time teaching at the University of Istanbul). In the same letter, he expressed understanding that von Mises himself would not come due to distance and expense (RVM: HUG 4574.5, box 3). On 20 May 1936, von Mises had written to his mother in Vienna that he had sent a proposal for a paper to Oslo, but on 19 June he wrote to her the following: "I still do not know whether I will go to the congress in Oslo and I cannot really decide. The travel takes five days, on water and on land. It would have the attraction that it leads through Lemberg [then Polish Lwów, von Mises' birthplace] which I could thus see again for the first time after 50 years. But I believe the whole thing is not worth it. Probably I will be in Vienna in the first week of July." (RVM: HUG 4574.5.2. Letters and postcards to his mother (6 boxes))

[13]Private communication to R. Siegmund-Schultze, 24 February 2011.

at least, most presentations referred directly to these publications (as we shall see in the next chapter), and the congress served the purpose of making them better known. But this implies that in order for a historical analysis of the contributions as they appear in the proceedings to have full value, one should try to relate them to the detailed publications that lay behind them.

A final remark about the impact of the Oslo congress may be connected to a retrospective historical view, for instance one given 64 years after the congress for the mathematics in the second part of the 20th century in the collection (**Pier, 2000**). Of all the plenary speakers of 1936 only Siegel is quoted there frequently (according to the index, there are 18 pages in the collection where his name appears). This has of course to do with the primary focus on recent mathematical accomplishments. The strong representation in (**Pier, 2000**) of the Russians (with Kolmogorov having a towering 30 page-references) and of the then-young French Bourbaki mathematicians points back to the absence of these in Oslo and thus to a real deficit of the congress.

8.3. Languages and geographical distribution of speakers

In this section, we offer some very brief comments on the languages employed by the plenary speakers in Oslo, as well as their countries of origin or residence, in comparison with previous congresses. There is certainly scope for a wider study of language use at the ICMs, and of the developments in the engagement of different nations at the congresses, but we do not have the space to attempt this here.

As we discussed in Section 2.8, it was the task of the programme committee not only to build a programme of interesting and new mathematical lectures for the Oslo congress but also to balance the programme in various ways. One concrete way in which the programme committee sought to balance the great mathematical nations against each other had to do with language: it was apparently a concern to them that no language, and in particular not German, should become too dominant at the congress.[14] Therefore, the official welcome addresses and speeches were divided among English, French, and German. In 1932, the fact that Switzerland was itself multilingual had made it very natural to welcome contributions in many languages.[15] Thus, plenary lectures were delivered equally in French (nine) and German (nine), with a minority in English (two), and some lectures in the parallel sessions were given in Italian. However, in Oslo, the balance was shifting — or deliberately being shifted — towards English with seven plenary lectures, although German was the preferred language with ten lectures, not only by German mathematicians but also by some emigrants, and French had dropped to being preferred in a mere two plenary lectures (see Table 8.1).[16]

In comparing the distributions of plenary lecturers among nations for the ICMs since 1928, another feature, concerning the geographical and cultural proximity of the hosts, becomes visible (see Table 8.2). In Bologna, six of the plenary lectures

[14]According to a letter from AMS secretary R. G. D. Richardson to Hans Blichfeldt, the Norwegian organisers had been eager to attract American and British mathematicians to the conference to balance the Nazi contingent (**Siegmund-Schultze, 2002**, p. 343).

[15]See the welcoming address by the *Bundesrat* Dr. A. Meyer in (**ICM, 1932**, vol. I, pp. 64–65).

[16]Note that we cannot be absolutely certain that the lectures were delivered in the languages given here: what we deal with are the languages in which the *printed* version of the lecture or abstract was produced.

TABLE 8.1. Languages of the plenary lectures at the 1928, 1932, and 1936 ICMs (spellings as in proceedings)

Language	Bologna 1928		Zurich 1932		Oslo 1936	
English	2	Veblen, Young	2	Morse, Alexander	7	Størmer, Veblen, Bjerknes, Birkhoff, Mordell, Wiener, Ore
French	5	Hadamard, Borel, Fréchet, Lusin, Birkhoff	9	Julia, Carleman, Cartan, Severi, Wavre, Riesz, Valiron, Sierpinski, S. Bernstein	2	Cartan, Fréchet
German	3	Hilbert, Weyl, Kármán	9	Fueter, Carathéodory, Tschebotaröw, Bieberbach, Noether, Bohr, Nevanlinna, Menger, Stenzel	10	Fueter, Siegel, Nielsen, Hecke, Neugebauer, Oseen, Hasse, Ahlfors, van der Corput, Banach
Italian	6	Puppini, Castelnuovo, Volterra, Tonelli, Amoroso, Marcolongo				

TABLE 8.2. Plenary lectures at the 1928, 1932, and 1936 ICMs, distributed among nations (spellings as in proceedings). *Emigrant, i.e., not representing country of origin (regarding the emigrants from Germany, see the comments in Section 8.2).

Nation	Bologna 1928 (16)	Zurich 1932 (20)	Oslo 1936 (19)
Austria		Menger	
Denmark		Bohr	Neugebauer*, Nielsen
Finland		Nevanlinna	Ahlfors
France	Borel, Fréchet, Hadamard	Cartan, Julia, Valiron	Cartan, Fréchet
Germany	Hilbert, Kármán	Bieberbach, Carathéodory*, E. Noether, Stenzel	Hasse, Hecke, Siegel
Great Britain			Mordell
Hungary		Riesz	
Italy	Amoroso, Castelnuovo, Marcolongo, Puppini, Tonelli, Volterra	Severi	
Norway			Bjerknes, Størmer
Poland		Sierpinski	Banach
Soviet Union	Lusin	S. Bernstein, Tschebotaröw	
Sweden		Carleman	Oseen
Switzerland	Weyl*, Young*	Fueter, Wavre	Fueter
The Netherlands			van der Corput
United States	Birkhoff, Veblen	Alexander, Morse	Birkhoff, Ore*, Veblen, Wiener

were given by Italians (and all in Italian), with three by Frenchmen — a distant second. In Zurich, the diversity of the plenary lecturers was much greater, and this tendency was continued in Oslo with the additional feature that many more Scandinavians numbered among the plenary speakers. Some of these, of course, had either emigrated to the USA (Ore and Ahlfors), whereas others were immigrants from Germany (Neugebauer, who was now in Denmark). Thus, in line with the remarks made throughout Chapter 2, we see that the Norwegian programme committee certainly made sure to showcase Scandinavian mathematics in the prestigious and visible plenary lectures.

8.4. Analysis of content: thematic restrictions

The congress in Oslo, as is probably the case for such conferences in general, was strongly shaped by the plenary talks, of which there were 19 (roughly as many as in Zurich four years earlier),[17] which were given during the morning sessions from

[17]See the comments at the beginning of Chapter 9 on the number of Oslo plenary lectures.

Tuesday 14 to Saturday 18 July. As we saw in Section 2.8, they had been planned by the organising committee much in advance and were obviously chosen to represent general trends in mathematical research. Indeed, the preliminary programme[18] that was circulated with the call for registrations contains the following remark:

> these lectures should give as complete a picture as possible of the current state of mathematics. We have primarily approached scholars who have not given talks of this kind at recent congresses.[19]

An immediate impression upon looking at the list of plenary talks in Oslo (Table 8.3) is that topics close to number theory, represented by the lectures of Siegel, Hasse, Hecke, Mordell, van der Corput, and (partly) Fueter, i.e., by almost one third of the talks, were clearly dominant. The other 13 plenary lectures were devoted to separate subjects, like functional analysis (Banach), complex function theory (Ahlfors), hydrodynamics (Bjerknes), mathematical optics (Oseen), structural algebra (Ore), quantum mechanics (Birkhoff), numerical integration of equations in mathematical astronomy (Størmer), history of mathematics (Neugebauer), gap theorems in analysis (Wiener), topology (Nielsen), *mélanges mathématiques* (from functional analysis, analytical functions, and probability, by Fréchet), projective geometry (Veblen), and Lie groups (Cartan). One may safely say that among the 19 speakers, the internationally most prominent were Siegel, Cartan, Banach, Wiener, Neugebauer, and Hasse. Slightly less prominent, but still well known, were Birkhoff (for his work in dynamical systems) and Bjerknes (meteorology). However, these two did not present results from their main areas of research.

The dominance of number theory would have been even greater if the originally planned programme, as described in Section 2.8 (and found in Table 9.1 on p. 219), had been able to go ahead; as we have seen, the plan, as late as spring 1936, had been to include talks by both Gel'fond and Landau, neither of whom was able ultimately to attend.[20] As we have also seen, Mordell stepped in for one of the two cancelled number-theoretic talks, but the loss to the congress was in both cases significant. Landau, who was almost 60 years of age in 1936 and would have been (together with Cartan) one of the oldest plenary speakers at the congress, was probably the most famous (but perhaps no longer the most creative) number theorist of the time, having been a teacher of the three German speakers on number theory (Siegel, Hecke, Hasse); in particular, he directed Siegel's doctoral thesis.[21] On the context of Landau's proposed talk for Oslo, see Section 9.21.3.

Before summarising some of the highlights of the mathematics that was presented in Oslo, it seems reasonable to look first at the thematic restrictions that the congress exhibited, both with respect to the topics put forward by the organisers (as far as the plenary sessions were concerned) and with respect to the work in the smaller sections. We know from Section 2.8 that the organisers had a strong influence on the choice of speakers for the plenary sessions. As to the sections,

[18]See footnote 45 on p. 40, and also footnote 43 on p. 113.

[19]"In einer Anzahl von ungefähr zwanzig sollen diese Vortrage ein möglichst vollständiges Bild des gegenwärtigen Standes der Mathematik geben. Wir haben uns vorzugsweise an solche Gelehrten gewendet, die nicht auf den letzten Kongressen Vorträge dieser Art gehalten haben." (Bessel-Hagen Nachlass)

[20]See, respectively, Chapter 4 and Section 9.21.3.

[21]In addition, van der Corput was strongly influenced by Landau: see (**Duparc and Korevaar, 1982**), as well as the remarks in Section 9.16.

TABLE 8.3. The timetable of plenary lectures as it appeared in the final programme (**ICM, 1936**, vol. I, pp. 14–16) (NB. We do not know how the one-hour slots were to be divided between two speakers — see the remarks in footnote 44 on p. 40.)

Date and time	Speaker	Title
Tues 14 July 10:15–11:00am	C. Størmer	Programme for the quantitative discussion of electron orbits in the field of a magnetic dipole, with application to cosmic rays and kindred phenomena
Tues 14 July 11:15am–12:00pm	R. Fueter	Die Theorie der regulären Funktionen einer Quaternionenvariablen
Weds 15 July 9:00–10:00am	E. Cartan	Quelques aperçus sur le rôle de la théorie des groupes de Sophus Lie dans le développement de la géométrie moderne
Weds 15 July 10:15–11:00am	C. L. Siegel O. Veblen	Analytische Theorie der quadratischen Formen Spinors and projective geometry
Weds 15 July 11:15am–12:00pm	J. Nielsen	Topologie der Flächenabbildungen
Thurs 16 July 9:15–10:00am	E. Hecke	Neuere Fortschritte in der Theorie der elliptischen Modulfunktionen
Thurs 16 July 10:15–11:00am	O. Neugebauer C. W. Oseen	Über vorgriechische Mathematik und ihre Stellung zur griechischen Probleme der geometrischen Optik
Thurs 16 July 11:15am–12:00pm	V. Bjerknes H. Hasse	New lines in hydrodynamics Über die Riemannsche Vermutung in Funktionenkörpern
Fri 17 July 9:15–10:00am	G. D. Birkhoff	On the foundations of quantum mechanics
Fri 17 July 10:15–11:00am	L. J. Mordell	Minkowski's theorems and hypotheses on linear forms
Fri 17 July 11:15am–12:00pm	L. Ahlfors J. G. van der Corput	Geometrie der Riemannschen Flächen Diophantische Approximationen
Sat 18 July 9:15–10:00am	S. Banach	Die Theorie der Operationen und ihre Bedeutung für die Analysis
Sat 18 July 10:15–11:00am	M. Fréchet N. Wiener	Mélanges mathématiques Gap theorems
Sat 18 July 11:15am–12:00pm	Ø. Ore	The decomposition theorems of algebra

the organisers were largely dependent on the registration of participants. In both cases, however, there were political factors that reached beyond the influence and the decisions of the organisers. The three mathematical areas that were hardest hit by these were topology, probability, and algebraic geometry. The absence of the Russians mostly affected the first two; the absence of the Italians impacted the third. The talks by Severi on algebraic geometry and by Khinchin on probability, which had been planned by the organising committee, could not be given.

With respect to the three disciplines just named, we may make some immediate observations.[22] By 1936, topology was already developing in many different algebraic, geometric, and analytic directions, beyond the original divide between its combinatorial and set-theoretic aspects.[23] Therefore one cannot talk about a general absence of topology from Oslo, rather a very restricted representation. However, (**James, 1999b**, p. 839) gives a somewhat distorted picture when he asserts of the Oslo congress that

> a fairly representative gathering of topologists took part, including Aleksandroff, Borsuk, Dehn, Ehresmann, Eilenberg, Freudenthal, Lefschetz, Heegaard, Hurewicz, Morse, Nielsen, Seifert, Threlfal, Veblen and Whitehead, and this time topology was also better represented on the lecture programme.

However, he does not specify which of those mentioned gave talks on topological topics. In fact, Dehn, Ehresmann, Lefschetz, Seifert, and Threlfall did not speak at all; Aleksandrov did not attend. Veblen's plenary talk was purely differential geometric, while J. H. C. Whitehead's sectional talk (**ICM, 1936**, vol. II, p. 127) was more group-theoretic.

The two ageing founders of topology, Dehn and Heegaard, attended the congress, but did not speak on that topic. Heegaard had left the field and was now more interested in the history of mathematics, on which he gave a talk.[24] Another early pioneer of topology, the Dutchman Brouwer, whose relationship with the ICMs was strained anyway (dating from his uncompromising opposition to participation in the Bologna congress in 1928: see p. 23), had left topology too. He now favoured mathematical intuitionism (which flew in the face of his own non-constructive topological fixed point theorems), but foundational research was barely represented at the congress (see below).

As far as topology was concerned, there was the plenary talk by Nielsen, "Einige Methoden und Ergebnisse aus der Topologie der Flächenabbildungen" (see Section 9.7), which could be connected closely to members of the Issai Schur school of combinatorial group theorists. Several of the latter were in attendance and gave talks in the section "Geometry and Topology". However, leaders of the discipline, such as Kurt Reidemeister, Schur, and Richard Brauer were absent from Oslo. Relations between topology and group theory were also discussed in Oslo by the English topologists M. H. A. Newman and Whitehead (see above), who gave a joint talk (**ICM, 1936**, vol. II, p. 127), and the Polish mathematician Karol Borsuk (**ICM, 1936**, vol. II, p. 129).

In addition to the politically enforced absence of the Russians, the non-attendance of Heinz Hopf must also have had an impact on the more general topological aspects of the congress. Hopf and Aleksandrov had just published their seminal book (**Aleksandrov and Hopf, 1935**), which was quoted by Fueter in an algebraic context in his plenary lecture (**ICM, 1936**, vol. I, p. 88). As we have

[22] See also the comments in Sections 2.8.1.3 and 2.8.1.4.

[23] On the development of topology, see (**Dieudonné, 1984**), (**Dieudonné, 1989**), and (**James, 1999a**).

[24] See footnote 59 on p. 183.

noted, Pontryagin's proposed talk on the transformation of spheres into spheres, an abstract of which was printed in the congress proceedings (**ICM, 1936**, vol. II, p. 140), was in reality read by Lefschetz. The latter, however, a leading American topologist of Russian origin, did not give a talk of his own. After the congress, Lefschetz wrote to Hopf in Zurich from Fontaineblau (near Paris). Pointing to the absence of the Russians, Lefschetz said: "Topologically of course the Congress was weak".[25] Lefschetz' countryman and colleague from Princeton, J. W. Alexander, who had given a plenary talk in Zurich four years before (**ICM, 1932**, vol. I, pp. 249–257), did not attend the Oslo congress. However, Hans Freudenthal and J. L. Synge did attend and gave lectures (**ICM, 1936**, vol. II, pp. 138–139). There were other topologists in Oslo, such as those from the Polish set-theoretic school, with talks in the Geometry and Topology section, in particular Edward Szpilrajn and K. Zarankiewicz (**ICM, 1936**, vol. II, p. 125).

The co-authors of the influential German book *Lehrbuch der Topologie* (1934), Herbert Seifert and William Threlfall, were in Oslo but did not give talks. Hassler Whitney, the student of G. D. Birkhoff and rising star of American topology, who, as we have noted (p. 145), attended the topology congress in Moscow in 1935, did not, however, go to Oslo. Topological dynamics was missing from Oslo, in spite of the attendance of Birkhoff, who had recently proved the individual ergodic theorem (**Birkhoff, 1931**).[26] Moreover, neither John von Neumann, who had established the statistical ergodic theorem (**von Neumann, 1932a**), nor Eberhard Hopf, who would publish his *Ergodentheorie* the year after the congress ((**Hopf, 1937**), which marks the real birth of ergodic theory), attended. Of course the absence of the Russians and their work on stochastic processes was felt here, given the close connections between probability and ergodic theory dating from the beginning of the 1930s.

To turn next to probability, we recall from Section 2.6 that the plan to include actuarial mathematics in the congress programme had been a strategic move on the part of the organisers, but that in the end, following deliberations over the plenary lectures, only one probability speaker remained on the plenary programme: A. Ya. Khinchin, with a topic not directly connected to actuarial matters (see Section 9.21.4). Lectures in this direction did appear in section IV (Probability, Mathematical Statistics, Actuarial Theory and Econometrics), however.[27]

As we observed in Section 2.8.1.4, Khinchin's was the one remaining probabilistic name from a list of three that had also included A. N. Kolmogorov and Harald Cramér. Neither of the latter went on to give a plenary lecture in Oslo, but Cramér did attend the congress, and gave a sectional lecture entitled "Some theorems connected with the 'Central Limit Theorem' in probability" (**ICM, 1936**, vol. II, pp. 205–206). A number theorist by background, Cramér had become interested in probability in the 1920s, and was concerned with the need for a greater rigour in the study of the subject. The topic of his Oslo sectional lecture was a result, the Central Limit Theorem, which Cramér proved under increasingly general conditions, and which is regarded as one of the highlights of his probabilistic work.[28] It

[25]Hopf Papers ETH Zurich, no. 896. The letter is dated "6 Juli 1936", which is probably a typo for 6 August.

[26]On the theorem, see also (**Birkhoff, 1942**).

[27]For example, Nilos Sakellariou's "Sur une formule générale sur l'assurance sociale" (**ICM, 1936**, vol. II, pp. 213–214).

[28]See, for example, (**Blom, 1987**, p. 1340).

is interesting to note that although the printed version of the talk is entirely theoretical, it was probably inspired by actuarial concerns, a topic of central importance to Cramér's work, for he had worked as an actuary in the 1920s:

> It was his work as an actuary that inspired him to study the behaviour of a large number of random variables when added and suitably standardized. In particular, he was interested in estimating the error of replacing the distribution function of a standardized sum by that of the standard normal distribution function. (**Blom, 1987**, p. 1340)

Indeed, Cramér had been professor of actuarial mathematics and mathematical statistics in Stockholm since 1929, so it is perhaps emblematic of the underrepresentation of actuarial science and the neglect of applied mathematics in Oslo that even someone in such a position should not speak explicitly on an actuarial topic.

The other originally proposed probabilistic speaker, Kolmogorov, was of course something of a star in this area, having provided a mathematical foundation for probability in his 1933 Springer *Ergebnisse* report, building upon the measure theory developed by Lebesgue and Borel in the preceding decades.[29] Even in Kolmogorov's absence from the congress, we nevertheless see the beginnings of his influence in probability, just three years on from his publication, in the few references to his work that appear in some of the sectional lectures.[30] In addition to Kolmogorov's axiomatic approach to probability, measure-theoretic ideas were very much in evidence in the section IV lectures in Oslo, as they were in other sections also. In view of its title (see Table 9.1 on p. 219), one wonders whether Khinchin's proposed plenary lecture might have been an overview of some of these new ideas in probability — we will return to this in Section 9.21.4.

Thinking of other probabilists of this period, we note that of course none of the Soviets Bernshteyn, Petrovskiy, or Slutskiy were present in Oslo, although of the three it was only Bernshteyn who had planned to attend. One prominent figure who *did* attend was Willy Feller,[31] who lectured in German on the theory of stochastic processes (**ICM, 1936**, pp. 194–196). From the French school, Paul Lévy was absent, while Émile Borel was present but did not speak; in addition to his plenary lecture (see Section 9.18), Fréchet gave a sectional talk entitled "Sur quelques idées modernes en théorie de probabilités" (**ICM, 1936**, vol. II, p. 219), which discussed critically certain ideas, in particular the notion of a "collective", that von Mises had presented as "generally recognised" to the Zurich congress four years earlier (**ICM, 1932**, vol. II, pp. 221–228). Scandinavian names were naturally very well represented in this area: we have already mentioned Cramér; also present were Herman Wold, who delivered two sectional lectures,[32] and Ragnar Frisch, who spoke on an economic theme.[33] "Big name" statisticians, such as R. A. Fisher and Jerzy Neyman, did not attend the congress, however — one wonders, in line with

[29] On the development of rigour in probability theory, see (**Doob, 1994**).

[30] For example, in Arthur H. Copeland's "Sequences with after-effect" (**ICM, 1936**, vol. II, pp. 218–219).

[31] Feller attended from Stockholm, where he had taken refuge after fleeing Germany; in 1939, he moved to the United States.

[32] "On multi-dimensional distributions" and "On the mean difference at random samples — a note on Prof. Bowley's lecture" (**ICM, 1936**, pp. 199–200, 212–213).

[33] "Price index comparisons between structurally different markets" (**ICM, 1936**, pp. 220–221).

comments made above, whether the existence of specialised statistical congresses made the ICM a less attractive event to attend.[34]

Overall, we might say that, with the restrictions noted above, the section on statistics and probability, which featured 26 short talks, fared well at the Oslo congress, even in the absence of a plenary lecture in this area. This is not something that we can say about the representation of algebraic geometry at the congress, however. Perhaps more so than any other discipline at this time, algebraic geometry was linked to a single national group. The absence of Italian geometers, such as Enriques, Castelnuovo,[35] and Severi from the ICM had a serious effect on its visibility there. Of course, algebraic geometry was not a *purely* Italian affair: new approaches were given by Zariski in the USA, by van der Waerden in Germany, and by Weil in France, but none of these were present in Oslo either. Émile Picard did not go, probably for reasons of age (he turned 80 a few days after the congress). Thus, while Khinchin's failure to deliver his probability plenary lecture had only a small impact on the representation of that discipline, the non-appearance of Severi, as part of the wider Italian boycott, meant that the algebraic geometry at the congress was severely limited. Nevertheless, it was not entirely absent: see, for instance, the talks of the Americans B. M. Walker and Virgil Snyder, as well as that of the Belgian Lucien Godeaux, all delivered in the algebra and geometry section.[36] Differential, homological, metrical, and topological themes were, however, much stronger with regard to the geometry in that section.

To turn to other topics, we saw in Section 2.8.1 that the organisers of the congress had not planned to have a section specifically on mathematical logic, although the title of the section "Philosophy and History of Mathematics; Education", which was in the first printed communication about the congress (**Anon, 1935**) did not expressly exclude logic. Indeed, the organisers had planned a plenary talk by the prominent Norwegian logician Thoralf Skolem (without specifying a title).[37] But for some reason the next communication of the organising committee announced a section on Logic, Philosophy and History of Mathematics, while Education got a separate section.[38] We conjecture that the organisers, using this new title, wanted to draw more logicians to the congress. In the end — maybe as a result of too few logicians registering — Skolem's plenary talk did not materialise. Instead, Skolem took part in the congress with a shorter contribution on the decision problem in the above-named section (**ICM, 1936**, vol. II, pp. 268–270), which altogether featured only twelve presentations. Fourteen years later, in Cambridge, Massachusetts, the organisers provided for a section on Logic and Philosophy, chaired by Tarski, with S. C. Kleene, Abraham Robinson, Skolem, and Tarski himself as invited speakers. While Robinson would have been too young in 1936 to have participated, neither Kleene nor Tarski had been in Oslo. The same is true of Alan Turing, Gentzen, and Gödel, who did not attend the Oslo congress. This is surprising because Oslo would have been a suitable occasion to celebrate the dramatic progress in the foundations

[34]The 23rd Session of the International Statistical Institute was held in Athens in 1936.

[35]Recall from Section 2.8.1.3 that Castelnuovo had declined an invitation to speak.

[36]Respectively: "The higher singularities of algebraic curves", "On a system of involutorial Cremona transformations defined by a pencil of quadric surfaces", and "Sur les involutions cycliques appartenant à une variété algébrique" (**ICM, 1936**, vol. II, pp. 82–83, pp. 150–151, pp. 151–152).

[37]We see this in the Autumn 1935 preliminary programme mentioned in footnote 43 on p. 113.

[38]See also the comments in Section 2.8.1.7.

of mathematics that had already been achieved by that point in the 1930s.[39] The only logicians to take part in Oslo with contributions in the section "Logic, Philosophy and History of Mathematics" were Skolem, Rózsa Péter, Jean Cavaillès,[40] and Alfred Errera. There were, however, further logicians in Oslo who did not give talks, among them Abraham Fraenkel,[41] Paul Finsler,[42] and Ludovic Zoretti.

As to Philosophy of Mathematics, there wasn't any in Oslo beyond the four contributions given by logicians, which, however, could more appropriately be described as belonging to mathematical logic and foundations of mathematics. The other talks in the section "Logic, Philosophy and History of Mathematics" were historical. Here again the organisation of parallel specialised conferences[43] was apparently a major factor militating against the participation of philosophers of mathematics. Indeed, the likes of Rudolf Carnap, Walter Dubislav, Hans Reichenbach, Gustav Bergmann, Herbert Feigl, Felix Kaufmann, Louis Rougier, Friedrich Waismann, Edgar Zilsel, Béla Juhos, Kurt Grelling, Olaf Helmer, Carl Gustav Hempel, Richard von Mises, or Paul Hertz, many of whom were refugees at the time, were not seen in Oslo. Consequently, even the Scandinavian allies of the Vienna Circle such as Arne Næss and Jørgen Jørgensen did not show up in Oslo either. It appears that the only members of the Vienna Circle who attended were the mathematician Karl Menger, who delivered lectures in the sections on analysis and geometry (**ICM, 1936**, pp. 45–46, 171–173), and the physicist and philosopher Philipp Frank, who apparently was there primarily[44] because he was already in Scandinavia, having given one of the main lectures, on the problem of causality in physics and biology (**Stadler, 2001**, p. 372), at the Second International Congress for the Unity of Science in Copenhagen just two weeks before.[45]

A further restriction on the participants has already been mentioned in passing (p. 148): the almost total absence of the young French mathematicians who would be known as the Bourbaki group in the decades to come. This is particularly noteworthy because most of them had been participants at the previous congress in Zurich (André Weil, Jean Dieudonné, Henri Cartan, Jean Delsarte, Claude Chevalley, René de Possel). However, much like the Vienna Circle, the Bourbaki group was also in some respects riding the wave of independent and specialised conferences; it had had its first internal "congress" in 1935 (**Corry, 2004**, p. 296). In 1936, they were preparing for their second congress in September at El Escorial in Spain

[39]In addition to the logicians mentioned above, the following prominent logicians also did not go to Oslo: von Neumann, Alonzo Church, Paul Bernays, Wilhelm Ackermann, Emil Post, Arend Heyting, Leopold Löwenheim, Kurt Grelling, Willard Quine, Heinrich Scholz, Ernst Zermelo, Stanisław Leśniewski, and Jan Łukasiewicz.

[40]Cavaillès' talk is only recorded by title ("Formalisme et expression d'une structure mathématique" (**ICM, 1936**, vol. II, p. X)).

[41]While at the congress, Fraenkel gave a largely politics-free interview to the Norwegian newspaper *Tidens Tegn* about his working life at the University of Jerusalem; a translation of this interview appears as our Appendix A.4, with some additional commentary in Chapter 6 (p. 122).

[42]According to van der Waerden, Finsler was a forerunner of Gödel (**Burckhardt, 1980**, p. 37).

[43]Principally, the International Congresses for the Unity of Science: see (**Stadler, 2001**).

[44]A secondary reason may have been his political interests, as he was very much engaged in the cause of refugees in mathematics and physics, partly in collaboration with his friend Richard von Mises, who gave up his plan to participate in Oslo at rather short notice (see footnote 12 above). On Frank, see Chapter 6 (p. 120) and Appendix A.2.

[45]We can only speculate as to why more people didn't travel across to Oslo from the Copenhagen congress.

(**Beaulieu, 1989**, p. 310). In any case, the absence of that group from Oslo had obvious consequences for the discussion of several modern disciplines such as algebraic geometry (Weil), algebra (Chevalley), and the theory of complex functions in several variables (Cartan).

This raises another point that is somewhat related to the absence of Bourbaki: there was virtually no representative of the Noether school of abstract algebra in Oslo, which had had so much influence on the mathematicians of Bourbaki. Emmy Noether herself had died the year before in Princeton from complications during routine surgery; van der Waerden, her most influential student and author of the Springer volumes *Moderne Algebra* (1930/31) was a professor in Leipzig, and the German Nazi authorities forbade him to go to Oslo (see p. 71). Others, such as Max Deuring and Heinrich Grell, had political problems,[46] partly due to having been Noether's students, and were not included in the German delegation either. Mathematicians like Wolfgang Krull and Emil Artin who were influenced by Noether's algebra, did not show up. Artin, who was married to a Jew, and was therefore on the brink of emigration from Germany, would certainly not have been a first candidate for the German delegation, even if he had wished to go (unknown). All the algebraists mentioned, with the exception of Artin, had taken part in the Zurich congress four years before, at which Noether had given her celebrated talk on hypercomplex systems.[47] Of course there were foreigners (in addition to the Bourbaki group) who were strongly influenced by Noether's structural algebra and of whom several took part in Oslo: Ore, who gave a plenary talk, Lefschetz, Garrett Birkhoff, and others. In addition, Hasse was close to Noether in his mathematical approach and gave a plenary talk (on which, see Section 9.12).[48]

The general impression — given that algebraic geometry also hardly featured in Oslo — is that the algebraic research included in the congress had a strong leaning towards number theory and was only weakly represented beyond that.

[46]See, for example, (**Siegmund-Schultze, 2009**, p. 163).

[47]"Hyperkomplexe Systeme in ihren Beziehungen zur kommutativen Algebra und Zahlentheorie" (**ICM, 1932**, pp. 189–194).

[48]The absence of representatives of another of the main currents in modern algebraic research, the school of Issai Schur of Berlin — focusing on group representations, and in a certain sense more mathematically "concrete" than Noether's brand of algebra — was mentioned on p. 153.

CHAPTER 9

The Plenary Lectures

We come now to the most substantial chapter in the book: our summary and assessment of the plenary lectures delivered in Oslo. In Sections 9.2–9.20, we take each lecture in turn, locating them both within the wider work of the speaker, and also within the broader mathematics of the twentieth century. In Section 9.21, we attempt to do the same thing, but necessarily much more briefly, for those plenary lectures that were planned but did not take place. At this point, we must acknowledge a small discrepancy: in the present chapter, we list 19 plenary speakers, and yet, according to some sources (including the lecture list in the congress proceedings, if not the printed programme) there were in fact 20. This is because these sources count Carathéodory's lecture at the award of the Fields Medals as a plenary lecture. However, since the lecture did not concern Carathéodory's own work, we do not count it as a plenary, although we note that it is a useful resource for the assessment of the work of the Fields Medallists, a topic that we pick up in Section 9.22; the latter is also the place where we deal with the work of the Fields Medallist Jesse Douglas, who, unlike the other recipient Lars Ahlfors (see Section 9.15), did not deliver a lecture in Oslo.

9.1. Evaluating the plenary lectures

Historical analysis of the mathematics presented at the congress has to start with the plenary lectures. Indeed, the leading mathematical reviewing journals of the time, *Zentralblatt für Mathematik und ihre Grenzgebiete* (hereafter simply *Zentralblatt*) and *Jahrbuch über die Fortschritte der Mathematik* (*Jahrbuch*), restricted themselves to reviewing the 19 plenary talks of the congress. In no case did they publish reviews of the talks given in the eight sections, which were limited in time to 15 minutes each and for which there appeared only abstracts (some of these, however, were more than a page long) in volume II of the proceedings; in the reviewing journals, these presentations were merely recorded by their titles. The discussions following the talks, both plenary and sectional, are not recorded in the proceedings. As to the plenary sessions, they usually received detailed accounts in the reviewing journals, although the talks on applied topics by Bjerknes and Størmer seem to have fallen outside the scope of *Zentralblatt* in spite of its title including the word "Grenzgebiete" ("bordering subjects"). Most of the reviews of plenary talks refer to more detailed publications by the respective speakers. But the talks in the sections, which were only abstracted, also often referred to more detailed publications.[1]

[1]For example, the lengthy abstract of J. A. Schouten and J. Haantjes' contribution to the section "Geometry and Topology" ("Zur Theorie des geometrischen Objektes": (**ICM, 1936**, vol. II, pp. 155–159)) directs the reader to their paper of the same title (modulo language): (**Schouten and Haantjes, 1937**). The "abstract" of Samuel Eilenberg's contribution to the same section

Our historical analysis can thus build upon both reviews and upon later historical secondary literature such as (**Birkhoff, 1976a**), (**Birkhoff, 1976b**), (**Chandler and Magnus, 1982**), (**Corry, 2004**), (**Dieudonné, 1978**), (**Duren, 1988/1989**), (**Fischer *et al.*, 1990**), (**Gray, 2000**), (**Jacobs, 1986**), (**Lützen, 1982**), (**Weil, 1950**), (**Pier, 1994a**), (**Pier, 2000**), and (**Rassias, 1992**). In addition, we have used the biographies of the main speakers that appear in the *Dictionary of Scientific Biography*, as well as various further biographies. Moreover, we are in the happy possession of Lietzmann's report on the congress (as discussed in Chapter 3, and reproduced in full in Appendix D), which is based on the expertise of the German participants and will be used prominently in Chapter 11. Our own mathematical expertise as authors is unfortunately not equal to the wide variety of topics covered at the congress. We have therefore been heavily reliant on the above-mentioned contemporary and retrospective judgements of the relevant and secondary literature, including general surveys written by competent mathematicians such as Dieudonné, Weil, Weyl, and others. Inasmuch as these surveys are themselves dated, our account will not reflect the current state-of-the-art in the respective fields and will only provide glimpses of the past state in the 1930s of various rapidly developing mathematical topics. For this reason, our accounts of the various plenary lectures are a little uneven; we have endeavoured to choose and arrange details of the mathematics that are in accordance with the wider historical goals of the present book. The evaluation of the talks given in the smaller sections of the congress is, of course, even more complicated, so we deal with some of these only in passing and focus for the most part solely on the plenary lectures.

Even in the 1930s, mathematics was already too specialised for plenary lectures to be understandable to all mathematicians gathered at an ICM. Indeed, most plenary speakers, then and now, made an effort to reach broader circles of colleagues. Some plenary lectures have been and are able to stimulate communication between various disciplines or have independent value with respect to the clarity with which this or that new subject has been presented.[2]

However, few plenary talks, and publications arising therefrom, can be considered as original contributions to research. Even such a technical presentation as Siegel's Oslo talk on the analytical theory of quadratic forms (Section 9.5) was merely a summary of a series of near-contemporary papers in the *Annals of Mathematics*.

In any case, plenary lectures are a valuable source for historians precisely because of their broad intent. And therefore they provide an appropriate basis for us to give an impression of the important topics of mathematics in the 1930s.

We refer throughout the following to the two tables included in this chapter and the preceding: one listing the plenary lectures as they were recorded in the preliminary programme for the congress (Table 9.1 on p. 219), and another that reproduces the plenary lectures as they are listed in the congress proceedings (Table 8.3 on p. 152). In this chapter, we take each of the (delivered) plenary lectures, in chronological order of presentation, and assess its mathematical impact, before

(**ICM, 1936**, vol. II, p. 141) consists simply of one line, referring the reader to the published version of the talk (**Eilenberg, 1936**).

[2]This remark has been made, for instance, with respect to Hasse's plenary talk in Oslo (see Section 9.12) and the question has been raised as to why the talk was not included in Hasse's Collected Papers (see p. 190).

turning our somewhat more speculative attention in Section 9.21 to those plenary lectures that were planned but not delivered.

9.2. C. Størmer: "Programme for the quantitative discussion of electron orbits in the field of a magnetic dipole, with application to cosmic rays and kindred phenomena"

As we described in Chapter 2, Størmer was a major figure in the organisation of the congress in Oslo.[3] In particular, he was a member of the committee that pieced together the programme of plenary lectures, one of which (the first one after Carathéodory's talk about the Field Medallists) he delivered himself.

Since, as we see from Table 8.3, Størmer's lecture was one of four devoted to mechanics or mathematical physics (the others being provided by Bjerknes, Oseen, and Birkhoff), we begin with a few general remarks on the place of applied mathematics at the congress. Uncharacteristically of the congress as a whole, three of the four lectures on this broad topic were given in English, and Scandinavians gave three of the four.

All three Scandinavians had previously been very active in the International Congresses for Applied Mechanics (ICAMs), the first of which, as we noted on p. 107, had been held in Delft in 1924.[4] Oseen had even organised the 1930 ICAM in Stockholm. Throughout the 1920s and 1930s, the ICAMs (later the ICTAMs) were well attended, and provided an important forum for applied mechanics in which Scandinavian scientists took part. The ICAMs can therefore be seen as part of the move towards specialised congresses that we mentioned above; it was perhaps Størmer's status as a main organiser of the Oslo congress that nevertheless ensured the presence, though smaller than for the other sections, of mechanics and mathematical physics in 1936.

Størmer had started his research career in number theory but had become excited by the physical simulation of the Northern Lights in the terrella experiments of his countryman Kristian Birkeland (1867–1917) in the 1890s. In order to calculate the trajectories of the Northern Lights from the second order ODE for the assumed electromagnetic interaction in the field of a magnetic dipole, Størmer independently reinvented Adams' numerical multistep method and reached a good correspondence both with experiments and observations. Working in a pre-computer age, Størmer and his assistants spent thousands of hours on the calculation, adding to the equally long hours spent on observing and photographing the Northern Lights.[5] In his 1936 lecture, Størmer remarked that he and other human calculators had carried out over 18,000 hours of work since 1904. But he added that the precision was "quite sufficient" for the applications; as Størmer also noted in his lecture:

> One only needs enough time and sufficient money for paying the assistants. (**Størmer, 1936/7**, p. 62)

For several decades, Størmer had also been lobbying intensely among mathematicians for his research on the Northern Lights. He presented photographs of the models with calculated orbits at no less than three ICMs (Rome 1908, Toronto

[3]For biographies of Størmer, see (**Brun, 1958**), (**Chapman, 1958**), (**Spjeldnaes, 2008**), and (**Egeland and Burke, 2013**).

[4]See `http://iutam.org/history-2/` (accessed 31 May 2019) and (**Battimelli, 2016**).

[5]Størmer himself had contributed to the application of the new technology of photography to the study of such physical phenomena: see (**Egeland and Burke, 2013**).

1924, and Oslo 1936). The presentations in Toronto and Oslo were plenary talks and the one in Rome was a "quasi-plenary" because it was an evening presentation. Størmer's sectional speeches at the ICMs in Strasbourg (1920) and Zurich (1932), which were also devoted to the Northern Lights, went without photographs, and only the Strasbourg talk presented Størmer's numerical method in some detail.

Several things were new in Størmer's talk in Oslo, as compared with his previous presentations at ICMs. New discoveries about cosmic rays had revealed that Størmer's methods, originally developed for calculation of electron orbits for the Northern Lights, had broader, even more convincing fields of application. Second, Størmer could refer to new technical progress in the realm of analogue computers which allowed the calculation of orbits even without recourse to his numerical method. Acting as a spokesman for Norwegian science in general, not just for mathematics, Størmer could barely hide his pride that the University of Oslo would soon be the home of the world's most powerful computer: the "differential analyzer" as it was properly called. The Norwegian physicist Svein Rosseland (1894–1985), who was attending the ICM in Oslo, and who had been Niels Bohr's assistant in Copenhagen, had been commissioned to build a "Bush machine", funded by money from the Rockefeller Foundation. The machine in Oslo was going to be twice as powerful as Vannevar Bush's differential analyzer in the USA; the machine was eventually put into service in 1938, and was operational until 1954 (**Holst, 1996**). However, biographies of Størmer do not say of how much actual use the machine was in his calculations. In any case, Størmer was personally located at the Astrophysical Institute in Oslo where the machine would also be placed.

Finally, Størmer applied some modern technology in his Oslo plenary talk, improving his photographic presentations of previous years. Based on numerical computations, his group had even produced a moving picture that was shown after his presentation. Based on some of the pictures from that film, stereographic images were then produced which were included in the congress proceedings (see Figure 9.1). And as H. von Schelling, the reviewer of Størmer's paper for *Jahrbuch*, noted, the calculated predictions corresponded extremely well to recent experimental data obtained by Ernst Brüche (1900–1985) at the physical laboratories of Allgemeine Elektrizitäts-Gesellschaft (AEG) in Berlin (JFM 63.0736.01).

Although Størmer represented the applications of mathematics in a broad sense in Oslo, he showed at the end of his talk that he was aware of new demands on mathematics to develop specific methods of numerical analysis and of computing, implicitly criticising the persistent lack of recognition of these methods among pure mathematicians:

> I hope you will agree with me, that the methods of numerical and mechanical integration of differential equations are very important not only in the applications but also in pure mathematics. (**Størmer, 1936/7**, p. 73)

Since the development of digital computers still lay a decade or so in the future, the full implications of Størmer's numerical method and of some error estimations also developed by him — topics to which he alluded only slightly in his talk — were not yet visible.

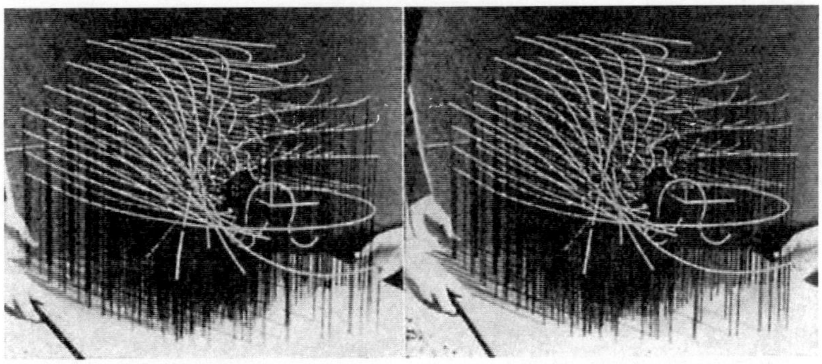

FIGURE 9.1. Stereographic image from Størmer's plenary lecture, taken from a movie which was shown after the talk. Original caption: "Stereoscopic picture of a bundle of trajectories from infinity towards the dipole and with asymptotes parallel to the magnetic equatorial plane." (Source: (**Størmer, 1936/7**, p. 65))

9.3. R. Fueter: "The theory of regular functions of a quaternion variable"

The Swiss mathematician Rudolf Fueter (see Figure 9.2) took his PhD in 1903 under Hilbert in Göttingen on a topic from class field theory (*Der Klassenkörper der quadratischen Körper und die komplexe Multiplikation*). His main mathematical achievements were in number theory and, in connection with it, function theory.[6]

In his doctoral dissertation, Fueter had already dealt with complex multiplication in its relation to number theory and elliptic modular functions. From the late 1920s onwards, he had turned to non-commutative number systems and worked on them first with number-theoretic, and later with function-theoretic, methods. Fueter's plenary address in Oslo was on this topic (**Fueter, 1936/7**); in his review of the lecture for *Jahrbuch*, P. W. Ketchum (a complex function theorist at the University of Illinois at Urbana-Champaign) observed that the talk was in large measure a summary of some of Fueter's earlier published papers on this topic (JFM 63.0310.04).[7] In his autobiography, which was published shortly after his death in 1950, Fueter gave special space to this theory which had also been closely related to his presentations at the 1928 and 1932 ICMs:[8]

> *Scientifically* I was privileged to discover the importance of hypercomplex functions, which brought me closer to my life's goal of developing class field theory with function-theoretic methods. I doubt that I will be able to complete this huge theory which requires totally new points of view, but I am confident that I have created the foundations for it. Foremost in *scientific* life I have to mention the International Mathematical Congress

[6]For biographical material on Fueter, see (**Speiser, 1950**) and (**Burckhardt, 1980**), (**Burckhardt, 2008**), as well as Fueter's own autobiographical notes (**Fueter, 1950**).

[7]Ketchum pointed in particular to (**Fueter, 1934**), (**Fueter, 1935**), and (**Fueter, 1936**).

[8]The titles of these presentations were, respectively, "Über Funktionen einer Quaternionenvariablen" (**ICM, 1928**, vol. II, p. 145) and "Idealtheorie und Funktionentheorie" (**ICM, 1932**, vol. I, pp. 83–92).

FIGURE 9.2. Karl Rudolph Fueter (1880–1950) (Public domain, via Wikimedia Commons)

in Zurich in 1932, which I had the honour to chair. This was certainly a success which earned me, at the following International Congress in Oslo 1936, an extensive presentation of my quaternion theory, which made this theory more widely known. The University of Oslo was particularly close to my heart because it had awarded to me an Abel honorary doctorate — I had always admired the genius of Abel.[9]

[9]Translated from the German in (**Fueter, 1950**, p. 103). On the honorary doctorates awarded by the University of Oslo in 1929, see Section 2.4.

According to his Zurich colleague, Andreas Speiser, Fueter did not achieve his goal, nor were any of his students able or willing to continue these difficult investigations in the direction of class field theory.[10]

In addition to his hopes for class field theory, Fueter had drawn particular satisfaction from his theory because, in his opinion, it comprised the theory of functions of two complex variables in its state of development as it existed in the 1930s. Fueter also articulated this claim in his presentation in Oslo in 1936, where he even spoke about a "conquest of function theory by hypercomplex systems",[11] and where he found that his theory was in a similar relationship to that of complex functions in two variables to "the general Galois theory of algebraic equations to the theory of Abelian equations".[12] This claim of Fueter's was not explicitly supported by Heinrich Behnke, the main specialist in multidimensional function theory at that time. In his *Zentralblatt* review of Fueter's plenary talk, he quoted Fueter's claims without commentary (Zbl 0018.31601). However, Behnke admitted retrospectively in 1978 that there was a "light tension" ("leichte Spannung") between his own presentation in section II (Analysis) in Oslo on "Der Kontinuitätssatz und die Regulärkonvexität" (**ICM, 1936**, vol. II, pp. 79–82) and Fueter's plenary presentation. At this later date, however, Behnke could rightly claim that Fueter's more general hope that multidimensional complex function theory would develop according to his theories had not come true:

> What will prevail as a function theory in higher-dimensional spaces? This question is now — as I write these lines forty years later — long decided. But now I would have to say that one should not overlook Fueter completely. This is injustice![13]

Indeed, Fueter's theory has not been "completely" overlooked, as the articles (**Deavours, 1973**) and (**Sudbery, 1979**) show. However, the former author admits (in reference to the USA) that it was "little known in this country" (**Deavours, 1973**, p. 996). And Sudbery says that

> the theory developed by Fueter and his school is incomplete in some ways, and many of their theorems are neither so general nor rigorously proved as present-day standards of exposition in complex analysis would require. (**Sudbery, 1979**, p. 199)

However these deficiencies could be — according to Sudbery — remedied by using the exterior differential calculus, i.e., methods that had been developed during the 1930s, by, among others, Élie Cartan, another plenary speaker in Oslo.

9.4. É. Cartan: "Some insights into the role of Sophus Lie's theory of groups in the development of modern geometry"

The plenary talk by the 67-year-old Élie Cartan (see Figure 9.3)[14] in Oslo was in French and was printed with the title "Le rôle de la théorie des groupes de Lie dans l'évolution de la géométrie moderne". It was his third (and final) plenary talk at an ICM, having already addressed this forum in 1924 and 1932 with lectures

[10] As reported by (**Burckhardt, 1980**, p. 8).

[11] Translated from the German in (**Fueter, 1936/7**, p. 91).

[12] *Ibid.*

[13] Translated from the German in (**Behnke, 1978**, p. 248).

[14] For biographies of Cartan, see (**Javillier, 1951**), (**Chern and Chevalley, 1952**), (**Whitehead, 1952**), (**Hodge, 1953**), and (**Dieudonné, 2008a**).

entitled, respectively, "La théorie des groupes et les recherches récentes de géométrie différentielle" (**ICM, 1924**, vol. I, pp. 85–94) and "Sur les espaces riemanniens symétriques" (**ICM, 1932**, vol. I, pp. 152–161).[15] Thus, on all three occasions, Cartan presented his seminal work on differential geometry in close connection to group theory, interest in which continued to grow among his scientific peers, particularly after the rise of Einstein's General Theory of Relativity of 1915 and the role of generalised geometries revealed therein. According to his biographer Jean Dieudonné (1906–1992), one of the founding members (around 1934) of the Bourbaki group:

> Cartan's recognition as a first rate mathematician came to him only in his old age. Before 1930 Poincaré and Weyl were probably the only prominent mathematicians who correctly assessed his uncommon power and depth. (**Dieudonné, 2008a**, p. 96)

Cartan's mathematical work can be roughly classified under three main headings: group theory, systems of differential equations, and geometry. These themes were, however, constantly interwoven with each other in his work. Almost everything Cartan did was more or less connected with the theory of Lie groups (**Chern and Chevalley, 1952**, p. 219).[16] Drawing on originally very different parts of mathematics, Cartan's work was, according to the younger Bourbaki member Roger Godement, exceptional in France at a time (the 1920s) when "theories of an essentially new character were born" (**Godement, 1948/1971**, pp. 337–338) outside France — in Germany, Poland, and the Soviet Union. However, being part of an older generation of mathematicians and having a strongly intuitive mathematical mind, Cartan was never interested in formalising — in the sense of the modern axiomatic method — the relationship between the various mathematical disciplines on which he worked. Not being the head of a "mathematical school", Cartan's approach was not everywhere easily understood, which led at times to neglect of his work by the younger generation, at least until it was revisited by André Weil, who later characterised the challenge that Cartan's heritage provided for the next generation with the following words:

> The work of Cartan alone contains enough material to keep busy several generations of geometers. ... The modern theory of Lie groups proper, studied by a combination of Cartan's methods with those of modern topology, is far from complete. ... But ... it is principally in the topological theory of fibre spaces, in the theorems of de Rham and in the notion of homotopy group, that we now find the tools best suited to the global study of the generalized geometries of Cartan. (**Weil, 1950**, pp. 301–302)

The broad and overall connection between the work of Siegel on modular functions of several variables and Cartan's geometries — although not discussed in Siegel's plenary lecture in Oslo (on which, see Section 9.5) — as well as the influence of algebraic geometry was mentioned by Weil too:

> In the work of Siegel, the local and global geometrical study of fundamental domains, in effect of manifolds with a complex

[15]The "sur" was dropped from the title in the printed version of the lecture in the congress proceedings.

[16]For a comprehensive survey of Cartan's contributions to the theory of Lie groups, see (**Cogliati, 2018**).

FIGURE 9.3. Élie Joseph Cartan (1869–1951) arriving in Oslo by boat (Source: (**Pólya, 1987**, p. 98))

analytic structure, tends to occupy a dominating role. Along this road, connection is made with the immense work accomplished by E. Cartan and its various extensions; at the same time one gets into the center of modern topology, the theory of fibre spaces; and the invariants of Stiefel–Whitney appear, along with their generalizations. The intimate connections between these two domains had been suspected for some time, but their actual merging was made possible only through the recent discoveries of Chern, stimulated, at least in part, by considerations of algebraic geometry. (**Weil, 1950**, p. 299)

These are results that were about to appear at the time of the Oslo congress without being made explicit there — not least due to the absence, as noted in Section 8.4, of leading topologists and algebraic geometers such as Whitney and Weil.

Like most of the other plenary presentations, Cartan's talk in Oslo cannot be considered as an original contribution to research. It is rather a summary, with a partly historical structure, and is not cited among his influential papers, for instance in (**Chern and Chevalley, 1952**). Van der Waerden reported on Cartan's talk in *Zentralblatt*:

> The development of geometry can be considered from the viewpoint of group theory in three different ways. The first is the perspective of Klein's Erlanger Program. The second originates from Veblen's definition of a geometric object, which is given with respect to each coordinate system by a certain number of components. The third perspective unites the notion of 'Übertragung' from differential geometry with Lie's notion of the infinitesimal transformation.[17]

Indeed, Cartan had commented explicitly on the third perspective, developed basically by himself, in the lecture:

> The principle of creating new spaces, the genesis of which I will describe here very cursorily, has been called "Übertragungstheorie" by some authors.
>
> Each of the generalised spaces in the sense outlined above has as its base a Lie group.[18]

The importance of Cartan's work and his creativity even this late in his life has to be seen in connection with this "third perspective", developed in particular in his books of 1935 and 1937. This and parallel work on differential manifolds, for instance by Whitney, caused an "explosion" in mathematics, as described many years later by (**Dieudonné, 1985a**, p. 545):

> After Poincaré died in 1912 and until H. Weyl inaugurated the global theory of Lie groups in 1925, only Cartan worked in Lie's style. These two people dominated the scene totally until 1935, when an explosion occurred. The Lie theory was found to relate to practically everything in mathematics. This is one of the most surprising phenomena in the last 50 years. As far as geometry is concerned, it is not too surprising since Lie theory is geometrical in nature, but its first main connections with geometry were far from obvious. Both were discovered by Cartan.

The years between the Oslo congress and the end of the war, saw, however, an increased strain on international relations between mathematicians, including some bitter personal experiences and disappointments too. Cartan's son Henri (1904–2008) succeeded to some extent in maintaining his relationship with German

[17]Translated from the German in Zbl 0018.27202. The corresponding *Jahrbuch* review, written by the Polish mathematician W. Ślebodziński (JFM 63.1155.04) is almost identical to van der Waerden's review, which, however, probably appeared much earlier.

[18]Translated from the French in (**Cartan, 1936/7**, p. 98).

mathematicians such as Behnke beyond his expulsion from Strasbourg to the temporary university campus of Clermont-Ferrand.[19] Nevertheless, both Élie and Henri Cartan received an unwelcome visit from another Oslo plenary speaker, the number theorist Helmut Hasse, in December 1940 in German-occupied Paris.[20] Hasse, who appeared in the uniform of a naval officer, invited the Cartans to collaborate with the Germans, apparently meaning in the first instance that they contribute to the German reviewing journals *Zentralblatt* and *Jahrbuch*.[21] Élie Cartan responded that this was impossible as long as Polish mathematicians were not involved on a collegial and impartial basis also. Hasse is reported to have said that the "Führer" had deemed the Poles "incapable and unworthy of co-operation".[22] In addition, Hasse allegedly opened up the possibility of Henri Cartan's wife visiting Germany at any time because she was born in Alsace, which Germany claimed as originally being its territory, while her husband could not expect treatment on the same basis. This offer, of course, was declined by Henri Cartan's wife. Hasse left the apartment "very angrily" ("très fâché"). Behnke told Henri Cartan much later that Hasse had told him that he considered Élie Cartan "disqualified to renew French mathematics after the war" ("disqualifié de renouveler les mathématiques françaises après la guerre").[23] Before the war ended, Cartan went through yet another and even more serious ordeal. His third son, the physicist Louis Cartan (1909–1943), was executed by the Germans for his actions in the resistance.

Élie Cartan published only a handful of further papers, plus a book (**Cartan, 1945**), between his wartime encounter with Hasse and his death in 1951. The vast majority of these related to the geometrical topics upon which Cartan had built much of his career. In an obituary notice of November 1952, fellow-geometer J. H. C. Whitehead described Cartan as "one of the great architects of contemporary mathematics" (**Whitehead, 1952**, p. 71).

9.5. C. L. Siegel: "Analytic theory of quadratic forms"

One of the most prominent and (for several decades to come) most creative mathematicians among the main speakers in Oslo was Carl Ludwig Siegel (1896–1981; see Figure 9.4), whose principal contributions to mathematics lay in number theory, partly connected with complex function theory, and in celestial mechanics.[24] Richard Courant, who was also in Oslo, referred to Siegel in 1934 *vis-à-vis* the Rockefeller Foundation as "the most able and promising mathematician alive".[25] Siegel had studied number theory with F. G. Frobenius (1849–1917) in Berlin, and took his doctorate with Frobenius' student Edmund Landau in Göttingen in 1920,

[19]Henri Cartan acknowledged after the war that Behnke had saved manuscripts and books from his deserted apartment in Strasbourg (**Jackson, 1999**, pp. 783–784).

[20]This is according to a personal interview that R. Siegmund-Schultze conducted with Henri Cartan on 22 January 1999, in the latter's apartment (95, boulevard Jourdan, 75014 Paris), which had also been Élie Cartan's apartment during the war.

[21]See (**Siegmund-Schultze, 1993**, p. 187).

[22]Cf. Lietzmann's remarks on Polish mathematicians on p. 82 (see also pp. 281–282).

[23]Hasse was apparently more "successful" in his contacts with Gaston Julia in occupied Paris in November 1940: see his report to the German Nazi authorities, published in (**Siegmund-Schultze, 1993**, p. 187). As we have seen (p. 59), Julia had attended the Oslo congress and had spoken there of his traumatising experiences from the First World War.

[24]For biographies of Siegel, see (**Deuring, 1985**), (**Hlawka, 1985**), and (**Dieudonné, 2008b**).

[25]Quoted in (**Siegmund-Schultze, 2009**, p. 159).

FIGURE 9.4. Carl Ludwig Siegel (1896–1981) (Author: Konrad Jacobs. Source: Archives of the Mathematisches Forschungsinstitut Oberwolfach)

the dissertation later being hailed as a "landmark in the history of Diophantine approximations" (**Dieudonné, 2008b**, p. 826).

Siegel held strong pacifist views which led to his discharge from the German army during the First World War. Although not "Jewish" by the Nazi definition, he could not tolerate the brutal methods of the new rulers and escaped to the United States: he spent 1935 at the Institute for Advanced Study in Princeton. However, Siegel, who held himself somewhat aloof from the world, struggled to adapt to American society. He decided to return to Frankfurt in 1936, apparently trying to support his Jewish colleague Max Dehn there, who had been dismissed — a decision which another participant in Oslo, the Dane Harald Bohr, called "unbelievably foolish".[26] In 1937, Siegel was appointed to a post in Göttingen, where he had occasional conflicts with Hasse. As we saw at the end of the preceding section, Hasse was more willing than Siegel to adapt to the political circumstances without being a fervent Nazi.[27] With the outbreak of the Second World War, Siegel fled, once again to the United States, this time via Oslo. He was, however, one of the few German refugees to return to Germany after the war.

Being temporarily back in Frankfurt in 1936, Siegel was named a member of the German delegation to Oslo, which would become one of his rare participations in congresses. His student Schneider later wrote:

> Siegel has never sought public recognition. Rather to the contrary he tried to evade it and has barely visited congresses. The International Congress in Oslo 1936 and the annual meetings of the German Mathematicians' Association in Jena 1921 and

[26] *Ibid.*, p. 160.

[27] See (**Segal, 1980**) — in particular p. 54, which features Siegel's remarks to Courant concerning Hasse's stance.

> Göttingen 1955, where he gave talks, seem to have been exceptions.[28]

One may speculate why Siegel made such an exception in 1936: maybe he found the maintenance of contacts with refugees and foreigners to be particularly important at that time. In addition, the invitation to deliver a plenary lecture was an honour that may have been difficult to decline.

In his German talk “Analytische Theorie der quadratischen Formen” in Oslo, Siegel presented results that were contained in essence in a paper of 1935 with the same title. It was written in German but published with two sequels in the American *Annals of Mathematics* (**Siegel, 1935–1937**). These results were connected with his recent move into the study of quadratic and modular forms. Siegel himself described the aims of his work at the interface of number theory and function theory in his talk in Oslo with the following general introductory remarks:

> It is well-known that there exist certain analogies between number theory and the theory of analytic functions. In particular some properties of algebraic number fields are mirrored in similar statements about algebraic functions. The function-theoretic theorems provide a connection between the local and global properties of analytic functions. ... If one wishes to find in number theory analogies with the integral theorems in analysis, one will ask to what extent number-theoretic functions in the Large, i.e. in the field of rational numbers, are determined by the corresponding number-theoretic functions in the Small, i.e. in the p-adic fields, or, what is the same, by the corresponding arithmetical problem modulo q, where q is an arbitrary natural number.[29]

General remarks about the importance of Siegel's investigation of links between number theory and function theory are contained in the account by (**Klingen, 1983**) of Siegel's work in function theory:

> Without any doubt Siegel considered himself to be a number theorist in the first line, and complex analysis was originally only a tool for him. But how can it then be explained that almost one third of Siegel's articles and, in addition, three books and various lecture scripts are devoted to function theory? The reason is the deep discoveries about automorphic functions in several complex variables that Siegel made in connection with his central investigations into quadratic forms in 1935. They were about the modular functions of degree n, as Siegel would call them, today mostly called Siegel modular functions. One has to be aware that back then the general theory of functions of several complex variables was still in its beginnings. That's why the discovery of new function classes was sensational. During the following decades these functions quickly obtained independent importance and found world-wide interest; just to name Princeton and Bombay. In this period of genuine pioneering work many

[28] Translated by the authors from the German in (**Schneider, 1983**, p. 147).

[29] Translated by the authors from the German in (**Siegel, 1936/7**, p. 104).

> further investigations by Siegel, but also ground-breaking work by other mathematicians, took place. The importance of these functions was regarded so highly, that applicability of general function-theoretic methods on Siegel modular functions was often considered a criterion for the worth or worthlessness of these methods.[30]

Siegel would always remain sceptical of modernist and abstractionist methods in mathematics and stressed historical traditions in mathematics.[31]

In order to gain a contemporary view of the content of Siegel's plenary lecture, we must of course turn to its reviews. However, as if to reinforce the point that Siegel's talk was closely linked to a published paper, we find that the reviews in *Jahrbuch* (JFM 63.0895.04) and *Zentralblatt* (Zbl 0018.24404) each consist of a single line, directing the reader elsewhere: to the above-mentioned papers, and to Siegel's own report on developments in this area (**Siegel, 1937**). The *Jahrbuch* reviews of the three main papers were written by Siegel's sometime-antagonist Hasse,[32] probably because the investigations of the former were linked to a paper of his of over a decade earlier (**Hasse, 1924**). Moreover, Siegel's paper made use of Hasse's local-global principle, to which we will return in Section 9.12. Hasse's reviews are mostly purely descriptive, although he described one of Siegel's results as being in an "extremely elegant and transparent form" (JFM 61.0140.01). Moreover, Hecke's *Zentralblatt* review of Siegel's first paper is also full (running to around two and a half pages) and complimentary, referring to "highly significant advances in the theory of quadratic forms".[33] Thus, it appears that contemporary number theorists, including fellow plenary lecturers, probably shared Courant's view of Siegel, as recorded at the beginning of this subsection, and we may speculate that they welcomed the attendance of this reclusive mathematician at the Oslo ICM.

9.6. O. Veblen: "Spinors and projective geometry"

Oswald Veblen (see Figure 9.5) was the nephew of the sociologist Thorstein Veblen and had Norwegian ancestors; it is not least for this reason that Veblen had

---

[30]Translated from the German in (**Klingen, 1983**, p. 158). In a message to the authors of March 2019, Norbert Schappacher commented on this quote by Klingen by stressing that "back in 1936 when 'Siegel modular functions' were not yet an established topic, it was above all Siegel's stroke of genius that was most admired by the experts: after all, his big 1935 paper starts rather modestly from what appears like a mere quantitative version of a result of Hasse's in the arithmetic of quadratic forms, but then takes off on a fantastic flight, defines higher-dimensional analogues of the complex upper half-plane in an arithmetic way, the points of the new space corresponding to symmetric matrices, i.e., to quadratic forms, and then introduces novel 'modular functions' on this new complex space of arithmetic origin."

[31]See, for example, (**Siegel, 1978/79**), in which Siegel noted his connection to the seminar on the history of mathematics in Frankfurt in the 1920s. A reviewer of Siegel's lectures on analytic number theory (**Siegel, 1965**), noted also that "the historic aspect is stressed well beyond the simple attribution of priorities" (MR0262150). In a letter to L. J. Mordell, another plenary speaker in Oslo (see Section 9.14), dated Göttingen, 3 March 1964, Siegel hoped to find in Mordell an ally against recent modernist books by Serge Lang and others. Siegel added there: "I am afraid that mathematics will perish before the end of this century if the present trend for senseless abstraction—I call it: Theory of the empty set—cannot be blocked up" (Mordell Papers, 28.17; see also (**Siegmund-Schultze, 2009**, p. 314)). For Lang's reaction to this criticism, see (**Lang, 1995**).

[32]JFM 61.0140.01, JFM 62.0136.03, and JFM 63.0120.01, respectively.

[33]Translated from the German in Zbl 0012.19703.

a special connection to, not to mention a special appeal for, Scandinavian mathematicians.[34] During the 1920s and 1930s he became one of the most influential American mathematicians. He was one of the foremost organisers of his science, in particular in connection with the Rockefeller philanthropy, and he was a founder of the Institute for Advanced Study in Princeton in 1932/33; as we saw on p. 28, he was at least partly responsible for the invitation of the ICM to Oslo in 1936. Politically a liberal, Veblen was a very good friend of the English mathematician G. H. Hardy. However, he was tolerant enough to remain on good and friendly terms with the politically conservative G. D. Birkhoff, a friend from their days of study in Chicago with E. H. Moore. Veblen was a key person in providing job opportunities for the refugees from Hitler's empire who were flocking to the United States; he was thereby supported by Hardy but sometimes in conflict with Birkhoff, who had reservations about immigration.[35]

Veblen's plenary talk in Oslo had the title "Spinors and projective geometry" (**Veblen, 1936/7**). It was his second and last plenary talk before an ICM and presented one of the main directions along which his research had developed after his more general, historically oriented, plenary talk in Bologna in 1928, entitled "Differential invariants and geometry" (**Veblen, 1929**). On the one hand, Veblen had gone in the direction of a general theory of "geometrical objects" and "invariance" in differential geometry and topology, a development which was referred to in Cartan's talk in Oslo[36] and had culminated in Veblen's joint book with J. H. C. Whitehead: *The foundations of differential geometry* (**Veblen and Whitehead, 1932**). According to Saunders Mac Lane:

> This monograph contained the first adequate definition of a global differentiable manifold. ... Soon afterward H. Whitney, starting from this Veblen–Whitehead definition, developed the simpler definition of a differentiable manifold, which has now become standard (and extended to other cases such as complex analytic manifolds). In this case the Veblen–Whitehead book had influence not through many readers, but essentially through one reformulation, that of Whitney. (**Mac Lane, 2008**, p. 600)

On the other hand, Veblen continued his research in "projective differential geometry", a discipline that had been founded by Ernest Julius Wilczynski (1876–1932), Guido Fubini (1879–1943), and Eduard Čech (1893–1960). Its main problem is to find a complete system of local invariants of a submanifold under the projective group and interpret them geometrically through osculation by simpler geometrical figures (**Chern, 1978**, p. xxii). "Projective relativity", in turn, aimed at the application of this theory to mathematical physics, in particular the new theories in relativity and quantum mechanics represented by the work of Einstein and Dirac.

Veblen's talk in Oslo was closely connected to his Springer *Ergebnisse* report (**Veblen, 1933a**), although he referred instead to his English publication of the same year (**Veblen, 1933b**). According to Chern:

---

[34]See (**Barrow-Green, 2011**) on Veblen's attendance of the 1913 Scandinavian Congress of Mathematicians in Kristiania (Oslo), and (**Monna, 1994**) on his Scandinavian ancestry. For biographies of Veblen more generally, see (**Montgomery, 1963**) and (**Mac Lane, 2008**).

[35]On attitudes to immigrations among American mathematicians, see (**Siegmund-Schultze, 2009**, Chapter 8); see also the comments on Birkhoff on p. 51 of the present book.

[36]He mentioned "the notion of geometric object introduced recently by M. Veblen" (translated from the French in (**Cartan, 1936/7**, p. 95)).

FIGURE 9.5. Oswald Veblen (1880–1960) (By unknown photographer circa 1915 (http://legacyrlmoore.org/photos/veblen_o.html) [Public domain], via Wikimedia Commons)

> Projective relativity (O. Veblen, J. A. Schouten) aims at singling out the projectively connected spaces whose paths are to be identified with the trajectories of a free particle in a unified field theory. They are defined by a system of "field equations". (**Chern, 1978**, p. xxiii)

Veblen began his talk with the following words, where he clearly described the derived and applied character of the notion of the spinor, a generalised tensor, which in the group-theoretic context had already been introduced by Cartan in 1913:[37]

[37]See (**Chern and Chevalley, 1952**, p. 219).

> The theory of spinors like so many other branches of mathematics received its first impulse from theoretical physics. Unlike some other branches it has not required the discovery of anything very fundamentally new on the mathematical side, but rather has called for a regrouping and restudy of known ideas. Most of these ideas can advantageously be clothed in the dress of projective geometry, and some of them suggest the use of devices which have not yet been exploited as much as they should be in projective geometry itself. (**Veblen, 1936/7**, p. 111)

The German booklet of 1933, which was based on Veblen's lectures in Göttingen in 1932, had met a rather cool reception, particularly with respect to its value for physics. The *Zentralblatt* reviewer, David van Dantzig, for example, noted that "[p]hysical questions in contrast to geometrical ones are very much in the background".[38] There was also scepticism in the USA; somewhat earlier, E. B. Wilson had written to R. G. D. Richardson:

> I do not believe there is the least bit of real physical interest in anything that Birkhoff and Veblen may say about space and time.[39]

Veblen's talk of 1936 focussed more on the geometrical side and developed into what the speaker called a "very sketchy account of what proves to be a rather extensive geometric theory" (**Veblen, 1936/7**, p. 127). Van der Waerden, who was mentioned by Veblen in his talk as the creator in 1929 of a formal theory of two-component spinors adequate to special relativity, said in his *Zentralblatt* review that

> [t]he n-dimensional generalization of this theory recurs to an investigation of linear families of involutions and is parallel to the algebraic theory of Brauer and Weyl.[40]

The latter theory (found in (**Brauer and Weyl, 1935**)) was also mentioned by (**Veblen, 1936/7**, p. 122).

While the future success of Veblen's projective relativity was limited inasmuch as the various unified field theories in physics remained controversial, projective differential geometry itself continued to stimulate research. As late as 1978, Chern, the most influential successor to Veblen and Cartan in differential geometry, said:

> I wish to say that I believe that projective differential geometry will be of increasing importance. In several complex variables and in the transcendental theory of algebraic varieties the importance of the Kähler metric cannot be over-emphasized. (**Chern, 1978**, p. xxiv)

[38] Translated from the German in Zbl 0006.41902.

[39] WP: HUG 4878, Box 4, 20 December 1922.

[40] Translated from the German in Zbl 0018.32604.

FIGURE 9.6. Jakob Nielsen (1890–1959) (Author: Konrad Jacobs. Source: Archives of the Mathematisches Forschungsinstitut Oberwolfach)

9.7. J. Nielsen: "Some methods and results from the topology of surface transformations"

The Dane Jakob Nielsen (1890–1959; see Figure 9.6)[41] gave a plenary talk in German entitled "Einige Methoden und Ergebnisse aus der Topologie der Flächenabbildungen" (**Nielsen, 1936/7**), which translates as "Some methods and results from the topology of surface transformations". Nielsen referred for details to his own publications in *Acta Mathematica* (**Nielsen, 1927–1932**).

[41]For biographies of Nielsen, see (**Fenchel, 1960a**), (**Fenchel, 1960b**), (**Hansen, 1993**), and (**Hansen, 1999**).

The 46-year-old Nielsen was a recognised researcher in topology and combinatorial group theory, a student of Dehn and Hecke, both of whom attended the Oslo congress. Born in a then-German part of Schleswig and mostly trained and employed at German universities, Nielsen moved to Denmark in 1920 and had worked in Copenhagen since 1921. Indeed, he remained in Copenhagen for the rest of his career, despite concerns about his safety during the Second World War:

> At the beginning of the German occupation of Denmark 1940–1945 ... some attempts were made to bring Jakob Nielsen to America, since it was feared that he might be assaulted by the Nazis. (**Hansen, 1999**, p. 984)

The content of Nielsen's plenary talk in Oslo is perhaps best summarised in the following review by Oscar Zariski in *Zentralblatt* (Zbl 0018.42503). Zariski, who was not in Oslo, stressed the possible use of Nielsen's work in connecting algebraic geometry and topology, the consolidation of which was typical for the 1930s:

> A report upon some recent results, methods and unsolved questions concerning topological automorphisms of a Riemann surface of an algebraic curve of genus $p > 1$. The search for invariants of transformation classes forms the main topic of this report. Particular attention is given to periodic transformations and to transformation classes of finite order. Quite remarkable are the topological proofs of the classical results by Hurwitz and Wiman on regular Riemann surfaces and the extension to general topological automorphisms, in particular those of finite order. This topologization of a classical topic in algebraic geometry can be put in the same class with Lefschetz's topological theory of algebraic correspondences, both being applications of general fixed point theorems in topology.

As Nielsen's biographer admitted, the link with Lefschetz' research does not seem to have resulted in active collaboration and Nielsen was a bit isolated in his work:

> When it really came down to details, Jakob Nielsen worked, however, very much on his own problems on surface transformations. His work was highly respected by his contemporaries, but it was not in the main stream of the then burgeoning field of algebraic topology. ... There is no evidence of any correspondence with Solomon Lefschetz on fixed point theory, as could have been expected, in the Nielsen archives discovered by Sigurd Elkjær at Mathematical Institute, University of Copenhagen. (**Hansen, 1999**, p. 984)

Nielsen is one of the few plenary speakers at Oslo who has not received a biographical article in the *Dictionary of Scientific Biography*; however his student V. L. Hansen organised the publication of Nielsen's collected papers, including English translations of those publications originally in Danish or German, more than two decades after his death (**Nielsen, 1986**). Writing for *Mathematical Reviews* (MR0865335, MR0865336), B. H. Neumann welcomed this belated publication of Nielsen's works, noting that

> Jakob Nielsen ... initiated much of the topology of surfaces and of combinatorial group theory, and for this reason alone he occupies an important place in the history of 20th century mathematics. However, his work is not merely of historical interest, but is still (or again) much used and quoted today.

9.8. E. Hecke: "Recent advances in the theory of elliptic modular functions"

Erich Hecke (1887–1947; see Figure 9.7) belonged to a somewhat older generation of German number theorists than Hasse and Siegel.[42] He had studied with Landau in Berlin from 1906 to 1908, and later thanked the latter for his considerable influence when he defended his doctoral dissertation (*Zur Theorie der Modulfunktionen von zwei Variablen und ihrer Anwendung auf die Zahlentheorie*) under Hilbert in Göttingen in 1910. Landau had in the meantime moved to Göttingen also; Hecke and Landau both became teachers there of Hasse.

Modular functions were at the heart of Hecke's Oslo lecture, which evidently contained at least some results that were surprising, to judge by a remark made by Hecke's former student Hans Petersson in his review for *Jahrbuch*:

> These propositions have hitherto neither been proved nor even suspected in arithmetic.[43]

Another former student, Bruno Schoeneberg (1906–1995), the author of a standard book on elliptic modular functions (**Schoeneberg, 1974**), described Hecke's work in this area in the following terms:

> Hecke turned in 1925 to elliptic modular functions. He systematically applied quadratic number fields to the construction of modular functions. ...
>
> In 1936, Hecke systematically investigated the connection ... of the Dirichlet series with a functional equation of the Riemannian type and functions belonging to a certain automorphic group. Through his "operator" T_n Hecke established a theory for the investigation of relations of modular functions to Dirichlet series with Euler's product development. He discovered new connections between prime numbers and analytic functions and new rules for the representation of natural numbers through positive quadratic forms of an even number of variables. (**Schoeneberg, 2008a**, p. 214)

Moreover, the number theorist Samuel Patterson has said that Hecke's work on modular functions was

> technically much easier [than his previous work] and was pointing towards the future. It is ... a connecting link between the mathematics of the nineteenth and twentieth centuries.[44]

[42]For biographies of Hecke, see (**Perron, 1944/48**), (**Petersson, 1949**), (**Schoeneberg, 1989**), and (**Schoeneberg, 2008a**).

[43]Translated from the German in JFM 63.0339.02.

[44]Translated from the German in (**Patterson, 1990**, p. 629).

FIGURE 9.7. Erich Hecke (1887–1947) (Sources: Archives of P. Roquette, Heidelberg, and the Archives of the Mathematisches Forschungsinstitut Oberwolfach)

In the printed version of his talk, Hecke pointed his readers towards papers published in *Mathematische Annalen*, (**Hecke, 1936**), (**Hecke, 1937**), for a detailed discussion of the results that he had presented. As sources for his material, he quoted another plenary speaker, namely Mordell,[45] and the Jewish refugee Hans Hamburger (1889–1956), who was not at the congress; a connection to the work of Siegel is also clear (cf. Section 9.5). Like Siegel, Hecke was careful to maintain contact with the classical mathematics of the nineteenth century.[46]

[45]Indeed, the connection between the work of the two was remarked upon elsewhere; in connection with Mordell's interest in modular forms (in particular, his proof of the multiplicativity of the Ramanujan tau function), (**Cassels, 2008**, p. 653) noted: "The central argument (the Hecke operator) was rediscovered by Erich Hecke in 1937". Cassells was referring here either to Hecke's talk at the 1936 ICM or else to the related *Mathematische Annalen* paper (**Hecke, 1937**).

[46]A letter from Hermann Weyl to his friend Hecke in April 1936, written shortly before the Oslo congress, shows that Weyl considered Hecke and Siegel as allies in a struggle against

Hecke was a clear opponent of the Nazis and for this reason was in permanent conflict with his colleague at Hamburg, Wilhelm Blaschke (1885–1962), the noted geometer with ambivalent relations to the regime, who also attended the Oslo congress. His former student Heinrich Behnke, also in Oslo, wrote in this connection:

> As Hecke's assistant I had much freedom. In Hamburg we had many talks during extensive walks. But his radical scepticism I found increasingly nerve-racking. Probably he had serious problems with himself. He did not find a balance. He had his huge success too early in life. Although I recalled anti-Semitic remarks by him in earlier days, he behaved very bravely at the time of the brown guards. While editing the Mathematische Annalen together we went through difficult times.[47]

9.9. O. Neugebauer: "On pre-Greek mathematics and its position relative to the Greek"

As a talk on a topic from the history of ancient mathematics, the plenary lecture of Otto Neugebauer (see Figure 9.8), "Über vorgriechische Mathematik und ihre Stellung zur griechischen", was rather different from the other lectures delivered in Oslo. Nevertheless, Neugebauer was very much the "mathematician's historian": he was certainly well known and respected within the European and North American mathematical communities, perhaps because of his own mathematical background, as well as his intimate links with the mathematical institute in Göttingen and his founding editorship of *Zentralblatt*; note, for example, the following comment by David Eugene Smith in a review of (**Neugebauer, 1934a**):

> For several years Dr. Neugebauer has been known to readers of the history of mathematics as one of the most promising scholars in this field. (**Smith, 1935**, p. 162)

Moreover, in 1934, Neugebauer had left Nazi Germany to take up residence in Copenhagen, so would therefore have gained extra prominence in Scandinavian mathematical circles in particular, making him a natural person to invite to Oslo: as we recall from Table 2.5 (p. 41), his name had appeared in the earliest draft of the congress programme in Spring 1935.

Neugebauer was born in Austria, and saw active service during the First World War, though he spent the later parts as an Italian prisoner of war.[48] Upon his release in 1919, he enrolled at the University of Graz to study mathematical physics. Following a brief stint in Munich, he moved in 1922 to Göttingen, where he studied mathematics with Richard Courant, Edmund Landau, and Emmy Noether; Courant in particular would go on to have a strong influence over Neugebauer's life and career.[49] It was in Göttingen that Neugebauer first began to work on ancient

---

the "degeneration [of mathematics] into an axiomatic game" (see (**Siegmund-Schultze, 2009**, p. 308)).

[47] Translated from the German in (**Behnke, 1978**, p. 79).

[48] For biographies of Neugebauer, see (**Davis, 1994**), (**Gray, 1999**), (**Pingree, 1990**), (**Swerdlow, 1993a**), and (**Swerdlow, 1993b**), as well as the volume (**Jones *et al.*, 2016**). For (incomplete) bibliographies of Neugebauer's work, see (**Anon, 1962**) and (**Sachs and Toomer, 1978/79**).

[49] On Neugebauer's relationship with Courant (and on the former's early career more generally), see (**Rowe, 2016**).

FIGURE 9.8. Otto E. Neugebauer (1899–1990) (Source: Mathematisches Institut, Georg-August-Universität Goöttingen)

Egyptian mathematics, an interest that led ultimately to a 1926 dissertation on Egyptian fractions (**Neugebauer, 1926**).[50] In the years that followed, he also expanded his interests into Mesopotamian mathematics and astronomy.

With the rise to power of the Nazis in 1933, life became more difficult for Neugebauer in Göttingen owing to his liberal views. When Courant was removed as director of the mathematical institute in April 1933, Neugebauer was appointed acting director in his place, but soon withdrew after having been informed by the

[50]For a survey of Neugebauer's dissertation, see (**Ritter, 2016**, pp. 140–147).

dean that he did not enjoy the confidence of the (Nazified) students.[51] In 1934, he accepted an invitation from Bohr to take up a professorship (part-funded by Rockefeller money) in Copenhagen.[52] While there, he continued his editorship of *Zentralblatt*, defending it from Nazi interference, until the removal of Levi-Civita from the editorial board in 1938 (as mentioned on p. 131), at which point Neugebauer and the rest of the board resigned in protest.[53] Later that same year, Neugebauer was offered a professorship at Brown University in the USA, which he took up the following year; he remained at Brown until his retirement in 1969. During this time, he built up a department for the history of mathematics, and was once again the founding editor of a mathematical abstracting journal: *Mathematical Reviews*, published by the American Mathematical Society from 1940.[54]

The essence of Neugebauer's work has been characterised as a "technical cross-cultural approach" to the history of mathematics (**Swerdlow, 1993a**, p. 141), which stressed the universal nature of mathematics across cultures and centuries, thereby allowing us always to reinterpret ancient mathematics in modern (symbolic) terms.[55] This approach, very much at odds with current trends in mathematical historiography, is certainly in evidence in Neugebauer's plenary lecture in Oslo, which begins as follows, in the language of a mathematician:

> Every historical representation is more or less forced to refer a reality that depends on a large number of parameters in a very complicated manner to a merely one-dimensional coordinate system, the time axis. Nevertheless, I will seek in the following not only to limit myself to assigning certain events in the history of mathematics to certain points of this timeline, but also to try, as far as possible, to bring out the deeper connections which have conditioned the course of these events. The guiding point of view will be to show you that at least the history of ancient mathematics is essentially a history of *symbolism*.[56]

Neugebauer then proceeded to argue a point of view that appears elsewhere within his work:[57] the idea that the thinking of ancient Mesopotamian mathematical scribes was in some sense already "algebraic", even without any form of concrete symbolism in the modern style, and that a more geometrical point of view was adopted by the Greeks when they took on the ideas of Babylonian mathematics — that the precedence of geometry was a feature of classical, but not pre-classical, mathematics. Indeed, Neugebauer followed in the trend of "geometric algebra":

---

[51] On Neugebauer and Göttingen under the Nazis, see (**Schappacher, 1987**).

[52] On Neugebauer's connections with Copenhagen, see (**Brack-Bernsen, 2016**).

[53] On the effects of Nazi rule on reviewing in Germany, and the flight of Neugebauer from Göttingen, see (**Siegmund-Schultze, 1994**) and (**Siegmund-Schultze, 2016b**).

[54] On the foundation of *Mathematical Reviews*, again see (**Siegmund-Schultze, 1994**), and also (**Lehmer, 1988**).

[55] The main thesis of (**Siegmund-Schultze, 2016b**) is that Neugebauer's own rationalistic and internationalistic ideals provided him with the guiding principle for his approach to history, which contradicts his claim of "having no Weltanschauung". In an interview for an Oslo newspaper during the ICM (reproduced here in English translation in Appendix A.3), Neugebauer hinted at the importance of collaboration between peoples for the development of mathematics, a barb presumably aimed at the likes of Bieberbach — see also the remarks concerning race in Chapter 6.

[56] Translated from the German in (**ICM, 1936**, vol. I, p. 157).

[57] See, for example, (**Neugebauer, 1945**, pp. 11–12) on the academic side, or (**Neugebauer, 1934b**) for a more general readership.

the idea that even ostensibly geometric constructions in classical and pre-classical mathematics were the result of algebraic thinking.[58] In his Oslo paper, Neugebauer began with an astronomical example, interpreted this in modern symbolic terms, and then argued for the inherently algebraic nature of the subject matter. It is a sign of how acceptable a feature this was in the study of ancient mathematics (at least among mathematicians) that in his *Jahrbuch* review of Neugebauer's paper (JFM 63.0799.01), Kurt Vogel (1888–1985) did not feel the need to comment upon it.[59] Indeed, Neugebauer's influence on the field was such that it was not until near the end of his life that more works arguing against "geometric algebra"; began to appear.

Leaving aside the issue of "geometric algebra", there is another feature of Neugebauer's work, as reflected in his Oslo paper, that is worthy of comment, and that may in fact represent a turning point in his investigations. In his earlier work, Neugebauer viewed Egyptian and Babylonian mathematics simply as precursors of the much greater Greek contributions that came later; when he turned his attention to Babylonian mathematics, it was to study this as an example merely of "pre-Greek" ("vorgriechisch") science. However, it appears that as Neugebauer studied ancient Mesopotamian mathematics, his regard for it grew, and that he even began to view the development of ancient Greek mathematics as a retrograde step (**Rowe, 2016**, p. 47). As (**Swerdlow, 1993a**, p. 147) later commented:

> his respect for Babylonian mathematical science was far too great for him to treat it as preparatory to anything.

He began thus to move the study of ancient mathematics and astronomy away from the traditionally Greco-centric view of European science. This also appears to have marked a turning point in Neugebauer's research — the terms "pre-Greek" and "vorgriechisch" disappeared from his published work (at least as far as the titles are concerned) from the mid-1930s onwards: his Oslo paper features the last use of the term (in either language). In accordance with Neugebauer's newspaper interview in Oslo, which we reproduce in Appendix A.3, the renouncement of Greek mathematics as the towering point of reference went on to make it easier to stress the equal value of cultures and the benefit of various styles of mathematics for the development of this science. This could at the same time be used as an argument against racist theories of separation between peoples, such as in Bieberbach's "Deutsche Mathematik".[60]

9.10. C. W. Oseen: "Problems of geometric optics"

Carl Wilhelm Oseen (1879–1944; see Figure 9.9) was a Swedish mathematician and theoretical physicist with a prolific scientific and popular publication record.[61] Early in his career he was much influenced by Hilbert in Göttingen, where he studied

[58] On the development of and debate over "geometric algebra", see (**Høyrup, 2017**).

[59] We note that Vogel gave a talk on Babylonian mathematics in Oslo in the section on the history of mathematics: "Bemerkungen zum Nachleben der babylonischen Mathematik" (**ICM, 1936**, vol. II, pp. 277–278). Poul Heegaard, on the other hand, spoke on an Egyptian theme: "Zahlen auf einem Papyrusfetzen in der Osloer-Papyrussammlung" (**ICM, 1936**, vol. II, pp. 276–277). In its written-up form, Vogel's talk refers to Neugebauer's plenary lecture, while the congress proceedings record remarks made by Neugebauer in relation to Heegaard's presentation.

[60] See footnote 55 above.

[61] For a biography of Oseen, see his entry in the *Svenskt biografiskt lexikon*: `https://sok.riksarkivet.se/Sbl/Presentation.aspx?id=7816` (accessed 18 August 2018).

partial differential equations in 1900–1901. He took his PhD from Lund University in 1901 with a thesis on finite contact transformations, connected to the work of the recently deceased Norwegian mathematician Sophus Lie. Oseen worked his way up the academic ranks there until 1909 when he became professor of mechanics and mathematical physics in Uppsala. He held a number of organisational and honorary roles. Most of Oseen's research was devoted to hydrodynamics, in particular the development of the theory of Navier–Stokes equations for viscous fluids. Oseen was a member of the committee of the Swedish Academy of Sciences which in 1922 retrospectively awarded Albert Einstein the Nobel Prize in Physics for 1921. The delay had to do with discussions about the eligibility of Einstein's work on relativity, something which was put into doubt by, among others, the ophthalmologist and Nobel laureate of 1911 for Physiology or Medicine, Allvar Gullstrand (1862–1930), who was also a committee member. As a compromise Einstein was awarded the Nobel Prize for his work on the photoelectric effect, which could be more easily connected to applied domains than relativity and thus fulfilled Alfred Nobel's original intentions the best.

Both Oseen's conflict with Gullstrand, meanwhile deceased, and the former's wish to make some broad points about the need for a general mathematical theory even in applied contexts, seem to be the background to the surprising fact that Oseen devoted his plenary lecture of 1936 to a topic such as geometrical optics, which lay somewhat on the margins of his research interests.

Gullstrand's reputation relied mostly on the invention of optical instruments such as the slit lamp and the reflexless ophtalmoscope, while his mathematical theory of optics had less influence. In 1935, Oseen wrote an obituary of Gullstrand in Swedish (**Oseen, 1935**), which biographers of the latter call "the most insightful". These authors refer to Oseen as a "17 years younger friend [*sic*]" of Gullstrand's and continue by saying of Oseen that

> [h]e clearly admired Gullstrand's achievements and stature, but had the position and strength to point out flaws in his mathematical work and his academic judgments. (**Ehinger and Grzybowski, 2011**, p. 707)

Indeed it seems to us that the following passages in Oseen's plenary talk point in this direction, when he recommended the use of "covariant differentiation, as introduced by Ricci and Levi-Civita" and asked the rhetorical question of whether Gullstrand would have admitted the mathematical shortcomings of his theory:

> He would surely say that his theory had the purpose of promoting ophthalmology and the optics industry, and that he had no concern with objects that are meaningless for this purpose. One can approve of this answer, and yet believe that it is best for the applications of the theory that the theory should develop according to its own laws.[62]

We restrict our discussion of Oseen's plenary talk to these short remarks and leave a more detailed analysis to historians of geometrical optics.

It is important for us to note how the Scandinavian mathematical physicist Oseen both propagandised and criticised another Scandinavian scientist before the

[62]Translated from the German in (**Oseen, 1936/7**, pp. 180, 183).

FIGURE 9.9. Carl Wilhelm Oseen (1879–1944) (Source: https://alchetron.com/Carl-Wilhelm-Oseen-1249537-W. Licensed under CC BY-SA 3.0, https://creativecommons.org/licenses/by-sa/3.0/)

international public of the ICM in Oslo, pointing to the awareness of the Scandinavians of the most modern mathematical tools of physics.

9.11. V. Bjerknes: "New lines in hydrodynamics"

Vilhelm Bjerknes (1862–1951; see Figure 9.10) was the son of Carl Anton Bjerknes (1825–1903), professor of mathematics at the University of Kristiania.[63] After studying in Oslo and working as his father's assistant, V. Bjerknes studied in Paris and Bonn before accepting a position in Stockholm (1893–1907). There he soon began to expand upon his father's work in hydrodynamics and around 1900 realised

[63]This section is based upon a draft supplied by Henrik Kragh Sørensen. For biographies of V. Bjerknes, see (**Gold, 1951**) or (**Pihl, 2008b**). For his work in meteorology, see (**Friedman, 1989**). On C. A. Bjerknes, see, for example, (**Pihl, 2008a**).

that his research could help to explain land and sea breezes, as well as cyclone formation, by referring these phenomena to rising warm air. In 1904, he outlined a plan for applying such ideas to the problem of weather forecasting that involved frequent observations at different heights of the atmosphere and the graphical approximate integration of the hydrodynamical equations. Since observations were necessary from a number of places, Bjerknes engaged himself in an international collaboration of exchange and standardisation of measurements. In 1907, after the dissolution of the political and administrative union between Sweden and Norway, Bjerknes was called to Oslo, where he stayed for five years before moving to Leipzig to set up the Geophysical Institute there. While in Germany, he trained a number of Norwegian scientists, and eventually relocated to Bergen in 1917 to begin setting up a weather service for Western Norway. There he again employed a number of mathematically skilled assistants, and Bergen became a significant centre for meteorologists from all over Europe from which the new approach to meteorology was spread. Indeed, (**Friedman, 1989**, p. xi) notes that

> [t]he formulation of the so-called Bergen meteorology, based on the concepts of fronts and air masses, has long been recognized as one of the principal turning points of this science.

In 1926, Bjerknes relocated again, this time to a professorship in Oslo. He retired from this chair in 1932, devoting himself even more to international collaboration and directing the meeting of the International Union of Geodesy and Geophysics held in Edinburgh during September 1936.

Bjerknes' approach to meteorology, which he had initiated in Leipzig and developed into the so-called Bergen School, relied on physical theories for atmospheric conditions — in particular for masses of air — and mathematical studies of their behaviour. It thus relied on both mathematical theory and on numerical computation that could be said to be exact. This programme of combining hydrodynamics and meteorology had also figured prominently at the International Congresses for Applied Mechanics, which Bjerknes and his disciples frequented.[64] And in 1936, it was time to present it also at an ICM.

As we have seen (Section 2.8.3), Bjerknes' lecture at the ICM had only been added to the programme at the last minute, and perhaps therefore consisted mainly of a presentation of his forthcoming paper "Application of line integral theorems to the hydrodynamics of terrestrial and cosmic vortices" (**Bjerknes, 1937**).[65] The printed version of the talk spanned less than three pages, in which Bjerknes sketched the novel approach that he was pursuing. The first part was summarised well in his own opening sentence:

> My aim with this address is not so much to present solutions of hydrodynamic problems as to direct the attention of my colleagues mathematicians and physicists [*sic*] to an enormous field of *unsolved problems*, problems belonging to a generalized hydrodynamics. (**Bjerknes, 1936/7**, p. 186)

---

[64]See, for example, the summary of Bjerknes' address to the Delft congress of 1924: (**ICAM, 1924**, pp. 411–413).

[65]At the beginning of this paper, Bjerknes noted that it also incorporated aspects of an address, "De l'application des théorèmes de circulation à la discussion des phénomènes atmosphériques", delivered at the Edinburgh meeting of the geophysical union mentioned above.

FIGURE 9.10. Vilhelm Bjerknes (1862–1951) (By unknown photographer, CC BY-SA 4.0 (`http://creativecommons.org/licenses/by-sa/4.0`), via Wikimedia Commons)

Bjerknes described how modern hydrodynamics often did not attempt to approach the "*true equation of state*" (Bjerknes' italics: (**Bjerknes, 1936/7**, p. 186)) which expresses the density of a fluid as a function of both pressure and temperature, $q = f(p, \vartheta)$. Instead, he lamented, they work from a simplified form of this equation in which temperature has disappeared, $q = f(p)$. Nevertheless, the hydrodynamic mathematician has good reasons for the simplification, although it is a serious one; the simplified state equation $q = f(p)$ allows for a

> model academic science ... characterized not least by the wonderful theorems of *Helmholtz* and *Kelvin* of the conservation of vortices and circulation (**Bjerknes, 1936/7**, p. 186)

but at the same time

> [w]e know that every *real* fluid can serve as working substance in a thermodynamical engine. But the unreal fluid defined by [the]

> equation [$q = f(p)$] can not. The science of thermodynamics can be built on no other equation than the true equation of state [$q = f(p, \vartheta)$]. The simplified equation of state keeps the two sciences of hydrodynamics and thermodynamics carefully separated. (*Ibid.*)

Bjerknes called therefore for "a generalized *physical hydrodynamics* which is a synthesis of the two separate sciences", and he pleaded for "cooperation for the creation of this combined science" (**Bjerknes, 1936/7**, p. 187). He had himself begun to explore — both quantitatively and otherwise — the consequences of these state equations, which have widespread applications in fluid dynamics, the circulation of the atmosphere, and many other areas, some of which he outlined in his lecture.

Looking at Bjerknes' work more broadly, we find that it is in meteorology that his influence has been felt the most. He argued that by combining theories of thermodynamics and hydrodynamics, one could use knowledge of the current state of the atmosphere and the oceans to determine their future conditions. On a trip to the USA in 1905, he was able to obtain Carnegie funds to employ assistants to handle the large amounts of data involved in such predictions. This funding continued until 1941.[66] Here again, we see Bjerknes' efforts to build meteorological research at an international level in order to tackle the inherently complicated nature of mathematical weather prediction. Indeed, in a speech in Leipzig in 1938, on the occasion of the silver jubilee of the Leibniz Geophysical Institute, which he had once directed, Bjerknes

> exhorted his listeners not to be discouraged by the fact that the problem of calculated weather forecasts, in spite of all the progress made, had not been solved and might perhaps never be satisfactorily solved ... (**Gold, 1951**, p. 312)

An indication of Bjerknes' status within the development of weather forecasting is given by the title of the book (**Friedman, 1989**), which goes some way towards explaining one biographer's description of him, at the time of his death, as the "doyen of meteorologists" (**Gold, 1951**, p. 303).

9.12. H. Hasse: "On the Riemann hypothesis in function fields"

As is the case for so many of the mathematicians profiled here, when one turns to biographies of the number theorist Helmut Hasse (1898–1979; see Figure 9.11), one finds superlative remarks about his work and abilities, as, for example, in the first paragraph of his entry in the *Dictionary of Scientific Biography*:

> One of the most important mathematicians of the twentieth century, Helmut Hasse was a man whose accomplishments spanned research, mathematical exposition, teaching and editorial work. In research his contributions permeate modern number theory; particularly noteworthy are his 'local-global principle' which established Kurt Hensel's p-adic numbers as indispensible tools of number theory, and his proof of the 'Riemann hypothesis' for elliptic curves. (**Edwards, 2008**, p. 385)

---

[66](**Friedman, 1989**, p. 237) refers to "Bjerknes the strategist", who was able to define research agendas and identify potential audiences, thereby securing funding.

FIGURE 9.11. Helmut Hasse (1898–1979) (Sources: Archives of P. Roquette, Heidelberg, and the Archives of the Mathematisches Forschungsinstitut Oberwolfach)

Hasse had studied briefly in Göttingen in 1918–1920 with, among others, Hecke, Landau, and Emmy Noether. Then he went to Marburg to work with Kurt Hensel, and it was there that he established the local-global principle (also known as the Hasse principle) in number theory in his dissertation of 1921.[67] As we have already noted (p. 172), the local-global principle also played a role in Siegel's talk in Oslo.

A pinnacle in the use of this principle was reached in the determination of the Brauer group of a number field, which led to the short but celebrated paper of Richard Brauer, Hasse, and Noether (**Brauer *et al.*, 1932**) on the "Hauptsatz" in the theory of algebras. This paper also demonstrates the close collaboration and points of contact at that time between clearly distinct streams within German algebra and number theory (group theory and the Schur school in Berlin, to which Brauer originally belonged, number theory in a more specific sense, and abstract,

[67]The local-global principle enables one to find an integer solution of a given equation by using the Chinese Remainder Theorem to find solutions modulo powers of different primes. The name "local-global" stems from the fact that the principle allows for the construction of a rational ("global") solution to an equation from ("local") solutions obtained over the reals or the p-adics; the converse construction is of course always possible. On the development of the principle, see (**Schwermer, 2007**).

structural algebra, respectively).[68] The collaboration of the three German mathematicians ended abruptly in 1933, however, with the emigration of Noether and Brauer.

At the same time, in the early 1930s, Hasse had begun work in yet another area that was to become one of the central subjects of modern number theory and that found its reflection in Hasse's Oslo talk. In his entry on Hasse for the *Dictionary of Scientific Biography*, Edwards introduces Hasse's work in this direction as follows:

> Challenged by his English colleague Harold Davenport [1907–1969] to bring his algebraic methods to bear on a problem Davenport and Louis Joel Mordell had been studying concerning the number of solutions of a congruence of the form $y^2 \equiv f(x) \bmod p$ for large primes p when $f(x)$ is a fixed cubic polynomial, Hasse succeeded magnificently. He first perceived that the problem Davenport and Mordell had been considering was in fact equivalent to a problem that Artin had formulated in an altogether different form as an analogue of the Riemann hypothesis, and he then proved this analogue in the case Davenport and Mordell considered, which was the 'Riemann hypothesis' in the case of a function field of an elliptic curve over a finite field. To establish connections between such apparently disparate subjects in mathematics and, more than that, to solve a problem using techniques that were developed for altogether different purposes is an achievement of the highest sort in pure mathematics. Great progress has since been made in generalizing Hasse's solution. It is now known that the 'Riemann hypothesis' is true in a vast range of cases, provided that algebraic techniques apply: the actual Riemann hypothesis, which is transcendental rather than algebraic, is still an unsolved problem. (**Edwards, 2008**, p. 386)

According to Schappacher, Hasse's Oslo lecture, "Über die Riemannsche Vermutung in Funktionenkörpern" (**Hasse, 1936/7**),

> contains what is arguably the clearest account of his proof for the elliptic case that Hasse ever wrote, and one wonders why it was not included in his Collected Papers. (**Schappacher, 2006**, p. 12)

At the end of his talk, Hasse also gave a short explanation of Deuring's ideas, based on a general theory of algebraic "correspondences", for the proof of the Riemann Hypothesis for curves of any genus over finite fields. However, the talk contains nothing on connections to the existing theory of geometric correspondences, which would play a major role in the reorganisation of algebraic geometry at the hands of Weil in the years to come.[69] Generally speaking, Hasse showed a certain zeal for the

[68]For a historical account of the development of the theorem (the Hauptsatz), see (**Roquette, 2013**, Chapter 1).

[69]As (**Schappacher, 2006**, pp. 12–13) explains, Weil, who was not at the congress, wrote to Hasse to this effect on 17 July 1936, the day that the congress ended. Moreover, the American algebraic geometer Solomon Lefschetz seems to have used the occasion of the Oslo congress to draw Hasse's attention to geometric correspondences. The 1938 *Zentralblatt* review of Hasse's talk by O. F. G. Schilling (Zbl 0018.34302) refers to the Italian geometer Severi, who was neither in Oslo (see Chapter 5) nor mentioned by Hasse in his talk. The review therefore already reaches

purity of algebraic-number-theoretic methodology, as also revealed in remarks with respect to both Siegel's and Hecke's function-theoretic approaches. With regard to the former, Hasse said at the very beginning of his talk:

> Siegel's finiteness theorem can be called an analytic result, as his proof also shows, appealing to deep tools from analysis. The number theorist would like to have more precise [genauere] and, if possible, purely arithmetically founded arithmetic insights.[70]

With respect to Hecke's efforts to solve a related famous Hilbert problem in class field theory, Hasse said at the very end of his talk that

> a purely algebraic approach avoids the difficulties of Hecke's efforts so far, which result from the occurrence of analytical functions in several variables.[71]

As to the algebraic method used, Hasse's biographer Leopoldt recognised in his proof "one of the most beautiful examples for the power of the [modern algebraic] method"[72] which was not the invention of a single individual, although Emmy Noether in Göttingen can be seen as its "most determined representative".[73] According to Roquette, Davenport, who also attended the Oslo congress, and went on to a prestigious career in number theory,[74] had finally (after initial doubts) commented upon Hasse's original proof with the words: "Well, Helmut, you *are* a mathematician."[75] With Noether having died in 1935 and van der Waerden having been barred by the Nazis from attending the Oslo ICM, Hasse was probably the foremost representative of "modern algebra" at the congress.[76] Given that his compatriots Hecke and Siegel, with their strong anti-Nazi feelings, but also the younger "natural conservative"[77] Davenport, were more sceptical towards "modern algebra" than Hasse himself, one has to realise that any efforts to connect a political argument to the new theories and denounce them as "Jewish" and "un-German"[78] were just propaganda noise or academic policies under false pretence. However, mathematicians like Hasse had to take such prejudiced policies into account and tried, occasionally, to counteract them by emphasising the particularly "German" nature of abstract algebra.[79] Davenport was in Göttingen in February 1934 when

---

into the future: "Essential will be a closer investigation of the ring of meromorphisms as defined by Severi and Deuring."

[70]Translated from the German in (**Hasse, 1936/7**, p. 189).

[71]Translated from the German in (**Hasse, 1936/7**, p. 206).

[72]"eines der schönsten Beispeile für die Kraft dieser Methode" (**Leopoldt, 1982**, p. 18).

[73]"entschiedensten Vertreterin" (*ibid.*).

[74]Davenport gave a plenary address at the 1950 ICM on "Recent progress in the geometry of numbers" (**ICM, 1950**, pp. 166–174).

[75]According to (**Schappacher, 2006**, p. 10), the proof was finished in 1932, and it is certainly contained in (**Hasse, 1935**). Davenport's comment comes from an endnote, citing a talk given by Roquette in 1980, in the German edition (1985) of (**Ellison and Ellison, 1978/1985**, p. 350).

[76]There were of course other followers of that method there, several of them non-Germans, such as the Norwegian-American Ore, who gave a plenary talk (see Section 9.20).

[77]This according to an obituary of Davenport in *The Times* (10 June 1969, p. 12), and reaching much further than just mathematical convictions.

[78]As students of Noether such as Deuring, Grell, van der Waerden occasionally experienced under the Nazi regime in Germany. Prejudices of this kind were for instance spread by the racist theories of "Deutsche Mathematik" in the sense of Bieberbach. See also (**Siegmund-Schultze, 2012**).

[79]Students of the Jewish mathematician Emmy Noether tried to annul her dismissal in 1933 by stressing the "Aryan character" of her theories.

Hasse (then from Marburg) gave a probationary lecture before taking over a chair in Göttingen that June. He wrote on this occasion to his teacher Mordell in Manchester:

> The Rektor and the Studentenführer attended Hasse's sample lecture in Göttingen the other week and were not convinced by his arguments that abstract algebra etc. is the perfect expression of the Third Reich.[80]

Hasse did not just try to sell his abstract algebra to the Nazi rulers. He apparently shared some German nationalist resentments with the Nazis, in a somewhat similar manner to others in the German delegation in Oslo — less so than Blaschke and more so than Erhard Schmidt — and in stark contrast to Siegel and Hecke (**Segal, 1980**). But he remained devoted to his Jewish teacher Hensel. Although the Nazis took away the teaching permit from the 74-year-old Hensel in 1935, Hasse wrote an address on the occasion of Hensel's 75th birthday in the following year (**Hasse, 1936**). Hensel died five years later, before the holocaust began. Hasse, who on other occasions (1938) agreed with the Nazis and spoke about a fictitious "state of war between the Jews and the Germans",[81] was thus spared a most tantalising burden on his conscience.

9.13. G. D. Birkhoff: "On the foundations of quantum mechanics"

George David Birkhoff (1884–1944; see Figure 9.12) was a prominent American mathematician from a family of Dutch origin.[82] He had studied at Harvard and in Chicago and graduated from the latter in 1907 as one of the first promising mathematicians educated in America during the first decades of the twentieth century. In 1909, Birkhoff went to Princeton, and two years later he was promoted to a professorship. He remained at Princeton for only two years before taking up a position at Harvard, where he stayed for the rest of his active career. In 1919, he was promoted to professor, and he was named Perkins Professor in 1932. In 1936, he became the dean of the Faculty of Arts and Sciences at Harvard. He was a pivotal figure in American mathematics and held many offices of the AMS.[83] His son, Garrett Birkhoff (1911–1996), would also become an important American mathematician (initially pure, later also applied), and he was also present as a young man in Oslo for the 1936 ICM (see the discussion of Ore's plenary lecture in Section 9.20).

G. D. Birkhoff's mathematical interests spanned a number of fields from topology, where he achieved new results in the line of Poincaré, to dynamical systems, and he is perhaps best remembered by mathematicians for his individual (pointwise) ergodic theorem (**Birkhoff, 1931**), although, as we have already noted (p. 154), this was not the topic of his lecture in Oslo. Birkhoff's interests also included the application of mathematical thought to other disciplines such as physics, aesthetics,

---

[80]Davenport to Mordell, 14 January 1934. Davenport also reported in his letter that Witt, the student of Emmy Noether, had become a Nazi (Mordell Papers, 4.38).

[81]Quoted in (**Siegmund-Schultze, 1994**, p. 322).

[82]For biographies of Birkhoff, see, for example, (**Veblen, 1946**) and (**Morse, 1946**; **2008**).

[83]On Birkhoff's role in building mathematics at Harvard, see the article by his son: (**Birkhoff, 1989**). On the history of the AMS up to 1938, including Birkhoff's role, see (**Archibald, 1938b**).

FIGURE 9.12. George David Birkhoff (1884–1944) (Source: Rudolf Fritsch, *Der Vierfarbensatz: Geschichte, topologische Grundlagen und Beweisidee*, BI-Wissenschaftsverlag, Mannheim, 1994, p. 29. Public Domain, https://commons.wikimedia.org/w/index.php?curid=4032315)

and even ethics. Of these endeavours, his biographer and contemporary Veblen wrote:

> These studies, though Birkhoff took them quite seriously, seem to me to be definitely less likely than his purely mathematical work to survive. Something similar, I think, must be said about his efforts in physics. Like Goethe and Hilbert, he always remained an outsider. It may have been that the very strength of his faith in mathematical insight prevented him from properly appreciating the insight of the physicists. (**Veblen, 1946**, p. 9)

Birkhoff's paper at the ICM in 1936 probably belonged to the "forgettable" category suggested by Veblen. He did not have a lasting effect on quantum mechanics and his name is not mentioned in either of the two standard volumes on the subject ((**Jammer, 1966**) and (**Jammer, 1974**)), whereas his son Garrett is, having contributed to quantum logic. We will therefore not go into a detailed analysis of the talk.

That Birkhoff's plenary talk is not to be found on select lists of his publications (**Morse, 1946**), counts less, given the broader purpose and appeal of plenary talks. We have observed a similar fact also for Hasse's plenary talk, although the latter was much closer to Hasse's major research. What counts more is that Birkhoff in his role as a spokesman for American mathematics was keen on showing the importance of mathematics for the most exciting and publicly most acknowledged contemporary research in mathematical physics, which was still dominated by Europeans or European emigrants to the United States. Birkhoff assumed this role also in his intimate relationship with the American Association for the Advancement of Science which incidentally awarded its annual prize to him for a previous paper in the same line with the title "A mathematical critique of some physical theories" (**Birkhoff, 1927**). In the early 1940s, Birkhoff came into conflict with the German immigrant Hermann Weyl (then at the IAS in Princeton) over Birkhoff's alternative to Einstein's general theory of relativity.[84] Even Birkhoff's mathematically important ergodic theorem was occasionally disputed as to its importance for physics, compared with John von Neumann's statistical ergodic theorem of the same year.

Interestingly enough, Veblen, who was another leader of American mathematicians, was also occasionally criticised for an alleged dubious relevance of his work for mathematical physics (see the quote by Wilson on p. 175). In the quotation above, Veblen had also indirectly referred to Birkhoff's plenary talk at the Bologna ICM, entitled "Quelques éléments mathématiques de l'art", which had also been critically received (although not openly so). In 1928, Birkhoff had already been officially celebrated as a leader of American mathematicians. Therefore we cannot avoid pointing to the formal and symbolic role of plenary talks at the time which often put emphasis on the persons who gave the talks instead of focusing on the content. This gave noted lecturers such as Birkhoff, Bjerknes, Oseen, Fréchet, and even Fueter[85] the opportunity to report on topics that did not necessarily constitute their most important work.

9.14. L. J. Mordell: "Minkowski's theorems and hypotheses on linear forms"

Louis Joel Mordell (1888–1972; see Figure 9.13) was an American-born number theorist, who spent most of his life in the UK.[86] He studied mathematics at St John's College, Cambridge — the attraction of studying in the UK rather than at an American university being that he could specialise in mathematics immediately — before taking a lectureship at Birkbeck College, London. The middle part of his career was spent at the University of Manchester, before he returned to Cambridge in 1945 to take up the Sadleirian Chair of Pure Mathematics, in succession to G. H. Hardy.

Mordell was heavily influenced in his early number-theoretic work by the Norwegian Axel Thue. We mention in particular the result of Thue that states that for any given binary cubic form f, the equation $f(u, v) = 1$ has only finitely many

[84]See (**Rowe, 2003**, p. 12) or (**Siegmund-Schultze, 2016b**, p. 93).

[85]We recall (p. 164) that Fueter said in his autobiography that he earned the honour of giving a plenary talk in Oslo by his previous role at the ICM in Zurich in 1932.

[86]For biographies of Mordell, see (**Davenport, 1964**), (**Cassels, 1973**), (**Cassels, 1974**), (**Cassels, 2008**), and (**Fairbairn, 2017**). For some autobiographical remarks, see (**Mordell, 1971**).

FIGURE 9.13. Louis Joel Mordell (1888–1972) (Author: Konrad Jacobs. Source: Archives of the Mathematisches Forschungsinstitut Oberwolfach)

integer solutions u, v (**Cassels, 1974**, p. 73). It was through attempts to extend this result that Mordell was led to what has been hailed as "his most important individual theorem" (**Cassels, 1974**, p. 76): the "finite basis theorem". The latter concerns rational points (x, y) on a curve $y^2 = 4x^3 - g_2x - g_3$, for integers g_2, g_3. These rational points form an Abelian group in a natural way; the finite basis theorem states that this group is finitely generated.[87] The theorem confirmed a conjecture[88] of Poincaré and was later generalised by (**Weil, 1929**).

In 1922 Mordell also conjectured that there are only finitely many rational points on any curve of genus greater than unity, a claim that gained notoriety as

[87]The result first appeared in (**Mordell, 1922**); see also (**Cassels, 1986**).

[88]Or perhaps, more accurately, an assumption: (**Cassels, 1986**, p. 31).

"Mordell's conjecture", and was finally proved by the 1986 Fields Medallist Gerd Faltings (**Faltings, 1983**); as "Faltings' Theorem", it went on to form an important step in Andrew Wiles' proof of Fermat's Last Theorem. According to his biographer (**Cassels, 1974**, p. 85), Mordell was a "problem-solver" not a "system builder":

> It is now fashionable to classify mathematicians into "problem solvers" and "system builders". Mordell was an archetypal problem solver. Even when the system lay at his feet asking to be built (as in his early work on modular forms or in his proof of "Mordell's finite basis theorem") he passed on to other problems and left the building for other hands. His work has left an indelible impress on the subject he loved so well.

Indeed, in his work on modular forms between 1913 and 1920, Mordell had anticipated some results of Hecke's, on which the latter reported in Oslo (see Section 9.8).

Mordell's talk in Oslo, entitled "Minkowski's theorems and hypotheses on linear forms" (**Mordell, 1936/7**), was linked to his period of intense interest in the geometry of numbers. According to (**Cassels, 1974**, p. 81),

> in the late 1930s and early 1940s the work of Mordell, Mahler and Davenport in this subject saw the greatest development since its initiation by Minkowski.

Indeed, as we see from his title, the link to Minkowski was emphasised by Mordell: he began his talk by noting Minkowski's lecture on the geometry of numbers at the Heidelberg ICM of 1904 (**ICM, 1904**, pp. 164–173).[89] Mordell also stressed in particular the methods introduced by Siegel, and referred to "Koksma's most valuable 'Diophantische Approximationen'" (**Mordell, 1936/7**, p. 238), meaning the recent Springer *Ergebnisse* report by the Dutch number theorist J. F. Koksma (**Koksma, 1936**), which was also cited by van der Corput in his plenary lecture (Section 9.16). In fact, Koksma, who was not at the Oslo congress, reviewed Mordell's address for *Zentralblatt* and called it a

> very clear and instructive report on the problems which originate in Minkowski's theorems and conjectures on homogeneous and inhomogeneous linear forms to which the author and his students have contributed much.[90]

In 1929, Mordell had taken British citizenship. After the Nazis came to power in Germany, Mordell, who was himself of Jewish origin, helped immigrants to the UK, such as Kurt Mahler[91] and Richard Rado, who also attended the Oslo congress. Mordell expressly thanked Rado, a representative of the Schur school in Berlin, for reading the manuscript of his talk (**Mordell, 1936/7**, p. 238). After the congress, Rado, himself Jewish, wrote an ironical letter to Mordell, which seems to allude indirectly to the political connotations of meeting German mathematicians (although ostensibly pointing to the Norwegians), or at least expresses the ubiquity of the Jewish question in the 1930s:

---

[89] Though Mordell gave the year erroneously as 1902 (**Mordell, 1936/7**, p. 226).

[90] Translated from the German in Zbl 0018.29502.

[91] After the congress, Mahler wrote a letter to Størmer (3 November 1936) thanking him for the hospitality and explaining that he had had to leave early due to lack of money (Størmer Papers, box 10).

> I hope you have safely come back from the far North, a bit 'aufgenordet' by your stay among those Hyperaryans. I enclose three snaps which I took in Norheimsund [near Bergen].[92]

In the political context, we note that it was only in 1959 that Mordell was allowed to reenter the land of his birth: his help of refugees in Britain had earned him the suspicion by American authorities that he was a communist (**Cassels, 1974**, p. 83).

Mordell was not overly modernist in his methods and shared some of the concerns of Siegel and others about an exaggeration of axiomatics in mathematics — hence Siegel's appeal to him for support in opposing these trends (see Section 9.5).

According to (**Cassels, 1974**, p. 83), Mordell was "greatly in demand as a visiting speaker",[93] and it is perhaps this that earned him the last-minute invitation to deliver a plenary lecture in Oslo. In Mordell, we see a great contrast to some of the remarks made about Siegel in Section 9.5:

> With his wide circle of mathematical friends and acquaintances all over the globe, Mordell enjoyed conferences large and small. He was particularly in his element in the big four-yearly International Congresses. At each of these he organized a dinner. At first they were comparatively modest, but that at Nice was particularly memorable when he was host to about a hundred mathematicians and wives.[94]

9.15. L. V. Ahlfors: "Geometry of Riemann surfaces"

The plenary lecture given by Lars Ahlfors (1907–1996; see Figure 9.14) at the 1936 ICM must of course have a special place within our account of the mathematics of the congress, in light of his status as one of the first winners of the Fields Medal, perhaps most particularly because of the fact that his co-recipient, Jesse Douglas, did not speak there (but see Section 9.22). Indeed, Ahlfors' stature as a Fields Medallist has made our task rather easier than it has been for some of the other plenary speakers, since there has been much analysis and assessment of his work, both around the time of the congress and subsequently.

As we learn from Carathéodory's Oslo address about the Fields Medallists ((**Carathéodory, 1936/7**), which we will use again in Section 9.22),[95] Ahlfors was born in Helsinki (then Helsingfors) in 1907, the son of a professor of mechanical engineering. In 1924, he entered Helsinki University, where he studied mathematics under Ernst Lindelöf and Rolf Nevanlinna; he earned his degree in 1928. A PhD soon followed (defended in 1930, but not formally awarded until 1932), with a

[92]Rado to Mordell, 6 September 1936, Mordell Papers, folder 21, Richard Rado, 21.8.

[93](**Cassels, 1974**, p. 83) noted further that Mordell "kept a list, arranged geographically, of the institutions at which he had spoken. Ultimately it had about 190 entries. There was some sort of competition with Paul Erdős about this, and Mordell claimed to be still in the lead, though resigned to losing in the end."

[94](**Cassels, 1974**, p. 84). Taking the membership lists in the proceedings as our guide, we see that Mordell attended the 1912, 1928, 1932, 1936, 1950, 1954, 1962, and 1970 ICMs. We have no additional documentation for Cassels' (implicit) claim that Mordell organised a dinner in Oslo.

[95]For other biographies of Ahlfors, see (**Bott *et al.*, 1998**), (**Lehto, 1998b**), (**Gehring, 2005**), and the volume (**Lehto, 2013**). For autobiographies, see (**Ahlfors, 1989**) and (**Ahlfors, 1997a**).

FIGURE 9.14. Lars Ahlfors (1907–1996) (Author: Konrad Jacobs. Source: Archives of the Mathematisches Forschungsinstitut Oberwolfach)

dissertation containing a proof of Denjoy's Conjecture:[96] that if an entire function has p distinct asymptotic values, then its order is greater than or equal to $\frac{p}{2}$.

Early in his career, Ahlfors travelled around Europe, first accompanying Nevanlinna on a stay in Zurich and Paris in 1928–1929, and then making further trips to Paris, Göttingen, and Munich during the following years, thanks to Rockefeller funding. He taught for a time in Helsinki, before accepting an offer, made on the recommendation of Carathéodory, to go to Harvard in 1935; it was while attending the Oslo ICM from Harvard that Ahlfors received what he later described as the "surprise of [his] life" by being awarded the Fields Medal.[97] Ahlfors remained at Harvard until 1938 when Lindelöf retired and a sense of both duty and of homesickness led the former to accept the chair in mathematics at Helsinki University. At the outbreak of war, his wife and children were sent to live with relatives in the safety of Sweden, but Ahlfors himself remained in Helsinki; as he described it:

> The university was closed for lack of male students, but otherwise life went on, partly in air-raid shelters. Because of an earlier physical condition, I had never been called to military duty, and my only part in the war effort was as an insignificant link in a communications setup. (**Ahlfors, 1997a**, p. 5)

In 1944, Ahlfors was offered a teaching position in Zurich (**Ahlfors, 1997a**, p. 6). Wartime travel from Finland to Switzerland was accomplished only by a tortuous

[96]For an account of Ahlfors' proof, see (**Sodin, 2000**).

[97](**Ahlfors, 1997a**, p. 5). In an address fifty years later at the 1986 ICM, Ahlfors admitted that he had in fact been forewarned of the award the day before by someone who had accidentally congratulated him (**ICM, 1986**, vol. I, p. XXV).

and dangerous route via Sweden, Britain, and France,[98] but in Switzerland Ahlfors was reunited with his family. Although Zurich provided a temporary home, Ahlfors was never entirely comfortable there, so was pleased to return to Harvard in 1946,[99] where he remained until his retirement in 1977.

The topic of Ahlfors' doctoral dissertation was an early indication of the direction in which his subsequent work was to go, the majority of it being in complex analysis, a subject to which he was first introduced via lectures given by Nevanlinna in Zurich. Accounts of Ahlfors' work have divided it under three headings: conformal geometry, Kleinian groups, and quasiconformal mappings.[100] We focus our attention here in this short survey on the first of these, since this is the research strand to which Ahlfors' Fields Medal is linked.[101]

As Robert Osserman notes with regard to conformal geometry in his portion of an article on Ahlfors' mathematics (**Gehring *et al.*, 1998**, p. 233):

> There are two directions in which one can pursue the relations between Riemann surfaces and Riemannian manifolds. First, if a two-dimensional Riemannian manifold is given, then not only lengths but also angles are well defined, so that it inherits a conformal structure. Furthermore, there always exist local isothermal coordinates, which are local conformal maps from the plane into the surface. ... The set of all such local maps forms a complex structure for the manifold, which can then be thought of as a Riemann surface. One then has all of complex function theory to bring to bear in studying the geometry of the surface.

But Ahlfors' work was more closely linked to the other direction:

> In the other direction, given a Riemann surface, one can consider those metrics on the surface that induce the given conformal structure. By the Koebe uniformization theorem, such metrics always exist. In fact, for "classical Riemann surfaces" of the sort originally considered by Riemann, which are branched covering surfaces of the plane, there is the natural euclidean metric obtained by pulling back the standard metric on the plane under the projection map. One can also consider the Riemann surface to lie over the Riemann sphere model of the extended complex plane and to lift the spherical metric to the surface. Both of those metrics prove very useful for obtaining information about the complex structure of the surface.
>
> For simply connected Riemann surfaces the Koebe uniformization theorem tells us that they are all conformally equivalent to the sphere, the plane, or the unit disk. Since the first case is distinguished from the other two by the topological property of

[98]In an interview in 1998, Ahlfors revealed that he had smuggled his Fields Medal out of Finland and then been forced to pawn it temporarily in Sweden (**Albers, 1998**, p. 89).

[99]Ahlfors was replaced in Zurich by Nevanlinna.

[100]See, for example, (**Gehring *et al.*, 1998**).

[101]On the other strands, see (**Gehring *et al.*, 1998**), (**Lehto, 1998b**), or (**Gehring, 2005**). For other aspects of Ahlfors' work, see the articles in (**Brooks and Sodin, 2000**), and also those in volume 13, fascicule 3 (2000) of *Annales Academiæ Scientiarum Fennicæ* (Series A. I. Mathematica).

> compactness, the interesting question concerning complex structure is deciding in the noncompact case whether a given surface is conformally the plane or the disk, which became known as the parabolic and hyperbolic cases, respectively. In 1932 Andreas Speiser formulated the "problem of type", which was to find criteria that could be applied to various classes of Riemann surfaces to decide whether a given one was parabolic or hyperbolic. That problem and variants of it became a central focus of Ahlfors's work for several decades. (**Gehring *et al.*, 1998**, p. 233)

A key tool in Ahlfors' work was Nevanlinna Theory, as developed by his former teacher in the mid-1920s (**Nevanlinna, 1925**): this is concerned with the asymptotic distribution of solutions of the equation $f(z) = a$, for a meromorphic function f, as a varies. The appearance of Nevanlinna's paper on this subject was later described by Hermann Weyl and his son, with perhaps a touch of hyperbole, as "one of the few great mathematical events in our century" (**Weyl and Weyl, 1943**, p. 8). The Weyls immediately noted Ahlfors' contributions to the theory, describing them as being of a "half topological character", before remarking that

> [a] good part of the theory of meromorphic functions is better understood when looked upon as a special case of the theory of meromorphic curves. (**Weyl and Weyl, 1943**, p. 8)

Ahlfors' geometrical version of Nevanlinna Theory, phrased in terms of specially constructed conformal metrics on a Riemann surface (**Ahlfors, 1935a**), was specifically cited by Carathéodory in his Oslo address. Indeed, Ahlfors produced two further geometrical versions of Nevanlinna Theory. The second of these appeared in 1937, and involved the application of differential geometry to the study of covering surfaces (**Ahlfors, 1937**), but it is the first, which uses covering surfaces as its main tool, that is of greater interest to us here, since, according to (**Carathéodory, 1936/7**, p. 309), this was a paper (**Ahlfors, 1935b**) that was singled out during the deliberations of the Fields Medal committee; in his address, Carathéodory noted that this work could be

> regarded as the first chapter of a new branch of analysis, for which perhaps the name "metrical topology" is the appropriate one.[102]

As Osserman explains, this approach via covering surfaces is based upon the idea that the fundamental quantities associated by the original Nevanlinna Theory with a meromorphic function can be regarded instead as properties of the Riemann surface of the image of the function, itself viewed as a covering surface of the Riemann sphere (**Gehring *et al.*, 1998**, p. 234). Ahlfors himself is recorded as having described this approach as being a "radical departure from Nevanlinna's own methods" (**Gehring, 2005**, p. 10); he also noted its reach in a commentary that he gave of the paper some years later:

> My paper met with immediate recognition and earned me the Fields medal. It has been incorporated in the leading textbooks: R. Nevanlinna's *Eindeutige Analytische Funktionen*, W. Hayman's *Meromorphic Functions*, K. Noshiro's *The Modern Theory of Functions* (in Japanese). (**Ahlfors, 1997b**, p. 9)

[102] Translated from the German in (**Carathéodory, 1936/7**, p. 309).

Indeed, the importance of the paper was recognised early on: the *Jahrbuch* reviewer, W. Saxer, described it as "fundamental to the whole theory of meromorphic functions",[103] while the *Zentralblatt* reviewer noted the relative simplicity of the proofs in this version of Nevanlinna Theory (Zbl 0012.17204). The search for further extensions of Nevanlinna Theory within the broader value-distribution theory was one of the research strands that was to occupy Ahlfors for many years to come.[104]

Having gained an impression of the work that led to the award of his Fields Medal, let us turn now to the lecture that Ahlfors delivered in Oslo. Its title, "Geometrie der Riemannschen Flächen" (**Ahlfors, 1936/7**), gives a clear sign that it too fits into the broad theme of conformal geometry discussed above. Indeed, both the *Jahrbuch* and *Zentralblatt* reviews simply direct the reader to the works by Ahlfors that we have cited.

Ahlfors began his lecture by comparing the analytical and geometrical approaches to the treatment of analytic functions, with a clear indication that he would develop the latter:

> On the one hand, geometrical function theory serves to make the handling of analytical problems graphical and thereby simpler by introducing a geometrical element. On the other hand, it also happens that the independent consideration of the geometrical side leads to an extended problem, which then allows a uniform treatment of the whole theory. In this lecture, I want to shed light on this behaviour by using the example of the value distribution theory of analytic functions. The central problem here was formulated and treated purely analytically. Later, geometrical concepts and methods were introduced that allowed a substantial formal simplification. Going even further, by formulating the whole problem geometrically, one arrives at a *covering theory* [*Überdeckungstheorie*] which, I believe, also offers a new clarity and simplification in comparison to the analytical theory.[105]

The text then proceeds to outline the ideas indicated above, with reference to the papers (**Ahlfors, 1935b**) and (**Ahlfors, 1937**).

Ahlfors went on to deliver a further two plenary lectures at future ICMs: in Stockholm in 1962, and in Helsinki, the place of his birth, in 1978; the topics remained closely related to the research themes described above.[106] Moreover, on the fiftieth anniversary of his receipt of the Fields Medal, Ahlfors was named honorary president of the 1986 Berkeley ICM, where he handed out that year's Fields Medals, as well as the IMU's Nevanlinna Prize.[107]

---

[103] Translated from the German in JFM 61.0365.03.

[104] See, for example, (**Gehring *et al.*, 1998**, pp. 234–236), (**Osserman, 1999**), or (**Eremenko, 2000**).

[105] Translated from the German in (**Ahlfors, 1936/7**, p. 239).

[106] The lectures were "Teichmüller spaces" (**Ahlfors, 1963**) and "Quasiconformal mappings, Teichmüller spaces, and Kleinian groups" (**Ahlfors, 1980**).

[107] (**ICM, 1986**, vol. I, pp. XXIII, XXV). The Fields Medals were awarded that year to Simon Donaldson (b. 1957), Gerd Faltings (b. 1954), and Michael Freedman (b. 1951), while the Nevanlinna Prize went to the computational theorist Leslie Valiant (b. 1949).

9.16. J. G. van der Corput: "Diophantine approximation"

In contrast to the situation with the subject of the preceding subsection, biographical materials on the Dutch number theorist Johannes Gualtherus van der Corput (1890–1975; see Figure 9.15) are rather fewer,[108] although an effort does appear to have been made in recent years to revive his memory (among non-number theorists and outside the Netherlands): in 2015, a special issue of the journal *Indagationes Mathematicae* featured a biography of van der Corput, together with a collection of articles on aspects of his work.[109] It is the biography published there (**Korevaar, 2015**) that we take as our main source.

Van der Corput studied mathematics in Leiden from 1908, going on, with a break for military service and a stint as a secondary school teacher, to complete a PhD under the supervision of J. C. Kluyver in 1919. The topic of his thesis (**van der Corput, 1919**), whose title translates into English as *On lattice points in the plane*, derived from a series of articles by Edmund Landau, although the problem concerned was an older one that had previously been considered by Gauss: to determine the number of points with integer coordinates in a given planar domain. Shortly after completing his PhD, van der Corput spent a summer working with Landau in Göttingen. As well as a joint paper, one of the results of this stay in Germany was van der Corput's development of a new device for finding number-theoretic estimates: the method of exponential sums, namely sums of the form $\sum_{n\in A} e^{2\pi i f(n)}$, where A is a finite set of integers and f is a smooth real-valued function — "van der Corput's method" approximates such a sum by relating it to simpler sums that are easier to estimate. He spent the next ten years, successively in Utrecht, Fribourg, and Groningen, refining the method of exponential sums and expanding its applications, including one, developed in collaboration with his student Jurjen Koksma, to the Riemann zeta function (**van der Corput and Koksma, 1930**).

Van der Corput's Oslo talk was entitled simply "Diophantische Approximationen", and cited both Mordell's earlier lecture within the congress (Section 9.14), as well as the recent report on Diophantine approximations by his student Koksma:

> An excellent overview of the field of Diophantine approximations, from the oldest authors to the middle of last year, is given in the report of J. F. Koksma, published half a year ago, which contains a complete bibliography.[110]

In his own talk, to which he appended an extensive literature, spanning nearly three pages, van der Corput focused on more recent results, particularly by the Russian I. M. Vinogradov,[111] in what was then a fast-moving field. However, he also emphasised the importance of earlier results, including those of Landau in Göttingen.

---

[108]We have found four biographical sources: (**de Bruijn, 1977**), (**Korevaar, 1975**), (**Korevaar, 2015**), and (**Duparc and Korevaar, 1982**). As of this writing (July 2018), van der Corput does not have an entry in the MacTutor History of Mathematics Archive (`http://www-history.mcs.st-and.ac.uk/`), but a search for his name reveals passing references to him in the articles on his students Jurjen Koksma (1904–1964), Johannes de Groot (1914–1972), and Johannes Boersma (1937–2004), as well as occasional references to aspects of his mathematics.

[109]Volume 26, issue 5 (December 2015): "In memoriam J. G. van der Corput".

[110]Translated from the German in (**van der Corput, 1936/7**, p. 249). As we have seen, the report was referred to elsewhere by Mordell as "most valuable" (**Mordell, 1936/7**, p. 238).

[111]Recall from p. 98 that Vinogradov had been on the list of possible Soviet delegates for the Oslo congress, but, like all the others, he had been prevented from attending.

FIGURE 9.15. Johannes van der Corput (1890–1975) (Source: University of Groningen)

Another name that featured prominently in the lecture was Weyl for a paper of 1916, where he linked exponential sums to the problems of Diophantine approximation. Indeed, as with so many of the plenary lectures in Oslo, van der Corput's was not an account of new results, but a survey of existing material. Koksma, as *Zentralblatt* reviewer, described the published version of van der Corput's lecture as "a clear overview" ("eine klare Übersicht") of the topic at hand (Zbl 0018.29501).

From the late 1930s onwards, van der Corput's life underwent dramatic changes, including involvement in the resistance against German occupation during the Second World War. After the war, he became the leading adviser to the Dutch government on policies for mathematics. Following a position as a visiting scholar at Stanford during 1950–1952, van der Corput became somewhat disappointed with the state of Dutch institutions, and so accepted a permanent position at the University of California, Berkeley. In 1966, he returned to Europe, where he lived alternately in Antwerp and Amsterdam until his death in 1975.

Even if he is nowadays not one of the best-known mathematicians to have delivered a plenary lecture in Oslo in 1936, van der Corput's mathematical achievements were certainly recognised at the time: hence the invitation to deliver a lecture at a congress that was already well supplied with number-theoretic topics. His lecture, however, may be taken as an indication of the poor health, by the 1930s, of those parts of analytic number theory with which van der Corput had begun his career:

> In the early thirties, at a time when the analytic theory of numbers had come to a certain stand-still, his powerful analytic skills sought a new outlet in the field of approximations. (**Duparc and Korevaar, 1982**, p. 14)

However, it is in connection with a wider interest in asymptotics that van der Corput receives his one and only mention in the main text of (**Pier, 1994a**),[112] in an article on the Prime Number Theorem (**Schwarz, 1994**, p. 589).

9.17. S. Banach: "The theory of operations and their significance for analysis"

Passing reference has already been made to the plenary lecture delivered in Oslo by Stefan Banach (1892–1945; see Figure 9.16) on the theory of operators in the sense of the fledgling theory of functional analysis: recall from p. 83 that in his report on the congress Lietzmann had sounded a note of approval for the fact that Banach had given his talk in German.[113] As we will see, Banach's book (**Banach, 1932**) was to be very influential for the development of functional analysis. Indeed, the development of functional analysis and Banach's role as one of its principal creators are so well known in modern mathematics that we can refer to many historical and biographical articles and do not need to repeat much of what is said there.[114] We will therefore focus on the representation of functional analysis at the various early ICMs up to 1936.

Note that the term "functional analysis" was originally coined by Paul Lévy (later a leading probabilist) as "analyse fonctionnelle" (meaning "analysis of functionals")[115] in his book of 1922,[116] but had a different meaning from modern functional analysis in the sense of Banach. It came slowly into use for Banach's and other similar more abstract theories during the 1930s and 1940s. The name "functional analysis" was not used in Banach's talk at the ICM in Oslo, although he did repeatedly use the noun "Funktional".[117]

---

[112] His only other appearance in this edition is in the list of highlights of twentieth century mathematics that we mentioned on p. 146, where his name is recorded under 1922 for "Estimates of sums of exponentials" (**Dugac *et al.*, 1994**, p. 11).

[113] But see footnote 103 on p. 83.

[114] On the history of functional analysis, see (**Dieudonné, 1981**), (**Birkhoff and Kreyszig, 1984**), (**Fichera, 1994**), and (**Siegmund-Schultze, 2003**). For biographies of Banach, see (**Ulam, 1946**), (**Steinhaus, 1963**), (**Katětov, 2008**), and material throughout (**Duda, 2014**). Further biographical material, as well as accounts of Banach's work, can be found in the various papers in volume 1, issue 2 (1947–1948) of *Colloquium Mathematicum*, and in the conference papers published in a 1963 special series of *Studia Mathematica*.

[115] Grammatically, "fonctionnelle" is an attribute here, but the same word is also a noun and that is what Lévy meant.

[116] See (**Birkhoff and Kreyszig, 1984**, p. 295).

[117] By the same token, the word "operator" was not used then, although "operations" was, as in the title of Banach's talk.

FIGURE 9.16. Stefan Banach (1892–1945) (By Hussein Abdulhakim. CC BY-SA 3.0 (http://creativecommons.org/licenses/by-sa/3.0), via Wikimedia Commons)

If we look for the representation of functional analysis in the Banach style[118] at the ICMs of the 1920s, there is just a little to be found at the later end of the decade. In Bologna in 1928, three plenary talks on generalised analysis by M. Fréchet, J. Hadamard, and V. Volterra still referred to very different theories and almost ignored Polish work.[119] Of the three talks the one by Fréchet, the inventor of the metric space (1904), was closest to modern functional analysis. The tradition coming from integral equations in the Hilbert school around 1906, which also

[118]Or, indeed, that of Hahn or Wiener, who were pursuing similar ideas: a minor issue of priority arose in this connection thanks to the appearance of similar concepts in a 1921 paper of Norbert Wiener, another Oslo plenary speaker, although the latter freely acknowledged Banach's priority. A further question involving Hans Hahn was effectively settled by the latter's early death in 1934: see (**Duda, 2014**, pp. 89–90).

[119]Hadamard made a passing reference to the "deeper" work of Banach and Wiener on function spaces (**Hadamard, 1928/9**, p. 151).

influenced the Polish school, was only represented by one sectional talk (F. Riesz). Recent Polish work, in particular by Banach, was present in the broadly defined section I-C (differential, difference, integral and integro-differential equations, calculus of variations, functional analysis [Analisi funzionale]), in talks by N. Bari, H. Steinhaus, and S. Kaczmarz.[120] Banach himself appears to have delivered a talk entitled "Sur les systèmes d'équations linéaires fonctionnelles" (**ICM, 1928**, vol. I, p. 123), although his presentation did not appear in the proceedings.

The situation four years later in 1932 in Zurich was little different: neither Banach nor any of his close associates (except for Schauder) attended the congress, and his approach to functional analysis was not represented at all. This, however, was the year in which the appearance of three monographs by Banach, von Neumann, and Stone[121] finally marked the establishment of functional analysis as an important mathematical subdiscipline. Banach's book, according to (**Birkhoff and Kreyszig, 1984**, p. 314), demonstrated the "scope and power" of functional analysis, and helped to establish it as a major branch of mathematics. In comparing these three books, (**Birkhoff and Kreyszig, 1984**, p. 314) note that

> [w]hereas von Neumann's book was primarily concerned with quantum mechanics and Stone's book presented the spectral theory of symmetric linear operators on Hilbert space, with applications to classical analysis ..., Banach's emphasized intriguing theoretical questions involving linear operators and functionals on a wide range of "*espaces (B)*" [i.e., Banach spaces].

With regard to von Neumann and Stone, we recall from Table 2.5 (p. 41) that in fact both had been considered as possible plenary speakers for the Oslo congress, but that in the end, neither had been chosen: in the end, von Neumann did not attend at all (see p. 154), while Stone attended and delivered a sectional lecture that was linked closely to Banach's work: "Some remarks on non-separable Banach spaces" (**ICM, 1936**, vol. II, p. 120). If all three of the books by Banach, von Neumann, and Stone showed the power of functional analysis, we can only speculate as to why it was Banach who was ultimately invited to deliver a plenary lecture in Oslo. We might perhaps argue that, by the early 1930s, Banach's abstract approach to functional analysis was more in tune with the greater abstraction and structural thinking that was appearing in mathematics at this time, most notably in the approach to algebra that had been advocated by Emmy Noether, but that was otherwise barely represented at the Oslo congress.

Whatever the reasons for his being chosen, Banach delivered his plenary lecture, "Die Theorie der Operationen und ihre Bedeutung für die Analysis" (**Banach, 1936/7**), on the morning of Saturday 18 July. He began by justifying his topic with a historical perspective:

> The importance of the theory of operations for all mathematics is most evident when we reconstruct its historical evolution.[122]

In the early parts of his lecture, Banach set out to do just that, citing the work of Volterra, Hadamard, Hilbert, and Fréchet in the earlier phases of the development of functional analysis. In what was perhaps an indication of the probable unfamiliarity

[120]In his 1936 talk, Banach recommended the new book by Kaczmarz and Steinhaus: *Theorie der Orthogonalreihen* (1935) (**Banach, 1936/7**, p. 264).

[121]Namely, (**Banach, 1932**), (**von Neumann, 1932b**), and (**Stone, 1932**).

[122]Translated from the German in (**Banach, 1936/7**, p. 261).

of this topic for his audience, he then proceeded to outline the basic ideas of the theory of linear operators and of his "type (B) spaces". He concluded by looking to the future:

> The circle of mathematicians who deal with the theory of operations from different points of view is becoming ever larger; unfortunately in such a short time it was not possible for me to discuss all important works, including those relating to spaces of type (B). But I believe that the examples that I have chosen to show you demonstrate the scope and significance of the methods of operations. They connect apparently entirely separate areas in unfamiliar ways, they increase the power of our geometric intuition, they allow complicated relationships in analysis to be traced back to simpler ones, thereby opening up new properties.[123]

It is important to stress that Banach did not only talk about his spaces and linear operations but also about the beginnings of non-linear functional analysis, for instance in the work of his student Juliusz Schauder and the latter's collaborator the Frenchman Jean Leray.[124]

One suspects that Banach's summary may have been a welcome introduction to the theory for some readers; the *Zentralblatt* reviewer N. Dunford, not having been in Oslo, described it as

> [a]n excellent but brief resume of the fundamental concepts, results and applications associated with the theory of (I) linear, polynomial, and analytic operations on spaces of type (B) and (II) convex bodies and the fixed point theorem. (Zbl 0018.36402)

Another American, E. R. Lorch, a participant in Oslo, however, said in a German review for *Jahrbuch*:

> The talk reports on the historical development and the main results in the theory of operations, without giving new results. Therefore the details need not be discussed here.[125]

After his return from Oslo, from 1938 to 1941, Banach served as dean of the mathematics faculty in Lwów, but with the Nazi invasion all universities were closed. Along with other now-unemployed university professors, he found "work" and comparative safety though as a feeder of lice at Rudolf Weigl's Typhus Research Institute. Schauder, however, and many other Jewish and non-Jewish Polish mathematicians, were killed by the Nazis. Following the capture of the city by the Red Army in the summer of 1944, Banach began to help to reestablish the university there, but the war years had taken their toll on his health and he died a year later, aged just 53.

---

[123]Translated from the German in (**Banach, 1936/7**, p. 268).

[124]Schauder worked under miserable conditions as a school teacher in Lwów at the time. He described his situation in a letter to Leray of 8 July 1936, still hoping that he would meet Leray in Oslo (**Leray, 1980**). Schauder traveled to the ICM and gave two talks, while Leray apparently did not appear. On Schauder, see (**Forster, 2008**).

[125]Translated from the German in JFM 63.0349.01.

9.18. M. Fréchet: "Mathematical mélange"

Maurice Fréchet (see Figure 9.17), founder of the theory of metric spaces,[126] was an experienced participant in the ICMs, having attended the 1912, 1920, 1924, and 1928 meetings. Moreover, Fréchet had been deeply involved in the preparation of the disputed congress in Strasbourg in 1920. Norbert Wiener, another plenary speaker in Oslo (see Section 9.19), had been invited to Strasbourg by Fréchet and later deplored his participation in this politicised congress for principled reasons.[127] In his autobiography, Wiener wrote about his host Fréchet in 1920:

> I hoped that I would be able to use the period before the opening of the Congress in September in working with some European scholar in whose field I was interested. The scholar I chose was Maurice Fréchet. It was Fréchet more than anyone else who had seen what was implied in the new mathematics of curves rather than of points ... At that time we all had great expectations that his work was to mark the next great step towards the mathematics of the future.
>
> Let me say that at the present time [1956] Fréchet's work has proved to be very important but less central than we had expected. It is written in a spirit of abstract formalism which is fundamentally hostile to any deep physical application. However, hindsight is easier than foresight, and it would have been somewhat difficult when we were in Strasbourg to predict that Fréchet was not to be the absolute leader of the mathematicians of his generation. (**Wiener, 1956**, p. 50)

Fréchet's international reputation earned him an invitation to give a plenary lecture in Bologna in 1928, where he presented his very abstract functional analysis which, inspired by the American E. H. Moore, he called "analyse générale" (**ICM, 1928**, vol. I, pp. 267–274). As Wiener indicated in the above quotation, the main stream of functional analysis at the time went rather through the work of Fréchet's Polish colleague Banach, as we have just discussed in the preceding subsection. Although rather abstract in its approach too, it allowed for various deep applications to differential equations and numerical analysis, including non-linear equations, as in the work of Banach's countryman Juliusz Schauder, who was also at the congress.

Since his previous plenary talk in Bologna eight years before, Fréchet had strongly developed his research in probability theory. While his general axiomatic approach, as found both in his theory of metric spaces and also his 1915 theory of the abstract integral, apparently influenced Kolmogorov's axiomatics of probability,[128] his speciality, the theory of Markov chains, flourished more at the hands of his students, such as Wolfgang Doeblin, than by his own contributions.

---

[126] The notion of a metric space was introduced in his PhD thesis (**Fréchet, 1906**). For biographies of Fréchet, see (**Siegmund-Schultze, 2005**) and (**Taylor, 2008**); for surveys of his work, see (**Taylor, 1982–1987**). On the development of the theory of metric spaces, see (**Pier, 1994b**).

[127] See the remarks recorded here in footnote 21 on p. 10.

[128] See (**Shafer and Vovk, 2006**). Kolmogorov quoted Fréchet's integral in connection with his abstract theory of mathematical expectation in probability (**Kolmogoroff, 1933**, p. 33). In his talk in Oslo, Fréchet claimed priority for his notion of the abstract integral (**Fréchet, 1936/7**, p. 270).

FIGURE 9.17. Maurice Fréchet (1878–1973) (Public Domain, https://commons.wikimedia.org/w/index.php?curid=1980574)

In his plenary talk in Oslo, Fréchet reported on his work in three rather different domains: functional analysis, complex function theory, and probability. It is probably the mixed and inhomogeneous nature of his talk that is responsible for the fact that it did not get a review in *Zentralblatt*. It seems unfortunate that Fréchet chose this format because at least one third of his talk, that on probability theory, as his new research area, deserved broader circulation.

In the first part of the talk, entitled "Analyse générale", Fréchet referred immediately back to his talk in Bologna in 1928. But then he discussed various forms of the differential in functional analysis, stressing the form of Paul Lévy, but not mentioning his compatriot René Gâteaux. Finally he returned to the theory of general spaces, stressing von Neumann's axiomatic definition of Hilbert spaces and the use of the latter in quantum mechanics. In the following section, on the "theory of analytic functions" ("théorie des fonctions analytiques"), he stressed the recent work of his countryman Gaston Julia on quasi-analytic functions, but placed greater emphasis on the more abstract approach by Nachman Aronszajn on the decomposition of sets of those functions (**Aronszajn, 1935**), work that was based partly on prior investigations by Fréchet himself (**Fréchet, 1930**).

By far the longest and most interesting section of Fréchet's plenary talk is the one on probability theory (**Fréchet, 1936/7**, pp. 275–283). In fact it can still be read as an interesting contribution to the history of the theory of Markov chains. As to the discussion of the foundations of probability theory, (**Fréchet, 1936/7**, p. 275) referred to his own critical discussion of von Mises' notion of the collective in section IV of the Oslo congress (see our p. 155). Neither in his plenary talk nor in his section IV contribution did Fréchet allude to Kolmogorov's booklet of three years earlier on the axiomatics of probability, which had positively mentioned and used Fréchet's notion of the abstract integral. Instead he discussed, in connection with his own work in the theory of Markov chains, the recent results by Kolmogorov and Doeblin in the same field:

> Thus M. Kolmogoroff and M. Doeblin have succeeded in rediscovering [retrouver] recently and independently by algebraic methods the principal results, thus extending to the general Markov case the remarkably straightforward nonalgebraic method presented by M. Hadamard at the Bologna congress in the case of shuffling cards.[129]

It seems remarkable that Fréchet, who at the time of his plenary talk in Oslo hoped that the upcoming generation of his compatriots would acknowledge his vision for a modern, structural mathematics (**Siegmund-Schultze, 2005**), did not see the importance of the recent work on the foundations of functional analysis and probability theory by Banach and Kolmogorov.

In general, Fréchet felt that he was more appreciated abroad than at home, as revealed in his failure to be elected a member of the French Academy of Sciences at this time. This feeling seems to find its expression in his many talks in the sections of the various ICMs.[130] Fréchet had to wait quite a long time before finally in 1956, at the age of 76, he was elected a member of the Academy.

9.19. N. Wiener: "Gap theorems"

As the titles of his autobiographies (**Wiener, 1953**) and (**Wiener, 1956**) indicate,[131] Norbert Wiener (1894–1964; see Figure 9.18) was a child prodigy, who graduated from Tufts College (now Tufts University in Medford, Massachusetts) in 1909 at the age of just 14, having been educated at home by his father, a professor of Slavonic languages and literature. A Harvard PhD in mathematical logic, supervised by Karl Schmidt, followed in 1913 with a dissertation establishing a

---

[129]Translated from the French in (**Fréchet, 1936/7**, p. 277). On Hadamard's talk in Bologna, see also (**Bru, 2003**).

[130]We may take the Toronto congress as an example, at which Fréchet gave no less than four talks: "L'expression la plus générale de la 'distance' sur une droite", "Number of dimensions of an abstract set", "Sur une représentation paramétrique intrinsèque de la courbe continue la plus générale", and "Sur une formule générale pour le calcul des primes pures d'assurances sur la vie" (**ICM, 1924**, vol. I, pp. 75, 80).

[131]For other biographies of Wiener, and accounts of his mathematical work, see (**McMillan, 1989**), (**Masani, 1989**), (**Masani, 1990**), (**Jerison and Stroock, 1995**), (**Mandrekar, 1995**), and (**Freudenthal, 2008**); see also the series of articles published in volume 72 (1966), number 1, part 2 of *Bulletin of the American Mathematical Society*, including a bibliography of Wiener's work on pp. 135–145.

FIGURE 9.18. Norbert Wiener (1894–1964) (Author: Konrad Jacobs. Source: Archives of the Mathematisches Forschungsinstitut Oberwolfach)

set-theoretic basis for the study of relations.[132] A Harvard grant enabled Wiener to pursue his interest in mathematical logic still further by travelling first to Cambridge to work with Bertrand Russell, and then to Göttingen to study with David Hilbert. It was these two figures, along with G. H. Hardy, who encouraged Wiener's move towards the study of mathematics more generally. However, Wiener found himself at odds with Hardy over the latter's famous disdain for applied mathematics[133] — even the "purer" parts of Wiener's work were strongly motivated by possible applications.

[132]The thesis title was *A comparison between the treatment of the algebra of relatives by Schroeder and that by Whitehead and Russell*; the main ideas were subsequently published in (**Wiener, 1914**).

[133]Wiener called Hardy's position "pure escapism" (**Wiener, 1953**, p. 189).

Upon returning to the United States, Wiener took a number of jobs, including teaching philosophy at Harvard and writing for the *Encyclopedia Americana*. Following the USA's entry into the First World War, he attempted to enlist in the army, but was rejected on the grounds of poor eyesight. Instead, he aided the war effort by undertaking work on ballistics, at the invitation of Oswald Veblen. Wiener was finally accepted into the army in the later stages of the war, but never saw active service. In 1919, he took a job as an instructor in the mathematics department at MIT, an institution at which he remained until his retirement in 1960. Wiener's early work at MIT during the 1920s covered such interlinked fields as Brownian motion, potential theory, and harmonic analysis; we will return to his work in this period below.

(**Levinson, 1966**, p. 24) noted that

> Wiener had an international outlook. Not only did he feel at home in Western Europe but also in Mexico, China and India. While he had visited the Soviet Union only briefly, he had long admired their achievement in mathematics.

Wiener later recounted his many travels in his autobiographies. Indeed, he travelled to the Oslo ICM on his way back from a trip to China,[134] returning via Cairo, London, and Copenhagen, as well as Clermont-Ferrand, where he spent four days working with Szolem Mandelbrojt on a joint paper whose results were presented in Oslo (**Wiener, 1956**, p. 204). More generally, Wiener's account of his travels in 1936 reads as a sequence of encounters with old friends:

> The meeting was delightfully arranged, and as exciting and overstimulating as such meetings always are. We found ourselves among a host of friends of all nationalities. We joined up particularly with our colleagues the Vallartas (for Vallarta had married since our early days, and Mrs. Vallarta became a good friend of my wife). Canon Lemaitre, of Louvain, also belonged to our group. He had worked for years at the Harvard Observatory. We found him in Oslo, as we had known him before, a delightful friend and table companion. We would go with him on excursions and walks during the twilit midnight of an Oslo summer.[135]

During the Second World War, Wiener returned to the study of ballistics, specifically the problem of aiming a gun at a moving target. A classified report that he wrote on this subject in 1942 was subsequently published by the MIT Press (**Wiener, 1949**), thereby establishing Wiener, along with A. N. Kolmogorov, as a co-developer of the linear prediction of stationary time series.[136] With this work, communication theory[137] became a more prominent strand in Wiener's research, leading him in turn to his famous work on cybernetics, and his book *Cybernetics, or control and communication in the animal and the machine* (**Wiener, 1948**).

[134]Wiener and his wife had considered making the trip via the Trans-Siberian Railway, but eventually concluded that this would be too hard on their children (**Wiener, 1956**, p. 201). Someone who did take the Trans-Siberian Railway to Oslo, by necessity, was Fritz Noether: see Appendix A.9.

[135](**Wiener, 1956**, p. 206). On both Vallarta and Lemaître, see Appendix A.3 and the corresponding notes in Chapter 6.

[136]On the parallel work of Wiener and Kolmogorov in the context of cybernetics, see (**Gerovitch, 2002**).

[137]For a short survey of Wiener's work in communication theory, see (**Root, 1966**).

To return to Wiener's work of the 1920s, we note the three major strands identified by (**Jerison and Stroock, 1995**, p. 432): Brownian motion, potential theory, and harmonic analysis. Since it is the third of these that is most relevant to Wiener's lecture in Oslo in 1936, we deal here only with this one.[138]

Wiener's work in harmonic analysis was motivated by the problem of processing electrical signals — in particular, the desire to construct an appropriate theory, a *generalised* harmonic analysis, that would take account of those aspects of the nature of signals within electrical systems that mean that classical Fourier analysis does not apply. Wiener therefore developed a new approach within the framework of Tauberian theory, which he went on to illustrate by applying it in analytic number theory to provide a new proof of the Prime Number Theorem. In their discussion of Wiener's work, (**Jerison and Stroock, 1995**, p. 436) point to his work on the theory just described as having fully established his reputation as a mathematician: in 1932, he was promoted to full professor at MIT, and then in the following year he was both elected to the US National Academy of Sciences, and awarded the AMS's Bôcher Memorial Prize. He had thus reached a stature that made him a natural choice for a plenary lecture at an ICM.[139]

Wiener's lecture in Oslo was concerned with so-called "gap theorems" in Fourier analysis. He described these in his opening paragraph as follows:

> In general, a gap theorem is the assertion that a function $f(x)$ with the Fourier or Dirichlet series
>
> $$f(x) = \sum a_n e^{i\lambda_n x},$$
>
> or the Taylor series
>
> $$f(x) = \sum a_n x^{\lambda_n},$$
>
> has certain properties which depend merely on the distribution of the λ_n. In this sense, it is a gap theorem to assert that the function with the Fourier series
>
> $$f(x) = \sum_{-\infty}^{\infty} a_n e^{inx},$$
>
> is of period 2π. (**Wiener, 1936/7**, p. 284)

Of particular interest to Wiener were of course the coefficients a_n (*Ibid.*):

> Closely connected with the very notion of a gap theorem is the problem of determining the coefficients a_n in terms of the function $f(x)$. The determinations which are of most interest are the integral determinations of the form
>
> $$a_n = \int_C f(x) g_n(x) dx.$$

[138]On these three strands, see also (**Mandrekar, 1995**); on Wiener's work in potential theory, see (**Brelot, 1966**).

[139]As we noted on p. 208, Wiener had previously attended the 1920 ICM, where he had delivered a lecture titled "Certain iterative properties of bilinear operations" (**ICM, 1920**, pp. 176–178). We recorded Wiener's subsequent regret of his attendance of the 1920 ICM in footnote 21 on p. 10. In Zurich in 1932, Wiener had spoken on joint work with Paley ("Analytic properties of characters of infinite Abelian groups": (**ICM, 1932**, vol. II, p. 95)). He went on also to speak at the 1950 ICM with a "Comprehensive view of prediction theory" (**ICM, 1950**, vol. II, pp. 308–321).

Indeed, a large part of the paper is concerned with the determination of the g_n in terms of the λ_n: via what the *Jahrbuch* reviewer, the Oslo participant J. Karamata, referred to as "remarkable tricks" ("bemerkenswerte Kunstgriffe": JFM 63.0974.01), Wiener showed how the g_n could be determined in such a way that the prior gap theorems of E. Fabry and S. Mandelbrojt were reproduced.

Considered alongside his wider research, Wiener's Oslo paper appears to have been a fairly minor piece of work. Indeed, the topic of gap theorems (with this paper as a representative) is not specifically singled out by the authors of surveys of Wiener's work. As far as his interests in (generalised) harmonic analysis are concerned, it is his other Tauberian theorems that seem to attract the greatest attention.[140] We may draw a contrast here with the surveys that we have seen of the work of other Oslo plenary speakers, where the topic of their ICM lecture is usually mentioned, even if the lecture itself is not.

9.20. Ø. Ore: "The decomposition theorems of algebra"

In Section 8.4, we noted the almost total absence from the Oslo congress of the abstract, structural approach to algebra, as advocated by Emmy Noether and others; the early death of Noether and the inability of her prominent followers (such as van der Waerden) to attend the ICM are of course important factors here. Nevertheless, one of the plenary lectures, that of the Norwegian-born Øystein Ore (1899–1968; see Figure 9.19), did feature this new approach.

Ore was born in Kristiania in 1899,[141] and he completed his PhD (*Zur Theorie der algebraischen Körper*) at the university there in 1924, under the supervision of Skolem. In his approach to mathematics, he was greatly influenced by Emmy Noether and her school, both in his PhD thesis and in his subsequent work, having spent time in Göttingen in the mid-1920s. In 1927, Ore moved to Yale University (being unable to find a suitable position in his native Norway), where he remained until his retirement in 1968. Nevertheless, as we see for example from his participation in the Oslo ICM, he remained closely connected to his country of origin throughout his life. During the Second World War, he was a leading figure in the "American Relief for Norway" organisation, and later chaired the "Fritt Norge" ("Free Norway") movement in New York. In 1954, Ore published a successful and well-researched Norwegian biography of his compatriot Niels Henrik Abel, which was subsequently translated into English (**Ore, 1954/7**). Ore's closeness to Norway could not be better symbolised than by his death in 1968 during the 15th Scandinavian Congress of Mathematicians in Oslo, at which he was due to speak on the four-colour problem, a topic related to his later research interests in graph theory.

Much of Ore's early work was in algebraic number theory, but he later worked in other areas, such as group theory and combinatorics. This, however, was not the subject of his lecture in Oslo, which instead concerned the burgeoning abstract theory of lattices, a topic whose early development may be traced back to the work of Dedekind at the end of the nineteenth century,[142] with parallel ideas appearing

[140]For a survey of these, see (**Mandelbrojt, 1966**).

[141]Biographical references for Ore are (**Anon, 1970**) and (**Aubert, 1970**). Unlike most of the other mathematicians discussed in this section, Ore does not have an entry in the *Dictionary of Scientific Biography*.

[142]On the development of lattice theory, see (**Mehrtens, 1979**), (**Corry, 2004**), (**Rota, 1997**), and (**Decaens, 2018**).

FIGURE 9.19. Øystein Ore (1899–1968) (Source: Department of Mathematics, Yale University)

also in the work of Ernst Schröder, who, in contrast to Dedekind's number-theoretic inspiration, came at the subject from the direction of the algebra of logic. However, as (**Corry, 2004**, p. 121) observes, "[b]oth works were ignored by the mathematical community at large". It was only in the 1930s that the theory of lattices began to re-emerge, and Ore was one of the main driving forces behind it. Another major figure was Garrett Birkhoff, son of the Oslo plenary speaker G. D. Birkhoff.[143] The younger Birkhoff spent much of the first half of the 1930s in Cambridge, where he came under the influence of the group theorist Philip Hall, who advocated a general approach to algebra which emphasised the role of binary relations on algebraic

[143]Others who contributed to this (re)birth of lattice theory were Karl Menger (**Menger, 1928**) and Fritz Klein-Barmen (**Klein-Barmen, 1929**), (**Klein, 1931**). On the latter's (largely axiomatic) investigations of what he called a *Verband* (now the modern German term for "lattice"), see (**Mehrtens, 1979**, §3.2.2) and (**Hollings, 2014**, pp. 95–96).

structures, leading Birkhoff to the notions of universal algebras and lattices.[144] He attended the Oslo congress, delivering an order-theoretic sectional lecture: "Order and the inclusion relation" (**ICM, 1936**, vol. II, p. 37).[145]

Ore's influence on the early development of lattice theory may be a little harder to discern than that of Birkhoff, since it is the latter's terminology "lattice" that has been retained in the modern theory. Ore's term "structure" for the same notion was perhaps deemed too vague to be used in this context, but this same feature does seem to point towards the fundamental role that Ore envisaged for these constructions. Like Birkhoff, he sought to understand other algebraic systems by studying the lattices formed by certain of their subsystems.[146] Lattice theory might serve, for example, to unify the commutative and non-commutative cases that had arisen in such areas as ring theory. However, as (**Corry, 2004**, p. 261) observes, Ore's "own enthusiasm for his program had already waned considerably" by the end of the 1930s: lattice theory had certainly been established as a valid discipline,[147] but hopes for its fundamental nature had faded.

Chronologically, Ore's development of lattice theory came a little after that of Birkhoff, though was entirely independent. His first presentation appeared in a paper of 1935, tellingly titled "On the foundation of abstract algebra" (**Ore, 1935**), with a further instalment the following year. Explicitly building upon Dedekind's work, and making occasional reference to Birkhoff's, Ore set out to construct the basic components needed to turn lattice theory into an algebraic discipline (homomorphism, quotient, isomorphism, etc.), before arriving finally at a suitable version of the Jordan–Hölder–Schreier Theorem in the lattice context. Further such decomposition theorems appeared in the second part of Ore's work, published in the year of the Oslo congress. (**Corry, 2004**, p. 275) notes of Ore's invitation to Oslo that it "makes clear that, at least at that time, the theory aroused some interest".

The title of Ore's Oslo lecture, "On the decomposition theorems of algebra" (**Ore, 1936/7**), gives us a clear indication that it was closely related to his recent interests. Moreover, the descriptions of this area in his opening paragraphs remind us just how new this topic was, and therefore probably unfamiliar to many in his audience. Indeed, Ore began by justifying these ideas with an appeal to applications:

> We shall in the following give an account of certain new ideas regarding the foundation of the so-called abstract algebra. These ideas, which have been developed only in the last couple of years, throw new light upon several of the basic problems of algebra. It should, however, be mentioned already at the beginning, that their application is in no way limited to algebra. They have originated principally in connection with algebraic problems and

[144]On Hall's influence on the study of universal algebra, see the comments in (**Hollings, 2014**, pp. 152, 270).

[145]This in addition to two further sectional lectures: "Integration of differential equations" and "Generalized convergence" (**ICM, 1936**, vol. II, pp. 50–51, 152–153).

[146]In the interests of clarity, we will employ Birkhoff's term "lattice" throughout, apart from where Ore's "structure" appears in direct quotations.

[147]We may see this in the fact that 1938 also saw the first symposium on lattice theory, papers from which were printed in volume 44, number 12, part 1 of the *Bulletin of the American Mathematical Society*. We note that both Birkhoff and Ore spoke at this symposium, each retaining his own terminology for the meeting's central concept.

> in the following we shall stress the algebraic consequences, but there are already important applications to geometry, to point-set theory and to the foundation of quantum mechanics. Since the theory is a very recent one and since it is in a state of rapid development, the following lecture cannot be expected to give any complete evaluation of its importance for algebra or for its several other domains of applications. It should serve to demonstrate, however, that a very fertile field or [*sic*] mathematical investigation has been opened up. (**Ore, 1936/7**, p. 297)

Nevertheless, Ore identified for his audience what he saw as the two essential features of this new algebra, axiomatisation and the exact determination of the domain of validity of theorems in algebra:

> One may say that abstract algebra has been developed from ordinary algebra through the realization that the various algebraic theories may be derived from a small number of axiomatic rules, to a large extent the same for all algebraic theories. The axiomatic synthesis is, however, only one side of the abstract theory. Among its most notable achievements, I should prefer to mention the solution of what one may call the *completeness problem* in several important cases, namely the determination of all algebraic systems with given properties. As a classical example one may mention ... the determination of all fields for which the Galois theory of equations is valid. (*Ibid.*)

(**Corry, 2004**, p. 275) notes:

> The fact that Ore used the privileged stage offered to him to mention these two aspects of algebra indicates how far they were from being taken for granted by the mathematical community at large still in 1936 and, on the other hand, the high degree of importance that he himself conceded to them.

What follows in Ore's paper is a careful development of the basic features of lattice theory. Reproducing the appropriate notion of quotient from his earlier papers, Ore derived, for example, a lattice version of the second isomorphism theorem. With various fundamental elements of the theory thus established, Ore noted that it would next be possible to go on to produce lattice versions of results that are familar from other branches of algebra: he mentioned the theorems of Schmidt–Remak and Jordan–Hölder. In his concluding pages, Ore drew connections between lattice theory and topology and, in particular, projective geometry (with reference to recent axiomatic works by (**Birkhoff, 1935**) and (**Menger, 1936**)).

As we have noted, however, Ore's programme of placing lattice theory at the foundations of abstract algebra came to nought. The theory remains an active area of research, but as one branch of abstract algebra among many, rather than occupying the fundamental position that Ore had hoped for. Moreover, as Corry remarks, the period in which Ore was working was hardly an auspicious one for this type of mathematics:

> since the early 1940s many American mathematicians turned their attention towards the war effort, and, as a result, applied

> mathematics received much more emphasis than the pure, foundational studies embodied in Ore's program. The program was not only abandoned; it fell into oblivion and it is not even mentioned as an object of historical interest. (**Corry, 2004**, p. 288)

9.21. The plenary lectures that were planned but did not take place

As we have already described in detail, four of the planned plenary lectures in Oslo did not take place, owing to the last-minute absences of the speakers in question (Gel'fond, Severi, Landau, Khinchin) for a range of reasons that we have outlined (except in the case of Landau) in earlier chapters. Nevertheless, in each case we do have a title for the planned talk, as found in the preliminary version of the programme (Table 9.1). We therefore devote this section to some brief comments on the nature of the work of the relevant mathematicians, and offer some speculations as to how their lectures would have fit into the Oslo congress, had they taken place.

9.21.1. A. Gel'fond: "Theories of transcendental numbers". As we have already noted in Section 4.1, the invitation for Alexandr Osipovich Gel'fond (1906–1968) to deliver a plenary lecture in Oslo probably stemmed from the fact that he had, just two years earlier, published an affirmative solution to Hilbert's seventh problem on transcendental numbers (**Gel'fond, 1934**): to determine whether the expression α^β for algebraic α and irrational β is necessarily transcendental. The title that was advertised for his talk, "Théories des nombres transcendants", appears to indicate that this solution would indeed have been his topic.

Gel'fond was born in St Petersburg in 1906, and entered Moscow State University to study mathematics in 1924.[148] Upon completing his undergraduate studies there in 1927, he went on to pursue postgraduate research under the supervision of A. Ya. Khinchin and V. V. Stepanov. In 1930, Gel'fond travelled to Germany, where he worked in Berlin[149] and Göttingen with Hilbert, and also with two other proposed Oslo plenary lecturers: Landau and Siegel. Following his return to the USSR, he took up the chair in number theory at Moscow State University, which he occupied until the end of this life; from 1933, he held a post also at the Steklov Mathematical Institute of the Academy of Sciences, and in 1939 was elected a corresponding member of the Academy. According to V. I. Arnol'd, in an interview of 2007, Gel'fond went on to become the Soviet Navy's chief cryptographer during the Second World War.[150]

The two branches of mathematics in which Gel'fond did his best-known work were the theories of analytic functions and, what is most relevant for our purposes, of transcendental numbers. He had published in both areas even before the end of his postgraduate studies (**Gel'fond, 1929a**), (**Gel'fond, 1929b**). Indeed, his 1929 paper on transcendental numbers laid the foundations for his later work on Hilbert's

---

[148] For biographies of Gel'fond, see (**Linnik and Markushevich, 1956**), (**Pyatetskii-Shapiro and Shidlovskii, 1967**), (**Evgrafov *et al.*, 1969**), (**Levin *et al.*, 1971**), and (**Yushkevich, 2008a**).

[149] In his later account of his German travels (**Gel'fond, 1977**), Gel'fond identified I. Schur and E. Schmidt as having been the major figures in Berlin at this time, but he does not indicate with whom he worked there.

[150] `https://web.archive.org/web/20100610052252/http://www.gzt.ru/topnews/education/-vladimir-arnoljd-opasatjsya-kompetentnyh-/308825.html` (accessed 3 February 2020).

TABLE 9.1. The timetable of plenary lectures as it appeared in the preliminary programme; those lectures that are highlighted here in bold did not take place (cf. Table 8.3).

Date and time	**Speaker**	**Title**
Tues 14 July 10:15–11:00am	C. Størmer	Programme for the quantitative discussion of electron orbits in the field of a magnetic dipole, with application to cosmic rays and kindred phenomena
Tues 14 July 11:15am–12:00pm	R. Fueter	Die Theorie der regulären Funktionen einer Quaternionenvariablen
Weds 15 July 9:00–10:00am	E. Cartan	Quelques aperçus sur le rôle de la théorie des groupes de Sophus Lie dans le développement de la géométrie moderne
Weds 15 July 10:15–11:00am	C. L. Siegel O. Veblen	Analytische Theorie der quadratishen Formen Spinors and projective geometry
Weds 15 July 11:15am–12:00pm	**A. Gel'fond** J. Nielsen	**Théories des nombres transcendants** Topologie der Flächenabbildungen
Thurs 16 July 9:00–10:00am	E. Hecke	Neuere Fortschritte in der Theorie der elliptischen Modulfunktionen
Thurs 16 July 10:15–11:00am	C. W. Oseen **F. Severi**	Probleme der geometrischen Optik **Théories et questions nouvelles dans la géométrie algébrique**
Thurs 16 July 11:15am–12:00pm	V. Bjerknes H. Hasse	New lines in hydrodynamics Über die Riemannsche Vermutung in Funktionenkörpern
Fri 17 July 9:00–10:00am	G. D. Birkhoff	On the foundations of quantum mechanics
Fri 17 July 10:15–11:00am	S. Banach **E. Landau**	Le rôle de la théorie des opérations dans l'analyse **Über einige neuere Fortschritte der additiven Zahlentheorie**
Fri 17 July 11:15am–12:00pm	L. Ahlfors J. G. van der Corput	Geometrie der Riemannschen Flächen Diophantische Approximationen
Sat 18 July 9:00–10:00am	**A. Khinchin**	**Hauptzüge der modernen Wahrscheinlichkeitstheorie**
Sat 18 July 10:15–11:00am	M. Fréchet N. Wiener	Mélanges mathématiques Tauberian gap theorems
Sat 18 July 11:15am–12:00pm	O. Neugebauer Ø. Ore	Über vorgriechische Mathematik und ihre Stellung zur griechischen The structure of algebraic systems

seventh problem, published in French in 1934. Number theory, and the study of transcendental numbers in particular, remained an important part of Gel'fond's subsequent work; he went on to write a monograph on this subject (**Gel'fond, 1952**). At the end of his life, he wrote an account of the history and solution of Hilbert's seventh problem (**Gel'fond, 1969**). We say no more about the problem here; the interested reader should consult (**Gray, 2000**, pp. 174–175).

So far as we can tell, Gel'fond did not attend any subsequent ICMs, with the notable exception of the 1966 Moscow congress in the year before his death — though he does not appear to have delivered a lecture there. The topic of transcendental numbers did, however, reappear at later congresses, most notably that in Nice in 1970, where a solution of a generalisation of Hilbert's seventh problem, often known as the Gel'fond Conjecture, earned a Fields Medal for the English number theorist Alan Baker (1939–2018).[151]

9.21.2. F. Severi: "Theories and new questions in algebraic geometry". Francesco Severi's name featured prominently in Chapter 5, both for mathematical and for political reasons.[152] In light of what we discussed there, it is thus no surprise that he should have wanted to attend the Oslo ICM, and that he appealed directly to the Italian Minister of Education to be allowed to travel — but to no avail, as we saw at the end of Section 5.3. The fact that he had originally been held in reserve for a plenary lecture in Oslo (see Table 2.5 on p. 41) is probably related simply to the desire of the organisers to prioritise people who had not previously spoken at an ICM (see p. 151).

To turn to Severi's mathematics, we note that although he contributed to other fields, such as complex function theory, the major theme of his work was algebraic geometry.[153] Indeed, this was the broad topic of his two prior ICM plenary lectures, and would also have been that of his Oslo lecture, which was advertised under the very general title "Théories et questions nouvelles dans la géométrie algébrique".

Severi's introduction to algebraic geometry came during his studies with Corrado Segre in Turin at the end of the nineteenth century. Turin was a centre for algebraic geometry at this time, with Luigi Cremona having founded the school some decades earlier. The major themes of the so-called "Italian school of algebraic geometry", which soon spread beyond Turin, were the study of birational geometry and algebraic surfaces. However, the rather intuitive nature of some of the studies of the school has been noted critically by subsequent commentators.[154] In later decades, these foundational issues were addressed by Chevalley, Weil, and Zariski.

As Severi's biographer Beniamino Segre (a student and great-nephew of Corrado) has noted, Severi's subsequent time was spent working with Enriques in

[151] The other Fields Medallists that year were Heisuke Hironaka (b. 1931), John G. Thompson (b. 1932), and S. P. Novikov (b. 1938); the Soviet authorities did not permit Novikov to travel to the ICM to collect his medal in light of his letters of support for dissidents who had been arrested and confined in mental institutions (see, for example, (**Buchstaber, 2001**)).

[152] For biographies of Severi, see (**Roth, 1963**), (**Segre, 2008**), (**Babbitt and Goodstein, 2009**), and (**Goodstein and Babbitt, 2012**). For his work in the foundations of algebraic geometry, see (**van der Waerden, 1986**).

[153] On the development of algebraic geometry, see (**van der Waerden, 1971**), (**Dieudonné, 1972**), (**Dieudonné, 1985a**), and (**Dieudonné, 1986**).

[154] See, for example, (**van der Waerden, 1971**, p. 171).

Bologna and with Bertini in Pisa, which reinforced his interests in algebraic geometry:

> He attempted, with great success, to explain important and still unsolved problems along with work in new areas. He perfected the theory of birational invariants of algebraic surfaces and created an analogous (but more complex) theory for algebraic varieties of arbitrary dimension. The completion of this work was to take him another fifty years. (**Segre, 2008**, p. 330)

We note in passing that Severi's algebraic approach to geometry led him to seek greater links with German algebraists, for reasons both mathematical and political — he concluded a talk at a conference in Germany in 1938 with the following words:

> I hope that the important progress that Germany has realized in modern algebra, will allow her magnificent mathematicians to penetrate deeper and deeper into algebraic geometry which has been cultivated in Italy over the past 40 years; and that the connections between German science and Italian science, which have already been so close in this domain at the time of our masters, become more intimate every day, as they are today in the political and general cultural realm.[155]

Despite his losing some influence after the Second World War, owing to his alignment with the fascists, the post-war biographical and review literature contains some very positive assessments of Severi's work, which serve perhaps to transcend political views and focus merely on mathematical content.

9.21.3. E. Landau: "On some recent advances in additive number theory". Unlike the other mathematicians dealt with in this section, we have yet to provide an explanation as to why the number theorist Edmund Landau (1877–1938; see Figure 9.20) did not, in the end, attend the Oslo congress. Indeed, the reasons are rather less clear in his case, although there seems to have been a little more forewarning of his absence, since he is not included in the list of members of the congress that was printed in the proceedings. Nor, apart from a few passing references to his work, does he appear elsewhere in the proceedings. It seems likely, therefore, that he did not attend. In the programme for the congress in Spring 1936, Landau was still listed as giving a plenary lecture, entitled "On recent progress in the additive theory of numbers" ("Über einige neuere Fortschritte der additiven Zahlentheorie": see (**Anon, 1936**)). On 5 August 1936, i.e., shortly after the congress, Harald Bohr wrote to Richard Courant, alluding to the complicated situation in Landau's home in Berlin ("Zustände bei Landaus"), in particular for his wife (Marianne, born Ehrlich), and that Landau allegedly had fallen "in his tendency to absolutism into absolute egotism".[156] One could speculate that Landau had refused to go to Oslo for some principled reason and thus also spoiled the occasion for his wife. We speculate further, however, that Landau was afraid of meeting people from Germany who had supported the Nazis and his dismissal (see

[155] Quoted in English translation in (**Schappacher, 2006**, p. 15). See also the comments in (**Siegmund-Schultze, 2014**, p. 1240).

[156] "Es ist nur traurig dass er in seinem Hang zu Absolutismus sich nur dadurch zu retten glaubt, dass er auf eine Art absoluten Egotismus verfällt." (RCP, Series I, box 2, folder 4)

FIGURE 9.20. Edmund Landau (1877–1938) (Copyright: Mathematische Gesellschaft (Hamburg) `https://www.math.uni-hamburg.de/home/grothkopf/fotos/math-ges/`. Source: Mathematisches Forschungsinstitut Oberwolfach)

below). He certainly would not have wanted to be at the mercy of a German delegation.

Landau had studied in Berlin in the 1890s.[157] Supervised by F. G. Frobenius, he had obtained a PhD in number theory in 1899. Landau taught in Berlin until 1909, when he moved to Göttingen as successor to Hermann Minkowski. From 1914, he was a full member of the Akademie der Wissenschaften zu Göttingen.

An observant Jew, Landau had hastily tried to improve his Hebrew in the mid-1920s in order to be able to lecture in that language at the opening of the Hebrew University in Jerusalem in 1925.[158] During 1927–1928, he taught in Jerusalem, but, facing some bureaucratic obstacles, he soon returned to Göttingen. Anecdotes told of Landau suggest that he was somewhat slow to recognise the Nazi threat.[159] Nevertheless, he was the victim of a student boycott of his lectures, and was forced eventually to resign his position in 1933, whereupon he retired to Berlin.[160] In the years that followed, he continued to lecture outside Germany (for example, in Cambridge in 1935, and in Brussels in 1937), but died in 1938.[161]

The bulk of Landau's mathematical work was in analytic number theory, of which he gave the first systematic account (**Landau, 1909**). Among his earliest works was a new (simpler) proof of the Prime Number Theorem (based upon properties of the Riemann zeta function), which allowed its application also to the distribution of ideal primes in algebraic number fields (**Landau, 1903**). Landau's systematisation of analytic number theory continued with his three-volume

[157]For biographies of Landau, see (**Hardy and Heilbronn, 1938**), (**Knopp, 1951**), and (**Schoeneberg, 2008b**); on his political stance, see (**Siegmund-Schultze, 2011**).

[158]On Landau's lecture, see (**Corry and Schappacher, 2010**). On the university in Jerusalem, see also the newspaper interview with Fraenkel that is reproduced in our Appendix A.4.

[159]See, for example, (**Segal, 2003**, pp. 454–455).

[160]In light of his wealth, Landau was deemed ineligible for the Rockefeller funding that was being used to help other Jewish mathematicians to escape (**Siegmund-Schultze, 2009**, p. 199).

[161]On Landau during this period, see also (**Siegmund-Schultze, 2012**).

Vorlesungen über Zahlentheorie (**Landau, 1927**). Many of Landau's books are characterised by a terse, clean, heavily symbolic style.[162]

Even before 1936, Landau had been invited to deliver a plenary lecture at an ICM: that of 1912, where he lectured with the title 'Gelöste und ungelöste Probleme aus der Theorie der Primzahlverteilung und der Riemannschen Zetafunktion' (**ICM, 1912**, vol. 1, pp. 93–108). The printed version directs the reader towards a further paper by Landau with the same title as his congress address (**Landau, 1912**).

The title advertised for Landau's proposed lecture in Oslo in 1936 was the very general-sounding "Über einige neuere Fortschritte der additiven Zahlentheorie". Nevertheless, we may begin to guess at its content by noting that there exists a published work by Landau with the same title as his Oslo lecture, namely the monograph (**Landau, 1937**), which was the outcome of the lectures that Landau delivered in Cambridge in 1935. One would assume therefore that the topics contained therein would have been at the forefront of Landau's mind when he was planning a lecture for Oslo — although it is worth noting that *additive* number theory had not previously been a major interest for Landau. The book was reviewed for *The Mathematical Gazette* very shortly after Landau's death in 1938, and was described as "a connected account of some of the remarkable advances which have been made in the additive theory of numbers since 1930" (**Ingham, 1938**, p. 190). Indeed, in a more general setting, Landau's biographer observed:

> Written with the greatest care, Landau's books are characterized by argumentation which is complete, and as simple as possible. The necessary prerequisite knowledge is provided, and the reader is led securely, step by step, to the goal. ... Through his books and his more than 250 papers Landau exercised a great influence on the whole development of number theory in his time. (**Schoeneberg, 2008b**, p. 616)

9.21.4. A. Khinchin: "Main features of modern probability theory". In Section 8.4, we made some brief comments on the new developments in probability theory, especially in its rigour, that had taken place in the years prior to the Oslo congress. We now pick up this story once more and consider the place of the work of Aleksandr Yakovlevich Khinchin (1894–1959) within it.

Khinchin studied mathematics in Moscow, and taught there for the whole of his professional life;[163] he was elected to associate membership of the Soviet Academy of Sciences in 1939 and also to full membership of the Academy of Pedagogical Sciences of the RSFSR (now Russia) in 1944.

Khinchin began his career in Luzin's school of function theory. However, during the 1920s, he turned to work in analytic number theory, probability theory, and the intersection of the two. For example, he studied metric problems related to Diophantine approximation, and established the law of the iterated logarithm, concerning the magnitude of the fluctuations in a random walk (**Khinchin, 1924**). As (**Yushkevich, 2008b**, p. 351) observes:

[162](**Hardy and Heilbronn, 1938**, p. 307) refer to "an incessant striving for compression".

[163]For biographies of Khinchin, see (**Anon, 1946**), (**Anon, 1955**), (**Gnedenko and Kolmogorov, 1960**), (**Gnedenko, 1960**), (**Gnedenko, 1961**), (**Doob, 1961**), (**Gnedenko and Markushevich, 1968**), and (**Yushkevich, 2008b**); on his work in probability, see (**Cramér, 1962**).

> Probability theory proved to be an auspicious field for the application of the methods of the metric theory of functions, and Khinchin was drawn more and more into the problems of the summation of independent random variables. During the 1920's and 1930's this classical branch of probability theory assumed its present form in the closely related works of Kolmogorov, P. Lévy, Khinchin, and others. Khinchin's contribution included results on the applicability of the law of large numbers to equally distributed random variables with finite mathematical expectations, on the coincidence of the class of all limit distributions with the class of all infinitely divisible laws, on the convergence of series of random variables (jointly with Kolmogorov), and on the structure of stable laws (jointly with Lévy).[164]

In 1933, Khinchin published the Springer *Ergebnisse* report, *Asymptotische Gesetze der Wahrscheinlichkeitsrechnung*, that we have mentioned previously (for example, in Section 8.2). There was certainly no shortage of probabilistic topics for Khinchin to cover in his proposed Oslo lecture, but we will never know exactly which "Hauptzüge" ("main features") he had in mind when he submitted his title: "Hauptzüge der modernen Wahrscheinlichkeitstheorie".

Khinchin's influence on the Soviet mathematical community is clear: we need only look to the language employed by his biographers and obituarists (principally his own students and collaborators, such as B. V. Gnedenko) — for example:

> He was without doubt known to all readers ... as one of the most eminent of Soviet mathematicians, a man who made substantial contributions to the theory of functions, to the theory of numbers and above all to the theory of probability and its applications to statistical mechanics, physics and technology. (**Gnedenko and Kolmogorov, 1960**, English transl., p. 93)

We conclude this section by recording a Western perspective of Khinchin's mathematical work: that of the Swede Harald Cramér, whose own interests in probability theory were noted on p. 154. In an article entitled "A. I. Khinchin's work in mathematical probability" (**Cramér, 1962**), Cramér sought to locate Khinchin's work within the growth in probability theory that we have noted above:

> The tremendous development which ... took place in the twenty years from 1920 to 1940 was, no doubt, a joint effect of the efforts of a large number of mathematicians and statisticians. However, it does not seem unlikely that future historians will ascribe this development, as far as the mathematical side of the subject is concerned, above all to the creative powers of three men: A. I. Khinchin, A. N. Kolmogorov, and P. Lévy. In fact, it may be said that the real turning point came with the publication in 1933 of the two famous monographs by Khinchin and Kolmogorov, and the appearance in 1934 of Lévy's paper on an important class of stochastic processes. (**Cramér, 1962**, p. 1228)

[164] For greater detail on these developments, see (**von Plato, 1994**).

FIGURE 9.21. *Dagbladet*, 14 July 1936, p. 3. The figure shows Ahlfors and Wiener in the front row, Størmer at the top right, and at the bottom Ahlfors receiving the prize from the hands of É. Cartan (the Norwegian text confuses the latter with Fueter).

9.22. The award of the first Fields Medals to Ahlfors and Douglas

As we indicated at the beginning of this chapter, there is one further Oslo lecture that needs to be dealt with here: that of Constantin Carathéodory (1873–1950; see Figure 9.22), which took place during the opening ceremony of the congress on the occasion of the award of the first Fields Medals (see Figure 9.21). Some sources include this talk on the list of Oslo plenary lectures, but we do not do so, for the reasons given above.

As we noted in Section 1.2.1, an international mathematical prize had been awarded, but just once, at the Rome ICM of 1908 in the form of the Guccia medal. It seems reasonable to speculate that the one-time award of such a prize had some influence on the subsequent creation of the Fields Medal by J. C. Fields. As (**Riehm and Hoffman, 2011**) relate,[165] plans for the award of the latter had been put in place some years before the Oslo congress.[166] The committee appointed to

[165]In particular, their Chapter 11; see also (**Lehto, 1998a**, §3.1). For an attempt to describe the major trends of twentieth-century mathematics through the lens of the work of the various Fields Medallists, see (**Monastyrsky, 1997**).

[166]Fields' original plans for the medal were somewhat vague, so it was left to subsequent generations of mathematicians to define the conditions of the award, a process that of course

consider the award of the first medals in Oslo in 1936 consisted of G. D. Birkhoff, É. Cartan, Carathéodory, Severi, and Takagi (**ICM, 1936**, vol. I, p. 39). Although the deliberations of the Fields committee were not made public, we can gain some insight into the decision-making process from a letter that Birkhoff wrote to his American colleague G. C. Evans at Berkeley three years after the Oslo congress, during the preparations for the proposed ICM of 1940:[167]

> I am writing to ask if you would be willing to undertake the Chairmanship of the Committee on the Award of the Fields Medals. ...[168]
>
> It may be well to give you something of the recent history of the Committee. I believe that I was the only regular member of the Committee who came to Oslo.[169] The awards which our Committee made did not seem to me to be the best possible ones perhaps, but our Chairman, Professor Severi, handled the affairs of the Committee in such a dilatory fashion that he and Professor Carathéodory had to practically settle the matter by themselves, and so the award went to Ahlfors and Douglass [*sic*] who doubtless were good candidates. My own candidates would have been Marston Morse and Siegel but I never had a fair opportunity to express my opinion for the reason above.[170]

It is thus perhaps because of the absence of Severi and Takagi, and the quiet opposition of Birkhoff (though we don't know Cartan's position), that it fell to Carathéodory to deliver a lecture connected with the work of the Fields Medallists. This talk went on to be published in the congress proceedings with the status of a plenary lecture (**Carathéodory, 1936/7**). However, this appears to have been a duty that Carathéodory was rather reluctant to perform: as we noted in Section 3.2, as late as June 1936, he still did not intend to go to the ICM, although he had been called upon to write a report on the work of the Fields Medallists, a task that he described as "not very rewarding".[171] Carathéodory's change of mind with respect to travel to Oslo was apparently prompted by the expected absence there of Severi, chair of the Fields committee, who probably would have presented the medals. Thus, it appears that Carathéodory was called upon to deliver his report in person.

---

began with the first deliberations over the recipients for 1936: see (**Curbera, 2009**, pp. 113–114), (**Riehm and Hoffman, 2011**, pp. 194–195), and (**Barany, 2018**).

[167]On the background to the eventual award of these Fields Medals in 1950 (to Laurent Schwartz and Atle Selberg), see (**Barany, 2016**, pp. 237–238), (**Barany *et al.*, 2017**, §8) and (**Barany, 2018**).

[168]The committee was eventually chaired by H. Bohr (**ICM, 1950**, vol. I, p. 126); on the circumstances surrounding his appointment, see (**Barany *et al.*, 2017**).

[169]Birkhoff was mistaken here: although Severi and Takagi did not attend the congress, Cartan and Carathéodory of course did.

[170]Birkhoff to Evans, 2 March 1939 (Birkhoff Papers, HUG 4213.4.5. box 1, file personal (1938/39)). The Oslo congress decided on the (absent) Hardy as the chairman of the new Fields committee for 1940, with Aleksandrov, Hecke, Julia, and Levi-Civita as members, and Lefschetz and Nevanlinna as reserves. The Oslo proceedings reveal that Hardy did not accept and was replaced by Lefschetz (**ICM, 1936**, vol. I, p. 48). Further, in his letter to Evans, Birkhoff explained that Lefschetz had agreed to transfer the chairmanship to Evans. This follows from another part of Birkhoff's 1938 letter to Evans that is reproduced in Chapter 7.

[171]See footnote 71 on p. 76.

FIGURE 9.22. Constantin Carathéodory (1873–1950) (Public domain, via Wikimedia Commons)

As one would expect in the context, Carathéodory's Oslo lecture was a survey of the mathematical work of the two Fields Medallists, with a particular focus on the research that had earned them their awards. We have already drawn upon the talk briefly in our account of Ahlfors' work in Section 9.15. In the remainder of this section, we give a parallel account of the work of Douglas. Carathéodory's address will once again have a role to play.

It is all the more important that we give an account of Douglas' work here because, unlike Ahlfors, he did not deliver a lecture in Oslo. Indeed, as far as the congress proceedings are concerned, we have much less information on Douglas than we have on Ahlfors: Douglas did not deliver a sectional lecture either, and appears to have had a rather low-key presence at the congress. He did not even attend the Fields Medal award ceremony, pleading tiredness;[172] his medal was collected on his behalf by Norbert Wiener (**ICM, 1936**, vol. 1, p. 45). Apart from the citation in Carathéodory's Fields Medal lecture (**Carathéodory, 1936/7**), Douglas is largely absent from the Oslo proceedings.

Perhaps surprisingly for such a prizewinning mathematician, and certainly in stark contrast to the case with Ahlfors, biographical material on Douglas is rather limited: the only substantial published biography that we have been able to find is that in the *Dictionary of Scientific Biography* (**Schaumberger, 2008**).[173] The limited detail that we have paints a picture of a rather nervous man, who was perhaps not particularly sociable: this may explain his "tiredness" in Oslo, and the fact that he does not appear to have attended any ICMs thereafter. While Ahlfors went on to become the "grand old man" of the ICMs, occupying a position of honour at the 1986 congress (see Section 9.15 above), there was never any possibility of Douglas attaining such a place, due to both temperament and a comparatively early death. Indeed, Douglas' impact on the mathematical community at large may have been rather limited: the Mathematics Genealogy Project does not list any students for him,[174] while zbMATH (accessed 5 August 2018) indicates that he only ever had one co-author (Philip Franklin, on a single paper of 1933), in contrast to the 23 listed for Ahlfors. Nevertheless, there are some brief accounts of Douglas' mathematical research, all of which focus upon the work that won him the Fields Medal: his solution of Plateau's Problem.[175]

We may take an elementary description of the problem from an expository article written by Douglas himself for the journal *Scripta Mathematica*:

> The problem of Plateau is to prove the existence of a surface of least area bounded by a given closed curve in space called a *contour*. Any surface having this minimum area property is called a *minimal surface*. If the contour is a plane curve, then the problem is trivial, the solution being merely the portion of its own plane limited by the curve; but when the contour is twisted, then to find the minimal surface which it bounds is a problem of a high order of difficulty. (**Douglas, 1938**, p. 159)

Douglas immediately noted the real-world manifestation of the problem:

> Nature, however, can solve this problem very easily. By immersing the given contour in a soap solution and then withdrawing it, we obtain a minimal surface, because the soap film, under its

[172] As a note of caution concerning the reliability of the book (**Reid, 1976**) (cited below in the main text), we mention that (**Reid, 1976**, p. 187) wrongly asserts that Douglas did not collect his Fields Medal because he could not afford the trip to Oslo.

[173] There also exist some fragmentary recollections in (**Rassias, 1992**) by people who knew Douglas, as well as some brief biographical notes, together with a bibliography up to 1936, in the "Biography" section of volume 4, number 1 (1936) of *Scripta Mathematica* (pp. 89–90). It seems that a biography and bibliography that were to have been produced by the US National Academy of Sciences (**Schaumberger, 2008**, p. 174) never appeared.

[174] `https://www.genealogy.math.ndsu.nodak.edu/id.php?id=37358` (accessed 5 August 2018).

[175] In a soundbite accompanying (and perhaps specifically solicited for) the biographical notes in *Scripta Mathematica* (see footnote 173 above), Douglas commented: "I think that most mathematicians would consider that my most significant work has been in Differential Geometry ... and on the Problem of Plateau."

> surface tension, takes the form of least area consistent with its constraints. (*Ibid.*)

Indeed, although the mathematical problem had been formulated in 1760 by Joseph-Louis Lagrange, it was named for Joseph Plateau (1801–1883), the Belgian physicist who had experimented with soap bubbles, ultimately formulating "Plateau's laws" governing their structure.

The problem had arisen for Lagrange in the context of a memoir on the calculus of variations for double integrals, where he formulated the problem of least area, and obtained a partial differential equation that is satisfied by all minimal surfaces. This PDE may be interpreted geometrically as a saddle surface whose principal curvatures are equal in magnitude but opposite in sign at each point: a minimal surface is therefore one of zero mean curvature. As Douglas noted:

> The difficulty lies in showing the existence of a surface obeying this condition and bounded by a given closed curve. (**Douglas, 1938**, p. 160)

During the second half of the nineteenth century, the problem was considered by Riemann, Weierstrass, and Darboux, among others, but they succeeded only in settling special cases. The general situation, for any Jordan curve in any number of dimensions, was resolved by Douglas in a series of abstracts published in the *Bulletin of the American Mathematical Society* in 1927–1930; a full treatment was published in the *Transactions* a year later (**Douglas, 1931**). The problem was approached around the same time via different methods, but with less generality, by René Garnier (1887–1984) and Tibor Radó (1895–1965): for Garnier, the curve in question had to have bounded curvature, or consist of arcs with this property, while for Radó it was necessary for the contour to bound at least one surface of finite area. In this connection, (**Gray and Micallef, 2008**, p. 299) note:

> If priority is assigned on the basis of published papers alone, then Radó was the first to put into print a comprehensible solution of the Plateau problem in anything like generality. Douglas's announcements give the impression that he was occasionally cavalier about what he could achieve. Most of the time he delivered on his claims, but it may not always be appropriate to use the timing of his claims to determine priority issues.

We note that a further contribution to the study of Plateau's problem was made by Richard Courant, and, indeed, presented in a sectional lecture in Oslo (**ICM, 1936**, vol. II, p. 143). (**Reid, 1976**, pp. 180–181) asserts that Courant found Douglas' method "roundabout", presumably making for a slightly critical sectional talk. The remarks of Heegaard to Engel concerning Courant's talk that appear in our Appendix B hint, moreover, of a certain intrigue surrounding the award of Douglas' Fields Medal.[176]

It appears, however, that the Fields Medal committee considered Douglas' work worthy of the prize,[177] perhaps because his name had already become well known even among non-specialists; after surveying other aspects of Douglas' work in his Fields Medal lecture, Carathéodory observed:

---

[176] A more detailed account of Douglas' work, and of the circumstances surrounding his receipt of the Fields Medal, will appear in (**Gray and Micallef, forthcoming**).

[177] Modulo the comments of G. D. Birkhoff at the beginning of this section.

> All these things, interesting as they may be, can only serve as a prelude to the series of great works that Jesse Douglas has dedicated to the Plateau problem, and through which his name is also known to other circles of colleagues.[178]

Carathéodory went on to provide a sketch of Douglas' solution of the problem in which the issue of priority, quite naturally, does not arise. Indeed, Douglas' name is inextricably linked with Plateau's problem, so we conclude this section with a final assessment of this work from (**Gray and Micallef, 2008**, p. 300):

> we conclude that Radó and Douglas share equal credit for solving the Plateau problem for disc-like surfaces spanning a single contour which bounds at least one disc-like surface of finite area. Radó deserves full credit for solving the least area problem. Douglas, on the other hand, was the first to solve the Plateau problem in complete generality, that is, for an arbitrary contour, including ones that bound only surfaces of infinite area. He was also the only one to consider more general types of surface than the disc, to which Radó's attentions were exclusively confined. Radó's method for solving the Plateau problem shifts almost all the difficulty onto problems in conformal mapping, whose solution for higher topological types was certainly not available at the time. By contrast, Douglas's method even helped solve some of these problems in conformal mapping. Thus, Douglas's contributions to the Plateau problem are more major, broader and deeper than those of Radó. Douglas's ideas, as developed later by Courant ... have remained important in the theory of minimal surfaces up to the present.

[178]Translated from the German in (**Carathéodory, 1936/7**, p. 311).

CHAPTER 10

ICMI

The activities of the International Commission on Mathematical Instruction (ICMI) at the Oslo congress, which formed a large part of section VIII (Pedagogy),[1] receive very little attention in the proceedings; as we noted on p. 16, most of the reports that were given in Oslo were subsequently published in the Swiss journal *L'enseignement mathématique.* What we find in the proceedings is largely a record of procedural matters pertaining to ICMI. For example, as part of the opening ceremony of the congress, the president of section VIII and ICMI secretary-general Fehr submitted the following resolution, which was adopted unanimously:

> *The Congress requests the International Commission on the Teaching of Mathematics to continue its work, prosecuting such investigations as shall be determined by the Central Committee.* (**ICM, 1936**, vol. I, p. 48)

As part of his short address in connection with the resolution, Fehr outlined a brief history of ICMI, from its founding at the Rome congress, remarking:

> At a time when international cooperation still faces obstacles in other areas, we are pleased to note that the work of the Commission has been able to continue in an excellent spirit of understanding and collaboration.[2]

In issuing an invitation in Zurich in 1932 for national delegates to present reports to the Oslo congress on the state of mathematics education in their respective countries, ICMI's central committee had sought not to impose too restrictive a questionnaire, but had suggested the following as useful headings under which to report (**ICM, 1936**, vol. II, p. 288):

I. School organisation; new types of schools.
II. Modern trends concerning the purpose of mathematics education; plans of study.
III. Exams; programmes of certification leading to higher studies (baccalaureate, certificate of maturity, etc.).
IV. Teaching methods; the links between different branches of mathematics; the place given to applied mathematics; textbooks.
V. The preparation of mathematics teachers.
VI. Various.

[1]The only section VIII talks that are printed in the proceedings are R. A. Fairthorne's "A method for demonstrating the qualitative properties of differential equations by means of cinematograph films" and Hans Przibram's "Beliebiges Wurzelziehen als Rechnungsart ohne Logarithmen" (**ICM, 1936**, vol. II, pp. 283–287). For Lietzmann's comments on these, see our p. 277.

[2]Translated from the French in (**ICM, 1936**, vol. I, p. 47).

TABLE 10.1. Presenters of national reports to ICMI at the Oslo congress

Germany	W. Lietzmann
Great Britain	E. H. Neville
Austria	G. Kantz, presenting a report prepared by E. Dintzl
Denmark	J. Mollerup
Hungary	J. Jelitai
Japan	M. Kuniyeda
Norway	K. Piene
Poland	S. Straszewicz
Romania	G. Tzitzeica
Switzerland	H. Fehr

Full reports were presented by the national delegates listed in Table 10.1, with further short reports received from P. Zervos (Greece)[3] and from J. Desforge and G. Iliovici (France), both of which were subsequently published as fuller versions alongside the other reports in *L'enseignement mathématique.*[4]

We do not have the space here to give full accounts of each of the reports presented to ICMI, so we confine ourselves instead to a few broad comments. The most prominent report to ICMI that we have already noted in Section 1.3 is of course that of Lietzmann: in particular, its remarks on changes to school mathematics teaching under Nazi rule (p. 84). Elsewhere (namely, in the unpublished report reproduced here in Appendix D), Lietzmann struck a rather pessimistic tone with regard to school mathematics education in Europe, based upon what he had heard in Oslo, but made recommendations as to how Germany might learn from other countries:

> It should be said here that surprisingly many changes in the organisation of higher education have been made in recent years. This is the case in Austria, Hungary, Norway, Poland. It may perhaps be recommended that the very interesting changes, notably in Austria, Hungary and perhaps also in Poland, be studied more closely through visits to schools and by consultation with teaching authorities. The situation of mathematics in secondary schools was repeatedly described as not very auspicious.[5]

In regarding secondary mathematics education as "not very auspicious", Lietzmann may have had Piene's report on Norway in mind — this was rather scathing in its criticism of the endangerment of school mathematics in Norway. Piene (see Figure 10.1) repeated some of his remarks in a newspaper interview during the congress (reproduced here in Appendix A.8), although his conclusion was that Norway was not lagging behind any other countries; Piene also commented on the effects of Nazism on mathematics education (see the comments on p. 125).

[3]We know from Heegaard's letter to Engel following the congress that Zervos was unable to attend: see Appendix B.

[4]The journal also contains (in vol. 37, pp. 205–211) a further report by A. Wahlgren (Sweden) that is not mentioned in the congress proceedings.

[5]See Appendix D, pp. 278, 284.

FIGURE 10.1. Kay Piene (1904–1968) (Author: Ragni Piene. Source: Archives of the Mathematisches Forschungsinstitut Oberwolfach)

The other reports to ICMI followed the above guidelines to a greater or lesser extent. For example, the reports for Austria and France are extremely detailed,[6] and outline the systems of school organisation in the respective countries, with a breakdown of the mathematical topics that were taught at different stages: the Austrian report placed a great emphasis on descriptive geometry, for instance. Other reports, such as those of Denmark and Greece (both presented in French),[7] are considerably shorter, consisting merely of a few pages of general comments. In line with the remarks made by Lietzmann, many of the reports focus on the nature and effects of recent reforms: for example, the Austrian report details a new teaching plan that had been put in place in 1935, while the Swedish contribution deals with reforms made between 1928 and 1933.

Let us conclude this brief outline of ICMI's activities in Oslo by returning to some remarks made in Chapter 1. We noted there[8] in passing Dunnington's positive comments concerning Lietzmann's ICMI report. Indeed, Dunnington's observations with regard to the ICMI session were as follows:

[6] *L'enseignement mathématique* **36** (1937), 237–244, 247–257.

[7] *L'enseignement mathématique* **36** (1937), 244–247, 257–259.

[8] In footnote 2 on p. 3.

> Unfortunately the reports read from the various countries were rather dry and perfunctory, and there was little enthusiasm or discussion. However, the report of Dr. Walter [*sic*] Lietzmann, professor in Göttingen and vice-chairman of the Commission, was exceedingly interesting and enlightening as to mathematics in the schools of present-day Germany. Much praise is also due the Japanese delegates who illustrated their report with actual textbooks used there, and distributed to those present a well printed summary in pamphlet form. (**Dunnington, 1936a**, p. 88)

That Dunnington would describe the reports from other countries generally as "rather dry and perfunctory" only underscores his preference for German matters — Piene's report, for instance, was certainly not dry, at least in its printed form, although some may have considered his criticism of the conditions of school mathematics in Norway as "perfunctory". The only report that could truly be described as "perfunctory", if we are to take its printed version as any indication, was that of the British differential geometer E. H. Neville, which consists of a single paragraph in which Neville noted that he had little to add to a report previously prepared by G. St. L. Carson in 1929, and which he ended with the following words:[9]

> It would not comport with the dignity of the Commission or with the seriousness of the reports to which you have listened that I should offer you a sheet of disconnected notes and emendations, and I prefer only to assure you of your [*sic* for 'our'?] continued adhesion.[10]

[9] *L'enseignement mathématique* **28** (1929), 269–278.

[10] *L'enseignement mathématique* **36** (1937), 237.

CHAPTER 11

Conclusions Regarding the Mathematics of the Congress

We must now begin somehow to assess the overall contribution of the Oslo ICM to the international mathematical community. We would certainly expect there to be some impact, given the sheer number of superlatives that we encountered in Chapter 9. These were rarely attached directly to accounts of the lectures delivered in Oslo, but they were certainly used in connection with topics closely related to the lectures, as well as in connection with the speakers themselves. The fact that most of the plenary speakers remain well known today even outside their specialisms indicates that the organisers of the Oslo congress did indeed manage to showcase at least some of the major mathematical talents of the decade. In counterpoint to this, however, we note the absence of other "stars", such as Turing, Kolmogorov, Tarski, and Weil, to name but a few. The congress certainly made an impression on the 25-year-old Garrett Birkhoff, who commented many years later:

> At the International Mathematical Congress in Oslo ... I was dazzled by the depth and erudition of the invited speakers, and the panorama of fascinating areas of research that their talks opened up. I was permitted to present three short talks (Marcel Riesz gave four!),[1] and there seemed to be an adequate supply of listeners for all talks presented. Paul Erdös gave one talk, and he must have been the only speaker who did not wear a necktie! (**Birkhoff, 1989**, p. 46)

One would hope that the inspiration of younger scholars in this way would have been one of the goals of the organisers, even if it was not explicitly stated.[2]

Having surveyed all the plenary lectures, a striking feature that emerges is the interconnectedness of several of them, either by immediate topic or by the further interests of the speakers. Thus, for example, Hasse's local-global principle was a key feature of the lectures given by both Siegel (Section 9.5) and Hasse himself (Section 9.12). The prior work of Mordell was heavily cited by Hecke (Section 9.8); moreover, Mordell's plenary lecture (Section 9.14) was cited by van der Corput in his (Section 9.16). We have noted also the links between the works of Banach (Section 9.17), Fréchet (Section 9.18), and Wiener (Section 9.19), if not between their Oslo lectures. In many cases, these interconnections were probably established via educational links: both Siegel and Hecke had studied with Landau, for instance,

[1] On Birkhoff's lectures, see the comments on p. 216; for Riesz's, see (**ICM, 1936**, vol. II, pp. 16, 36–37, 44–45, 62–63).

[2] Birkhoff had previously accompanied his father to the 1928 ICM, before attending the 1932 meeting, apparently in his own right, although he was then only 21 years old. He went on to attend at least the 1950 (as a co-organiser), 1954, and 1974 ICMs.

with Hasse then going on to study with Hecke and Landau in turn — the Göttingen connection is certainly relevant here. Moreover, the majority of the connections came about through shared interests in number theory, so the preponderance of this subject in Oslo that we noted in Section 8.4 is also relevant. Landau's proposed lecture would have fit in here too (Section 9.21.3). At any rate, it is tempting to look back at Dunnington's notion of "Oslo under the integral sign" (particularly in light of the analytical flavour of the common themes) and suggest that this image of integration is more appropriate to the mathematical side of the congress than to the political.

This is not to say, of course, that all of the speakers were interconnected. The talks on applied mathematics were all quite separate, for instance. As we have seen, Ore was the lone representative, at least among the plenary speakers, of the structural approach to algebra (Section 9.20). Neugebauer was certainly the only historian of mathematics (Section 9.9). Had it taken place, Khinchin's would have been the only lecture on probability (Section 9.21.4); likewise Severi's for algebraic geometry (Section 9.21.2). Thus, as we saw throughout Chapter 9, the congress featured a very broad range of mathematical topics, but the organisers may perhaps be accused of having permitted — or, more fairly, been forced by the difficulties over the speakers to allow — the formation of an analytic/number-theoretic clique within the broader congress. Recalling from the end of Section 8.1 that the organisers of the 1950 ICM went on to allow specialised meetings within that congress, we may speculate that, had such a route been taken in 1936, a special session on analytic number theory would probably have been the result. What we see in the Oslo congress is perhaps the organisers struggling to find a place for an ICM in a mathematical world in which specialised meetings were increasingly becoming the norm, a problem that was subsequently solved by their 1950 counterparts.

Throughout Chapter 9, we drew upon a range of sources in order to provide an assessment of the plenary lectures that were delivered in Oslo. But there is one major source that we have so far used only briefly for mathematical purposes, namely Lietzmann's congress report, as reproduced in Appendix D, and based upon partial reports that he had solicited — this is clearly documented in the case of Hasse.[3] Picking up from the discussion in Section 3.2, we make some comments here on Lietzmann's survey of the mathematical lectures at the congress, to see whether it tallies with the remarks made above.

As we can see from his account of the plenary lectures,[4] the visibility of number theory is something that Lietzmann remarked upon also, in his case as a point upon which to assert German dominance (in reference to the plenary lectures of Hasse, Hecke, and Siegel). Following his comment (quoted here on p. 82) about Germany's "leading position", he went on:

> Due to the wealth of new and deep results and the brilliant presentation, the three German talks were among the most impressive experiences of the congress, an impression that was confirmed to us by numerous foreign scholars.[5]

[3]See footnote 78 on p. 77.

[4]See Appendix D, pp. 273–274, 280–281.

[5]See Appendix D, pp. 273, 280.

The disproportionate representation of number theory was not a source of criticism for Lietzmann, but seems to have been regarded by him as quite natural, perhaps because of recent developments:

> Because of its beauty and the difficulty of its problems, number theory remains at the centre of general interest ...[6]

As far as the applied mathematics talks of Størmer, Oseen, and Bjerknes were concerned, Lietzmann was rather lukewarm, noting that they "met with much less response than those from areas of pure mathematics".[7] Lietzmann also included Birkhoff's quantum-mechanical lecture with the applied talks, but made no remark on the "forgettability" that we noted in Section 9.13 — perhaps because Lietzmann was not familiar enough with the topic, or could not solicit appropriate help, to be able to critique it fully.

Lietzmann's commentary on the remaining plenary lectures was quite neutral. He mentioned only in passing, for example, Cartan's "much-noted" talk at the unveiling of the bust of Lie, and recorded that the speaker "gave full credit to German contributions" to the study of Lie groups.[8] The lectures of Ahlfors and Banach were noted in one line each, although the fact that the latter spoke in German was mentioned specifically (see our p. 204). The emphasis on German contributions continued in Lietzmann's one sentence about Ore's lecture, "which is substantially based on German research"[9] — one wonders whether the reference here was to the prior work of Dedekind, or whether Lietzmann was making a reference to the contributions of Emmy Noether, which could be interpreted as either hypocritical or subversive. Referring to the lectures of Banach, Cartan, Siegel, and Hecke, Lietzmann asserted that the tendency, exhibited in each of them

> towards unification and convergence of different mathematical domains may be considered a chief characteristic of contemporary developments in mathematics.[10]

What Lietzmann also did, but that we have not, was to provide a survey of some of the sectional lectures. Naturally, in contrast to the general picture that we would have tried to present if space had permitted, he reported only on those sectional lectures that he had been able to attend. Again, the emphasis here was on German contributions, including even brief comments on expatriates such as Weyl. Lietzmann's assessments of sections in which there were no speakers from the official German delegation (for example, section (I) "Arithmetic and Algebra", and section (V) "Mathematical Physics, Astronomy and Geophysics") were, however, somewhat perfunctory. His account of the analysis section is rather fuller, running to seven paragraphs, and including remarks, for instance, on Behnke's "memorable and transparent lecture" on complex function theory, in connection with which he remarked upon the "friendly" collaborations between French and German mathematicians (see our p. 82). The geometry section, on the other hand, featured, in Lietzmann's view, "a good deal of incomprehensible exotica",[11]

---

[6] *Ibid.*
[7] See Appendix D, pp. 274, 280.
[8] See Appendix D, pp. 274, 281.
[9] *Ibid.*
[10] *Ibid.*
[11] See Appendix D, pp. 276, 282.

> while synthetic and algebraic geometry came up short due to the absence of the Italians ...[12]

Section IV was recorded by Lietzmann as having some significant German contributions.[13] Overall, Lietzmann's report accords German mathematicians a central role within the congress, and indeed, this may not simply be the skewed presentation of a politically minded author: German mathematics had already been decimated by the actions of the Nazis, but perhaps in the Oslo congress we see one of the last times that what remained of it was showcased prior to the Second World War.

An assessment of the mathematics of the Oslo congress must also take account of the fact that this was the first ICM at which the Fields Medals were awarded. We have already established the very different temperaments and careers of the two Oslo medallists (Section 9.22): while the medal was for Ahlfors a great boost to an already well-established career, in which he continued to be prominent at the international level for the rest of his life, it does not appear to have done the same for Douglas, who did not remain particularly visible thereafter, and worked for the later part of his career in relative obscurity. The same can probably be said of later winners of the prize: the case of Douglas shows that the award of a Fields Medal may simply be a passing honour that does not necessarily lead to sustained mathematical stardom. The award of the prize to Douglas, however, and indeed to Ahlfors, who was based at Harvard at the time, may have had a greater significance for American mathematics more generally. As (**Parshall and Rowe, 1994**), among others, have described, the American mathematical research community emerged at the end of the nineteenth century, and in the early decades of the twentieth gradually came out from beneath the shadow of European mathematics as more centres of research were established in North America. In the view of one of Douglas' biographers, the Fields Medals of 1936 had a role to play in this:

> Symbolically, the 1936 Award to Jesse Douglas and Lars Ahlfors signals the emergence of North America as a new force in the world of mathematics. (**Tropp, 1992**, p. 48)

Americans were certainly present at the congress in greater numbers than ever before.[14] And, indeed, the emergence of the United States as a leading mathematical nation was acknowledged at the congress: see Chapter 7, as well as the remarks in the newspaper interview with Harald Bohr that is reproduced in Appendix A.1.

We have observed several times the almost total lack of structural algebra at the Oslo congress. Nevertheless, this subject went on to thrive during the twentieth century. The same is true of algebraic geometry, although this had a longer history prior to the congress. We have noted also that probability theory enjoyed a very active section at the congress, even in the absence of a plenary lecture in that subject. We might therefore begin to question the value of such meetings for the large-scale organisation and development of mathematics. A branch of mathematics is not going to die away just because it has not been promoted by the organisers of an ICM, nor are the choices of the organisers, who are subject to their own biases, necessarily representative of the major strands of contemporary research. Trends in mathematical research are rather more "organic", shaped by the interests of all practicing mathematicians, rather than by those of an elite few (although the

---

[12] *Ibid.* Lietzmann's similar remark about the Russian absence appeared on our p. 82.

[13] As well as that of Gumbel, Lietzmann's comments concerning which we saw in Section 3.2.

[14] See Table 2.6 on p. 60, and indeed footnote 9 on p. 249.

influence of these may be somewhat stronger). Thus, although we have focused here on the plenary lectures, the general conclusion that we might reach from our analysis is that these do not give the full picture concerning the trends in the mathematics of the mid-1930s — to pick up on the comments made at the end of Section 9.13, such lectures perhaps tell us more about people than about mathematics. Although we do not have the space here to do so for the Oslo ICM, we believe that a survey of the contributed talks of the sections of a congress, taken as a whole, will provide a better picture of the topics that mathematicians consider as important at that moment in time. Moreover, the variety of topics, and consequent dynamism, that is a feature of such sectional meetings will in particular drive precisely that collaboration among mathematicians that the congress as a whole is intended to promote.

By way of bringing this chapter to a close, we link back to the remarks made in Section 8.1 concerning mathematical communication. It is well acknowledged that much of the value of conferences of all sizes, and indeed in all disciplines, lies not in the specific details of the presentations made there, but in the opportunities for personal interaction that such occasions afford. It is therefore very pleasing to note that these sentiments are indeed echoed, by its participants, in sources contemporary with the Oslo congress. For example, we find the following in an interview that A. A. Fraenkel gave to the Norwegian newspaper *Tidens Tegn* (see Appendix A.4):

> I don't think the greatest value of such a congress lies in its scientific lectures; mathematics is so broad that even for a professional it can prove difficult really to comprehend that which is outside his own special area. Instead, the importance of a meeting such as this one rests on the personal gatherings with colleagues whose papers you otherwise can only read or with whom you have perhaps only corresponded.

Moreover, in his report on the congress for *Neue Zürcher Zeitung* (Appendix C), Rudolf Fueter made special mention of the social events of the congress in precisely this connection:

> They allowed the mathematician the so uncommonly necessary "chatter", without which each Congress would miss its purpose. No other science needs the spoken word and personal contact as much as mathematics. In the shortest possible time, the mathematical conversation can give an account of mutual ideas and investigations, whereas to attempt the same with the printed word is troublesome and time-consuming.

The absence of various parties from the Oslo congress would therefore have left a gap in the personal interactions that are so necessary for academic life. Although we have done our best here to describe the impact of these absences upon the mathematics of the congress, there is nevertheless an immeasurable aspect arising from the lack of the face-to-face contacts that would otherwise have taken place.

Appendices

APPENDIX A

Norwegian Newspaper Items Relating to the Congress

We present here a selection of English translations of Norwegian newspaper items that relate to the Oslo congress. This is by no means an exhaustive collection of such pieces (see the references on p. 300, for example); we have selected those articles that shed interesting contemporary light on the congress beyond what can be found in the proceedings. See our commentaries above in Chapter 6.

FIGURE A.1. *Aftenposten*, 13 July 1936, p. 1 (from left to right: M. Fujiwara, I. Shimomura (both Japan), R. Wavre (Switzerland), M. Ahmed (Egypt), G. Kantz (Austria), R. Poiter (Spain), P. J. Daniell (England))

A.1. *Aftenposten*, 13 July 1936, no. 346, pp. 1+5 (signed by S.-L.)

We constantly encounter Abel (Abel møter vi stadig).

Princeton is the centre of the mathematical world but for the moment it is Oslo.

Professor Harald Bohr on the importance of the congress of mathematicians.

[Photograph with caption "Professor Harald Bohr photographed during the interview this morning"]

It is not to be expected that a man who can easily handle billions while at the lectern and who moves with ease in the infinite spheres of mathematics should be able to remember the number on his luggage-slip. It was with the utmost difficulty that the great Danish mathematician Harald [5][1] Bohr found his suitcases in the Sunday-school crowd at the train station this morning. Eventually they were loaded into a car and the famous scientist walked to his hotel together with a Danish colleague, Professor Steffensen. On their way up Karl Johan,[2] the two gentlemen discussed the architectural style of the 80s. Bohr would prefer to have a hammer and chisel all the ornaments off the facades. He prefers the modern clean faces. Steffensen viewed the confusion of styles with more forbearance. When hideous things are placed next to each other, a curious and beautiful effect can emerge as is witnessed by the characteristic patterns on the houses near Egertorget.[3] From that point to the castle, they were both fascinated by Karl Johans gate, but as Steffensen claimed: Oslo should not get too pretty; it would distract the eye from the surroundings which are the most important thing.

— The importance of a congress such as this one is closely connected to the general character of our discipline, Professor Bohr explains after he has finally been seated in an armchair in one of our overfull hotels. — That cannot be emphasised enough in times such as these when people from all sides — how should I express myself to avoid becoming too aggressive — when from all sides the general humanity is forgotten in favour of the narrow-minded nationalistic. Different scientific schools will always exist in different countries, founded around great and creative scientists. And the importance of having them meet at congresses lies not least in the mutual fertilisation that results from bringing different points of view into contact.

— From the crucible of mathematics, not a single drop of truth comes out that has not been put in there, an opponent[4] of the increasingly dominant use of mathematics in economics is reported to have said. What does that mean?

— Mathematics has advanced so far in its abstractions that there is a certain sense in which it can be said to be empty. But that does not at all mean that no applications for it can be found. For instance, all of natural science is dependent on mathematics. The freedom made possible by the vast abstractions is a condition for applying the concepts of mathematics in numerous mutually distant areas. The characteristic feature of mathematics is its absolute freedom from contradictions. You can add consequence to consequence in infinite number. The great English mathematician Hardy,[5] who regrettably will not attend the congress, has therefore in jest claimed that mathematics is the only science to deal with realities.

Harald Bohr is the brother of the physicist and Nobel laureate Niels Bohr. Therefore, it comes naturally to ask him whether he, by his mathematical studies, has been able to assist his brother in his famous studies of the structure of the universe.

— Unfortunately not. My brother's work is highly dependent on mathematics, but not of the kind which I pursue, which is pure mathematics.

---

[1]Numbers in square brackets indicate a page change in the original clipping.

[2]The main street in Oslo.

[3]Egertorget was the site of the first high-rise building in Oslo, the eight-storey Horngården built in 1929–1930 (`https://no.wikipedia.org/wiki/Egertorget` (Oslo), accessed 31 May 2019).

[4]Among opponents of the use of mathematics in economics at the time were Austrian economists such as Ludwig von Mises and Friedrich Hayek. The particular indirect quote could not be traced.

[5]Hardy was a good friend of Bohr's and we have here another confirmation of his non-participation in Oslo (see p. 43).

— What is the difference between that and other forms of mathematics?

— It is the part that has not yet been applied. By the way, numerous instances have occurred when mathematical studies of a purely speculative nature, pursued for no other reason than to erect a conceptual theory, have been of crucial importance in science. Also in the vast domain of practical technology, it is of the utmost importance that mathematicians plan and direct work. With their deeper insight they possess the prerequisites to present things more easily than others.

— Which continent is the leading one in the field of mathematics?

— America has had the good fortune of having a series of eminent German emigrants come to its universities. Just as Göttingen used to be the centre of mathematics, it is now to Princeton in the USA that the younger generation turn for their mathematical pilgrimage. There are also excellent schools of mathematicians in Russia and Japan. Here in Norway, proud traditions from Abel's days are still present. When together with Norwegian colleagues, you feel the influence on them of having been immersed in the world created by that genius. The great name of our times, Professor Hilbert, once said that a mathematician's importance can be measured by the time it takes before posterity has understood and solved the problems he posed. Concerning Abel, I can confidently say that we all constantly encounter his problems in our daily work. Also Lie, who lived in Leipzig, was one of the really great Norwegian mathematicians. In one of the plenary lectures of the congress, the famous French participant Cartan will account for the continued importance of Lie's ideas.

— To completely change the subject, Professor, how is football these days? Harald Bohr used to be one of the key players in Copenhagen, quick in the attack and a true cannon in front of the goal. The quadrature of the circle, which no mathematician has ever been able to solve on the blackboard, Bohr has repeatedly demonstrated on the green carpet of the stadium [Idretsparken = Idrætsparken, the traditional national stadium in Copenhagen].

— I only play football occasionally, and I don't attend internationals very often. I have become somewhat separated from the sport of my youth.

A.2. *Aftenposten*, 14 July 1936, no. 348, p. 3

"Mathematics and logic are one and the same".

There is no mystery therein.

The philosopher Dr. Frank takes part in the congress of mathematicians

[Photograph with caption "Dr. Frank"; see Figure A.2]

At the large congress of scientists in Copenhagen a fortnight ago, the Czechoslovak Dr. Frank delivered one of the plenary lectures. He spoke in continuation of Niels Bohr's opening lecture on the importance of quantum theory for philosophy. Yesterday, Dr. Frank arrived in Oslo to take part in the congress of mathematicians. He is equally well-known in the worlds of philosophy and numbers, highly respected by all scientists for his treatise on the law of causality and its limits and a number of other works.

— It has always been the subject of reflection, he explains as an answer to our question about the relations between philosophy and mathematics, how it is possible that Nature manifests itself everywhere through simple mathematical laws. It has occasioned mystical speculations. Research in logic has shown that we cannot really

„Matematikk og logikk er det samme".

Der er ingen mystikk i den.

Filosofen dr. Frank deltar i matematiker-kongressen

gen mellem filosofien og matematikken, hvordan det kan være at naturen overalt åpenbarer sig i enkle matematiske lover. Det har gitt anledning til mystiske spekulasjoner. Logisk forskning har vist, at vi i virkeligheten ikke kan si noget som helst sikkert om selve verden, bare om to beskrivelser av den er like eller forskjellige. Vi kan på intet punkt motsi erfaringen. Matematikkens store betydning ligger deri at den kan bringe naturbeskrivelser som vanskelig eller slett ikke lar sig prøve ad eksperimentell vei, i en form som gjør det lett å prøve dem på tavlen, f. eks. setningen at summen av vinklene i en trekant er to rette. Det kan ved målinger lett kontrolleres. Der er ingen mystisk sammenheng mellem naturen og matematikken, matematikken er bare et verktøi for naturbeskrivelser. De kan også gjøres med ord, men blir da så meget omstendeligere. Hvis jeg for

Dr. Frank.

FIGURE A.2. *Aftenposten*, 14 July 1936, p. 3 (extract from interview with Frank)

say anything certain about the world itself, but only whether two descriptions of it are the same or not. We can never contradict experience. The great importance of mathematics lies in its ability to bring descriptions of Nature that are difficult or outright impossible to test through experiments into a form that allows for them to be tested on the blackboard; for instance the proposition that the sum of the angles in a triangle is two right ones. That can be checked [kontrolleres] easily through measurements. There is no mystical connection between Nature and mathematics; mathematics is just a tool for describing Nature. Such descriptions can also be made in words but then they become so much more cumbersome. If, for instance, I were to describe in ordinary language where the planets will be tomorrow in relation to each other, it would take some time. Mathematics can describe it in a single equation. Mathematics simply shows us, if one were to express it quite popularly, whether two verbal claims have the same meaning; and from this perspective, it [mathematics] coincides with logic. Logic is itself also nothing but a theory about the reformulation of linguistic laws. I believe that many people will find this view too unsentimental but it protects against any mystical interpretation of science.

Tirsdag 14. juli 1936. Arbeiderbladet

700 berømte matematikere

Lyn-intervjuer om universet som utvider sig, om hvad kileskriften fra Ur kan fortelle og mange andre ting

Amerika det førende land på matematikens område -- takket være nazismen.

FIGURE A.3. *Arbeiderbladet*, 14 July 1936, p. 5 (extract and picture of Koht)

A.3. *Arbeiderbladet*, 14 July 1936, no. 161, p. 5

700 famous mathematicians[6]

Lightning-interviews about the expanding universe, lessons from cuneiform writings from Ur and many other issues

America [*USA*] *leads the world in the field of mathematics — thanks to Nazism.*

[Photograph with caption "The foreign minister Koht speaks at the opening of the congress" (see Figure A.3). Also: collage (Figure A.4) of photographs with caption "Some participants in the mathematical congress, pictures taken at the reception in the Aula. To the right the Norwegian-American Øystein Ore, in the middle top one of the six Japanese at the Congress, Schur, and below the Catholic priest and scientist Le Maitre. To the left the famous German Otto Neugebauer, who has become homeless in Nazi Germany but found a new opportunity to work in Denmark."]

Yesterday evening, the Aula was filled with speech in all tongues. Several hundred mathematicians from all parts of the world together with wives and even children were gathered to be welcomed in several languages by Professor Heegård [*sic*], enjoy tea and small round biscuits and — most importantly — greet each other, catch up on old acquaintances and be introduced to new people. For five days, the mathematicians will conduct their congress here in Oslo; time is short but must be exploited to the full.

Professor *Størmer*, who is the chairman of numerous committees in connection with this congress and who manages everything, is very pleased with the state of affairs. It has been hard work for a long time — this is, of course, the first time mathematicians have met for an international congress in Oslo[7] — but the participation has been extraordinary and the outcome will be great.

[6]This newspaper article is also reproduced as a figure in (**Siegmund-Schultze, 2016b**, p. 82).

[7]This means "international" in the sense of "beyond Scandinavia", and potentially including the whole of the rest of the world.

FIGURE A.4. *Arbeiderbladet*, 14 July 1936, p. 5 (Neugebauer, Schur (?), Lemaître, and Ore)

— The greatest importance of this congress consists in its aid to international collaboration, says the professor. It contributes to levelling the differences between nations and races;[8] and that is immensely important. But also from a purely scientific viewpoint, great results can come out of a congress such as this one.

We mingle about in the hall among all the foreign characters in order to find subjects for interviews. It is a truly rich field to choose from; small smiling Japanese, gesturing Frenchmen, people with long hair down to their shoulders and others completely bald, fair Nordic types and dark-skinned Indians, women scientists with glasses and chic young ladies.

In all the commotion, we catch a glimpse of an acquaintance from school, *Øystein Ore*, who has become a great mathematician with the title of professor in America.

Øystein Ore is still a young man in his mid-30s but he has already accomplished being professor for nine years and his title is "chairman of the head of department" [*sic*] at Yale University. That means that he directs all the mathematicians there, a total of 15, and has an excellent position — also in financial terms. A rich benefactor has donated a lot of money to Yale University and the library alone has been given 8 million dollars, Ore explains. Every year, the university there has 800–1000 mathematicians [mathematics students] whereas the maximum here at home is 150.

— You must then have a very large auditorium for your lectures?

— Me? No, I only lecture to the advanced students — for those who seek to become PhDs — and normally I have no more than 8 in the audience. I occupy myself mostly with scientific studies.

---

[8]On the ease of use of the term "race" (twice in this appendix), see the commentary in Chapter 6 (in particular, p. 121).

— You will be giving a lecture here?

— Yes, I will be presenting a survey of mathematics [see (**Ore, 1936/7**) and our Section 9.20]. In mathematics, as in other sciences, we have had to specialise. It is like having to map a country: you start in a corner but eventually you can join all the pieces together to see the mountain ranges and the valleys in the greater picture. Similarly, I shall now try to present a survey of the different branches of mathematics that seem to be separate but in reality belong together.

— Are there many Americans here?

— Very many,[9] and with the influx of German mathematicians we have received during the past years, it is fair to say that America has become the leading country in the field of mathematics.

— Presumably it is nice to see old familiar places again?

— I come here every year, and I have a house here as well. My family and I live in Stavern [south of Larvik] and had it not been for this congress I would have been [dressed up as] Haile Selassie for the party there on Saturday.[10] Now, somebody else will have to step in for me!

Øystein Ore says this almost with regret in his voice. But such are the terms: you will have to sacrifice much on the altar of science!

*

— Now, you must talk to *Le Maitre*! says Ore and drags us across to a Frenchman[11] in black robes and the characteristic round dog collar. Le Maitre is a catholic priest and famous scientist. His theories about the ongoing expansion of the Universe have received fierce attention in the world of science.

— Any news, Le Maitre? asks Ore.

— No, everything is still the same, answers the smiling priest. The Universe is still expanding!

*

Neugebauer is a small fair German who — like so many others in Germany — has become homeless; but he has been received with open arms in Copenhagen, where he works together with two such illustrious scientists as the Bohr brothers. Now he speaks Danish like a native.

Neugebauer was the first to begin investigating the history of pre-Greek mathematics[12] and he has shown that Greek mathematical science builds on Babylonian foundations. The Greeks introduced the modern perspectives into mathematics, but the foundation belongs to the Babylonians. This is a discovery that Neugebauer has made through studies of cuneiform texts on clay tablets taken from excavations

[9]Indeed, this was the first congress in Europe at which American mathematicians were represented in notably large numbers: the proceedings record 86 full participants (with 56 accompanying persons), making them the largest national group (**ICM, 1936**, vol. I, p. 39). In contrast, the only other delegations that numbered over 25 were the Norwegians (59 + 25), the British (48 + 18), the Germans (35 + 2), the French (28 + 13), and the Swedes (26 + 10). Although, as we know from Chapter 4 and elsewhere, the attendance figures given in the proceedings are not entirely to be trusted, they nevertheless give an indication of the force of numbers with which American mathematicians attended the congress. In Zurich four years earlier, for example, they had formed the fourth largest delegation, with 66 members (**ICM, 1932**, vol. I, p. 36).

[10]See the remark in Chapter 6 (p. 121) on the popularity of Haile Selassie at this time.

[11]In fact, Lemaître was Belgian.

[12]This is of course a journalistic exaggeration.

in Mesopotamia and now to be found in museums in Berlin, London, Paris, etc. The science of mathematics thus has a close connection to the science of language.

— One wonders why the Babylonians were so disposed to mathematics?

— Presumably through the specific blend of different types of people down there.[13] The Babylonians built on the Sumerian culture which the English have studied very extensively through excavations at Ur that you have probably heard about.

Neugebauer is not the only German emigrant mathematician attending the congress. All the way from Tomsk in Siberia comes Professor *Noether*, who three years ago lost his position as professor in Germany and was offered a small pension from which he could neither live nor die, and certainly insufficient to give his children an education. Then he was invited to the university in Tomsk where he has now learned Russian and lectured for two years. He has a three-year contract but has settled in so well up there that he intends to stay for longer.[14]

— How many students attend the university in Tomsk?

— 3–4000, and among them a few hundred mathematicians. There are many different races: Russians who for a large part descend from prisoners in Siberia, Tartars, Mongols, etc.

— The students receive financial support from the state?

— Yes, the vast majority. Previously, all were given a state stipend, but now parents who have means have to pay at least part of the expenses of their children's education.

— How is the student material?

— Exactly the same as elsewhere. Some are very gifted, others quite ordinary. The young people who are now taught in universities were born during war and famine and physically this is for some of them clearly visible.

— Are you a communist?

— No, not really a communist, certainly not what the Germans mean by a communist. The role of communism in the Soviet Union is to build up the new state and in that I would like to take part. I am leftist, but my political viewpoints play no role over there.

*

Vallarta is a small, thin and dark Mexican who is deeply interested in our own Professor Størmer's theory of the Northern Lights, in the trajectories of electrons, cosmic radiation and the rotation of the entire solar system. At the astrophysical department in Massachusetts,[15] he works on a calculating machine for Størmer's theory and he is very interested in the calculator which we shall have here at home in the new astrophysical institute at Blindern [see Section 9.2]. It will become the world's biggest, he tells us, making us very impressed with ourselves.

*

[13]See the commentary on this paragraph in Chapter 6.

[14]In general, the Soviets began to throw out foreigners at that time in an atmosphere of suspicion and hysteria. For Noether things ended much worse, namely in false accusations and finally in execution — see the commentary on pp. 122, 125–126.

[15]Meaning the Massachusetts Institute of Technology (MIT), near Boston.

Finally, we talk to a small Japanese, *Schur*, who with a broad smile assures us that it is a pleasure for him to meet so many famous men and women. There are six Japanese attending the congress — two of them study in Germany but four have come the long way from Japan directly to the congress. Among them is Fujiwara, who has become a famous and esteemed name in the international world of mathematics.

The opening of the congress in the Aula this morning.

The solemn opening of the congress of mathematicians took place as early as nine o'clock this morning. After the King had assumed his seat far down in the hall, Professor Størmer ascended the podium and delivered a rather long welcome, mainly in French, and including all the polite and nice words that are integral parts of this language in particular. Professor *Schieldrop* then introduced the Government's representative, Professor and Foreign Minister *Halfdan* [*sic*] *Koht*, who delivered a rather jesting speech in English and assured everybody of the joy and pride we Norwegians felt at having so many prominent scientists as our guests. Nothing expands the human intellect as mathematics does, said the minister, and he congratulated the mathematicians on their efforts. In his speech he emphasised the close connection between mathematics and music, making the audience regret even more the complete lack of any musical part to the opening festivities. A small philharmonic orchestra would have done well.

Professor *Fueter* from Switzerland proposed that Professor Størmer be elected president and the congress agreed. Størmer expressed his gratitude for the honour and in turn proposed a number of vice presidents who were also elected with acclamation.

A white-haired Frenchman, *Eli* [*sic*] *Cartan*, then announced the Fields gold medals which were awarded this year for the first time; and from an adjacent podium Professor *Carathéodory* from Munich read out the details of the two winners. One was the Finn Lars *Ahlfors*, who received the medal in person, and the other was the American Jesse *Douglas*, who had come all the long way to receive the honour but was so tired after the journey that he was represented by his colleague Norbert *Wiener*.[16] The audience congratulated the gold medal winners with enthusiastic acclamation.

Finally, the congress agreed to send a telegram to Crown Prince Olav who is the honorary president of the congress but was prevented from being present at the opening.

The opening ceremony lasted less than an hour and after a break of half an hour the lectures began for the mathematicians whereas wives and children left in a number of buses to look at the city and its surroundings.

[16]See Section 9.22.

TIDENS TEGN tirsdag 14. juli

500 STUDENTER ved Jerusalems universitet.

Professor Fraenkel fotografert av Tidens Tegn igår.

Studentenes ydelser gjennemgående høiere enn i Europa.

En samtale med den palestinske regjerings representant ved matematikerkongressen, professor Fraenkel.

iske institutt omfattende de forskjellige jødiske disipliner, som vår litteratur, historie, talmud, arkeologi og historisk topologi over Palestina. En annen avdeling beskjeftiger sig med mere orientalske emner som arabisk, arabisk- og islamitisk kunst, assyriologi og egyptologi, for å nevne noen av disiplinene. Alle de nevnte ting skulde jo ligge godt til rette for et universitet med vår beliggenhet og lette adgang til kontakt med de gamle kulturcentra. Men dessverre er politikken foreløbig kommet imellem.

FIGURE A.5. *Tidens Tegn*, 14 July 1936, p. 2 (extract from interview with Fraenkel)

A.4. *Tidens Tegn*, 14 July 1936, no. 160, p. 2

500 Students at the university of Jerusalem

[Photograph with caption "Professor Fraenkel photographed yesterday by Tidens Tegn"; see Figure A.5]

The efforts of the students are generally higher than in Europe.

A conversation with the representative of the Palestinian government at the congress of mathematicians, Professor Fraenkel.

Among the 700 participants of the congress of mathematicians, one also finds a representative of the world's youngest university, Professor Adolf Fraenkel from Jerusalem who is also the representative of the Palestinian government.

We meet the professor at his hotel and he immediately informs us that he will not touch upon anything that has to do with politics, and he says: I will say outright that I don't think the greatest value of such a congress lies in its scientific lectures; mathematics is so broad that even for a professional it can prove difficult really to comprehend that which is outside his own special area. Instead, the importance of a meeting such as this one rests on the personal gatherings with colleagues whose papers you otherwise can only read or with whom you have perhaps only corresponded. But for me and several others, the congress holds special importance because old colleagues who, owing to political conditions, are scattered all over the world, have the opportunity to see each other again. Before I went to Jerusalem, I was professor at two German universities, but I had already left Germany in 1929, so my departure from that country was entirely voluntary.[17]

— How is the scientific life at the university in Jerusalem?

— The university is only 11 years old and not yet completed. But it is, of course, an incredible extension from the two or three first institutes to the present. As of yet, we only have two faculties, one in the humanities and one in the natural sciences. Central to the faculty of humanities is the Judaic institute comprising the different Jewish disciplines such as our literature, history, Talmud, archaeology, and historical topology of Palestine. Another section is devoted to more oriental topics such as Arabic, Arabic and Islamic art, Assyriology and Egyptology, to name just some of the disciplines. All the things mentioned should be obvious for a university with our position and easy access to the old centres of culture. But unfortunately, this far politics have interfered.[18]

— The faculty of natural sciences?

— This includes among other things mathematics with four professors and docents, biological disciplines such as botany, zoology, bacteriology, etc. Tropical diseases are particularly relevant but we do not yet have a medical faculty and will not need one in the near future. At the moment, there is a surplus of physicians in Palestine.

The last-mentioned faculty altogether employs nearly 100 professors, assistants and lecturers, whereas the faculty of humanities has approximately 30–40 teachers including 10–12 professors.

— The number of students?

— Approximately 500 who in part originate in Palestine but also to a high degree from Eastern and Central Europe and partly from America, where a considerable portion have received their school education. If I were to compare the student material with that which I personally know from Europe, their interest is not less and their efforts are generally higher. This is perhaps even more remarkable as most of the students are very poor and can only make ends meet by working hard in the vacations to attend the university.

— Is there a demand for so many scientifically educated people?

— Some of them will be able to find work within the country as teachers, in industry and in the efforts against diseases such as malaria. But a considerable part will probably — for as long as large parts of the world are closed to us — go into

[17]Fraenkel's move from Germany to Jerusalem was in fact a little more complicated than he indicated here: see (**Fraenkel, 1967/2016**). See also the remarks in Chapter 6.

[18]Fraenkel apparently wanted to say here that the "other section" in the humanities faculty, the more secular one, had still to struggle in its development.

purely practical work and, we must then hope, continue their scientific interests as a hobby [*liebhaberi* for German *Liebhaberei*].

A.5. *Aftenposten*, 15 July 1936, no. 349 (morning issue), p. 3

Russians did not come to the congress of mathematicians.

It cannot be denied that one is disappointed that not a single one of the many Russian mathematicians who were registered for the congress has come. The list of participants showed a number of well-known names: Oguiewetzski, Pontragin, Kantorowitsch, Kolmogoroff, Bilimowitsch, Depoutatov[19] and others, but, as stated, they have not come.

Aftenposte

ris: Innenbys Oslo, onsdag 15. juli 1936 Abonn.pris:

Matematikerkongressen — hyldest til Sophus Lie.

De russiske matematikere som uteblev.

Var man redd for at de ikke vilde vende tilbake til paradiset?

FIGURE A.6. *Aftenposten*, 15 July 1936, p. 1 (article on Russian absence and picture of Cartan lecturing on Lie)

[19]On whom, see (**Demidov and Lëvshin, 1999/2016**, p. 225).

A.6. *Aftenposten*, 15 July 1936, no. 350 (regular issue), p. 1

[See Figure A.6.]

The Russian mathematicians who did not come.

Were [authorities] afraid they would not return to the paradise?

Earlier we published a note that the Russian mathematicians who were registered for the congress have not come. Inquiry at the Russian legation results in the statement that they have no knowledge about the case. The congress's office has from some of the registered [Russian mathematicians] short notes that they would not come, from others no information at all. The dean of the faculty of sciences, Professor Heegaard, could not give any information either. As he reports, a year ago one Russian scientist was allowed to travel abroad on his word of honour that he would return. However, this he did not do. I do not know, the professor says, but maybe this has some connection to the problem.

1500 tyske videnskapsmenn jaget av nazistene

Og nuller satt i deres sted. En av emigrantene på matematikerkongressen, professor Gumbel, forteller om kulturbarbariet i Tyskland.

Blandt de tyske matematikere som deltar i matematikerkongressen, er det hele ti emigranter. Likesom tyrkerne i sin tid jaget de lærde ut av Konstantinopel, har nazi spredt de tyske videnskapsmenn utover verden, og de ti på kongressen er kommet fra mange land, fra Sibir som fra Amerika, fra Danmark, England og Frankrike. Vi har tidligere hatt intervjuer med

Professor Gumbel, den første videnskapsmann som blev fordrevet av nazi.

FIGURE A.7. *Arbeiderbladet*, 16 July 1936, p. 13 (extract from interview with Gumbel)

A.7. *Arbeiderbladet*, 16 July 1936, no. 163, p. 13

1500 German scientists hounded by the Nazis.

And nobodies put in their place. One of the emigrants present for the congress of mathematicians, Professor Gumbel, talks about the cultural barbarism in Germany.

[Photograph with caption "Professor Gumbel: the first scientist to have been expelled by the Nazis"; see Figure A.7]

Among the German mathematicians participating in the congress of mathematicians, no fewer than ten are emigrants. Just as the Turks in their period drove out the scholars from Constantinople, the Nazis have scattered German scientists all over the world, and the ten [emigrants] at the congress have come from many different countries — from Siberia as well as America, from Denmark, England, and France. We have previously brought interviews with the emigrant Professor Neugebauer who works with Bohr in Copenhagen and with Noether who is a professor in Tomsk. Today, we have spoken with Professor Gumbel.

Professor Gumbel had served at the university in Heidelberg for ten years when the Nazis came into power, and he was the first scientist to be deprived of his position. Back then, the Nazis still proceeded formally, and a trial was staged before Professor Gumbel was ousted.[20] But the trial was of course only a sham, and soon thereafter even law and justice were flouted. The professors were simply told they had been ousted and others had taken their place.

— What did your crime consist of, we ask Professor Gumbel.

— I had spoken in a closed socialist student organisation and the Nazis produced two witnesses who could repeat sentences I was reported to have said. Eyewitnesses who could testify that I had not uttered those sentences were never examined.

— And what were you accused of having said?

— That is too complicated to explain in this interview, but among the things I really did say was that "our recollection of the war is a memory of hunger and misery, of malnutrition and turnips".[21] The Nazis found that I had insulted Germany's "military honour".

— How many professors have lost their positions in Nazi Germany?

— In total 1500. Just in Heidelberg, 56 professors out of 200 were removed.

— Were the 56 [professors] Jews and communists?

— No, not even half of them were Jews. They were not all radicals, either. Many of them were even conservatives, some were republicans, centrists, etc. But everybody who is not a Nazi is "politically unreliable" in the eyes of the Nazis and that is sufficient reason to oust them. Apart from the 56 who were ousted, 44 others have also ceased [in their positions]; some have died, others have retired, and the rest have opted out. So, in total, one hundred [the number is missing in the newspaper] new professors have been appointed in Heidelberg just over the last three years.

— And all of these 100 new professors are Nazis?

— Yes, they are veteran P. G., Parteigenossen (party members). All of them are nobodies instead of scientific researchers.

— What has happened to the 1500 ousted professors?

— 500 have left the country, the rest are starving in Germany. Some have a small pension and are shivering at the thought of losing it; others have not been

[20]This is either a misrepresentation or a misunderstanding: Gumbel was expelled for political reasons, but before the Nazis came to power (see p. 86). Indeed, this is one of several statements in this interview that seem unlikely and would bear further investigation: for example, the assertions that so few of those professors expelled from Heidelberg were Jewish, and that their replacements were Nazi Party members who lacked any scientific credentials. We are unable to decide which errors are Gumbel's and which are due to the journalist.

[21]See also the reference to "turnip Gumbel" in Appendix E.

given a pension. It is completely random who has been given a pension[22] and who has not been found worthy of it and very many are suffering terribly. One of the professors in Heidelberg was picked up directly from his lecture by students who took him to a concentration camp. He was a centrist and therefore "politically unreliable".

— What is the status of the mathematical science in Germany at the present?

— Göttingen, which used to be the centre for mathematical research, is now completely destroyed. Instead, America has become the leading nation in the field of mathematics — to some extent thanks to the German emigrants who have settled there. A journal such as "Deutsche Mathematik" has been published with a slogan by Adolf Hitler; apparently he has now also become a great mathematician. And the leading article in that journal was written by a student who reprimanded the professors for not having done their duty. And that duty was to prove the supremacy of German mathematics over the mathematics of the East, the South, and the Jews!

The victory of Nazism in Germany is a tragedy for the country itself, because the Nazis are not only enemies of the Jews but of all of European culture. A general discontent dominates the German universities at the present. And the students cannot just go abroad to study, because they cannot afford it.

— Where have you yourself settled?

— First, I gave guest lectures in Paris but now I am employed at the university in Lyon where I am doing fine. My colleagues are proper (*helt korrekte*) and the students are much more studious than in Germany.

— Will you be lecturing at the congress?

— Yes, tomorrow afternoon.[23] I am a mathematical statistician and I have tested my field, probability theory, on Swedish observations from the last 50 years. Of a hundred whales, one must necessarily be the biggest, and of any number of people, one must be the last to die.

I can calculate which age can be assumed to be the highest and this age is connected to the number of people who have died. I can tell you the age of the oldest man, but not his name.

— How old is the oldest person?

— In Sweden, he is 106 years old and in a hundred years, the oldest man will be 107 years old. The truth of my prophecy can thus only be proved in 100 years.

— I assumed that the duration of life would increase much faster?

— Yes, the average age for people to die has increased much faster because of the improvements in hygiene over the past years. But the maximum age increases much more slowly.

— It is strange, Professor Gumbel remarks finally, that here in the North where people are really Nordic, here they are not the least bit Nazi. Here, you have what the Nazis call a "Fluchwürdige Zeit der Schmach".[24] You have a *labour government.*[25]

[22]The related Nazi law made a distinction between dismissal due to at least one grandparent adhering to the Jewish religion (§3, not taking into account the religious denomination of the incriminated person him- or herself) and political "unreliability" (§4). The latter was handled arbitrarily and implied a 25% cut in pension. Nazi authorities often used it in order to save money and therefore dismissed Jewish scholars under §4.

[23]See (**ICM, 1936**, vol. II, pp. 200–205).

[24]This means roughly "A damnable time of humiliation" and was often applied by Nazis to the Weimar Republic and the Treaty of Versailles.

[25]This had been installed in 1935 — see the remarks on p. 49.

Norway is so sympathetic to me — and I say this not as a cordial conclusion of an interview but because I truly mean it — because it has nurtured a man like Nansen. I wish the League of Nations would follow his example and come to the aid of the German emigrants like Nansen did in the past for the Russian ones![26]

A.8. *Arbeiderbladet*, 18 July 1936, no. 165, p. 5

Teaching of mathematics must be reformed

International commission meets in Oslo.

Interview with the government's delegate, Lector Piene

[Photograph with caption "Kai Piene (on the left) in conversation with the Egyptian professor Bey Boulad"; see Figure A.8]

In connection with the congress of mathematicians, the International Commission for Mathematics Instruction has met.[27] The commission was set up 30 years ago and aims to further mathematics teaching in all countries. Professor Heegaard has been a member of the commission from the start and has devoted a great effort to the subject; he now holds an honorary position as the commission's vice president. Next to him, Norway has another delegate to the commission, Lector (*lektor*) Kay Piene, from the school in Vestheim. Lector Piene was appointed a few days ago but he is well informed in the field; thus, he is a member of the committee for mathematics education in this country and has studied teaching abroad.

We meet Lector Piene just before the first lecture at the congress of mathematicians this morning and ask him to tell a little about this international commission and its work.

— How many countries are listed as members of the commission?

— That is not easy to say. A series of countries want to take part but it costs a certain annual sum and there can be difficulties with the payment. Most European states are on board, as are America, Japan and a few countries outside Europe. Every member can name one delegate — a few countries have more delegates — and within every country a sub-commission is set up dealing with conditions within that country. The international commission meets every fourth year — in conjunction with the congress of mathematicians — and there the delegates present reports from their respective countries. Then the delegates are given the task of undertaking new studies on some question, and then they meet at the next congress with the results. In this way, you get a basis for comparison between the mathematics education in different countries and you stimulate the teachers.

— What question had been given to the commission ahead of this congress?

— To investigate the current tendencies within mathematics education in the different countries. The most interesting report was the German one. It becomes evident that even such a neutral subject as mathematics is influenced by politics, and you can indirectly trace Nazism also in the teaching of mathematics these days. Schools are now no longer supposed to be for the individual pupil but primarily serve the country. Germany now deliberately aims to strengthen applied mathematics in schools whereas the rest of us emphasise theory more. The fatherland must be

[26] The so-called Nansen passports were created in 1922 and mainly designed for Russian refugees by the Nansen International Office for Refugees within the League of Nations. The office was closed in 1938; passports were thereafter issued by a new agency, the Office of the High Commissioner for Refugees under the Protection of the League of Nations in London.

[27] See also Chapter 10.

FIGURE A.8. *Arbeiderbladet*, 18 July 1936, p. 5 (Piene and Egyptian Farid Boulad)

furthered through education, that is the purpose of the entire German teaching plan now.

The Japanese report was also very interesting. The teaching of mathematics in Japan has progressed remarkably in recent times.

— And the Norwegian report?

— From here, we mainly presented wishes and demands. There has been and will remain to be a heated discussion about mathematics teaching in the humanistic secondary education [gymnasium] in this country, and I pointed this problem out to the international commission. It appeared that other countries were dealing with the same problem. For the science-line [*reallinjen*], the situation is clear, but there is a widespread opinion that the mathematics teaching in the humanistic lines must be restricted or removed altogether or rearranged. It is clear that very many are struggling hard with mathematics without benefitting from it.

— What do you think about this problem?

— I think that the teaching should be rearranged, in particular the textbooks should be reworked so as to capture the students' interest better than at present.

The demand for a rearrangement is also bound to pass when we get the new legislation concerning schools. Among the teachers of philology there is a majority in favour of cancelling mathematics and strengthening the teaching of other subjects such as biology, and even among some teachers of mathematics there is an opinion to restrict the subject. Thus, there is a suggestion of cancelling written mathematics in favour of written French.

— But the teaching in primary schools?

— The primary schools are not affected by this question, because there only arithmetic is taught. But I mentioned principal Ribsskog's "Handbook in arithmetic teaching in primary schools" ["Håndbok i regneundervisningen i folkeskolen"], and this work attracted a lot of attention.

— How does the teaching of mathematics in Norway compare with other countries?

— The prerequisites here are lower than in, e.g., Central Europe. On the other hand, they are higher than in Sweden and roughly the same as in Denmark. Another matter is the effectiveness of the method of teaching, which it is very difficult to say anything about. On the one hand there is the curriculum and on the other hand is how to teach this curriculum. But I have studied mathematics teaching in Austria, Switzerland and Denmark, and I believe I can say that we are not behind any of these countries.

— Which tasks have the delegates of the international commission been given in advance of the next congress?

— For the next congress, we will take up the problem with the organisation of exams, and those reports will surely be met with great anticipation.

A.9. *Tidens Tegn*, 18 July 1936, no. 164, p. 1

A rail journey of six days to speak for 15 minutes in Oslo

[Photograph with caption "Professor Noether photographed yesterday by Tidens Tegn"; see Figure A.9]

Many of the participants of the congress of mathematicians come from the most remote corners of the world but surely only one has had to spend six days on the railway to come here and deliver a lecture of fifteen minutes, namely Professor Fritz Noether from the university in Tomsk in Siberia.

— Well, fortunately I did not complete the whole journey in one go, says the professor. I stopped in Moscow on the way, but six days would probably be the shortest time one could plausibly set aside for the journey from Tomsk to Oslo, and that even presupposes that all the connections work out which is not always to be expected on these vast distances. It happens that the trains are full and you cannot be allowed onboard and such calamities can occur even on the first leg of the journey. From Tomsk one has to go on a minor track to the Trans-Siberian Railway which originates in the Far East, Vladivostok and such places, and on these trains all seats could already have been taken by other passengers.

SEKS DØGNS JERNBANEREISE FOR AA SNAKKE 15 MINUTTER I OSLO.

Mange av matematikerkonferansens deltagere skriver sig jo fra jordens fjerneste egne, men det er sikkert bare én som har måttet tilbringe hele seks døgn på jernbanen for å kunne komme og avlevere et femten minutters foredrag, professor Fritz Noether fra universitetet i Tomsk i Sibir.

— Nå, heldigvis har jeg ikke gjort reisen i ett, sier professoren. Jeg har nemlig vært innom Moskva, men seks døgn er nok den korteste tid man må regne på turen Tomsk—Oslo, og det forutsetter enda at forbindelsene klaffer, hvad man ikke alltid kan vente på disse veldige distanser. Det hender nemlig at togene er fulle, så man ikke kommer med, og den kalamitet kan man komme ut for allerede ved reisens begynnelse. Fra Tomsk må man jo med en sidebane ta til den transsibiriske bane som kommer borte fra det fjerne Østen, Vladivostok og de steder, og på disse tog kan alle plasser på forhånd være optatt.

— Det må være en merkelig fornemmelse så å si å være forvist fra et europeisk sentrum som et tysk universitet og til Sibir.

— Som forvisning er den meget interessant. Forholdene i Tomsk er jo vidt forskjellige fra de man kjenner fra Europa. Selve byen er nokså liten, skal ha vært på 120,000 innbyggere, men nu er det jo lenge siden befolkningen blev tellet, så ingen vet med sikkerhet hvor mange den rummer. Landskapet omkring har det typiske tungsinn, som til en viss grad preger befolkningen og gir den dens efter vestlandske begreper langsomme tempo. Er man først innenfor bygrensen vil man finne en utpreget russisk by. Universitetet skriver sig fra 1888 og har et godt bibliotek på et par hundre tusen bind, men er ellers kanskje lit enkelt utstyrt. Av studenter er det 3—4000 stykker, og gjennemgående er de preget av den samme planbevissthet som utmerker hele universitetsarbeidet. For selv om russerne ikke forsømmer den rene forskning, så er det jo om å gjøre å skaffe folk til å føre opbygningsarbeidet. Forskning med et praktisk tilsnitt er derfor velsett, men på grunn av de veldige avstander må en utnyttelse ta lenger tid enn i Central-Europa.

— Hvad blir det av alle studentene?

— Efter avsluttede studier fortsetter mange ved universitetet som assistenter. Senere går de som lærere til de høiere skoler, ikke minst de tekniske. Det er i det hele stort behov for videnskapelig utdannet lærerpersonale. Før revolusjonen lå jo undervisningen nokså nede i Russland, og nu skal på kortest mulig tid imøtekommes et behov som ved kunst er holdt latent så altfor lenge. Det er veldige muligheter som skal utnyttes i dette land som nærmest er en hel verdensdel.

Professor Noether fotografert av Tidens Tegn igår.

— Faller professorens arbeid lett?

— Jeg kan ikke klage, og livet der borte er jo meget interessant. Selve naturen er også en oplevelse med de uendelige stepper i syd, de store elver og det kuperte landskap i nord. Klimaet er et utpreget innlandsklima med korte, men varme sommere og meget kolde vintre, slik at årets middeltemperatur ligger nesten en grad under frysepunktet. Elven som passerer byen er derfor isfri bare halvparten av året. Nedbøren er liten, bare 384 millimeter om året, og kommer fortrinsvis om sommeren. Det snefallet som regel er lite. Men vinteren begynner og slutter ofte med voldsomme snestormer, og på den strengeste vintertid kan man være forberedt på optil 50 graders kulde, en veldig forskjell fra de 30 grader man kan opleve på den varmeste sommertid.

i behold!

Dette så mere enn mystisk ut. Men lensmannen, som ikke kunde frigjøre sig for en mistanke om at 19-åringen fra Oslo allikevel var synderen, kom på den utmerkede idé, at skrinet var lagt der i urén for å hentes senere. Han satte vakt på stedet, og man fikk noen spennende ventetimer. Klokken 3 natt til igår så vakten en kar i skjorteermene, med jakken på armen, komme nedover helen fra Turtagrø. De gjemte sig og snart efter stod mannen der, hvor skrinet hadde vært gjemt under en bordstump og ganske riktig — det var Oslo-gutten. Han løftet på bordstumpen, fikk et sjokk da han så skrinet var borte, og satte sig i dyp eftertanke. Så blev han tatt, men erklærte da kort og godt, at han ingen forklaring vilde gi. Det viste sig at han efter tyveriet altså først hadde tatt veien over Elvesæter til Juvasshytten, derfra med lensmannen til Røisheim og så derfra tilbake og til Geiranger med båt inn Sognefjorden og så endelig over Turtagrø frem til „åstedet". Nu sitter han i hjelpefengslet i Sell.

Slik endte den historie, og så er det forhåpentlig atter ren luft i Jotunheimen.

FIGURE A.9. *Tidens Tegn*, 18 July 1936, p. 1 (article on Fritz Noether)

— It must be a strange experience to be — so to say — expelled from a European centre like a German university to Siberia.

— Considered as a place of exile, it is quite interesting. The conditions in Tomsk are obviously vastly different from those known in Europe. The city itself is quite small; it is said to house 120,000 inhabitants, but it has been a long time since the last census so nobody knows with any certainty how many people live there. The landscape surrounding the city displays the melancholy that is to a certain extent typical of the people and gives to them a slow tempo of life compared to western standards. Once inside the city limits you will find a typical Russian city. The university was founded in 1888 and has a well-equipped library of a few hundred thousand volumes but is otherwise somewhat simply equipped. There are 3–4000 students who in general are primed to the same attention to plans [*planbevissthet*] which distinguishes the whole effort of the university. Although Russians do not neglect pure research, it is of greatest importance to provide people to carry out material production. Research with a practical aim is therefore looked upon with favour but due to the vast distances, any application [of research] will take longer than in Central Europe.

— What happens to the students?

— After completing their studies, many of them continue at the university as assistants. Subsequently, they will move on to jobs as teachers in higher education [*de høiere skoler*], not least the technical ones. Throughout, there is an immense need for scientifically trained teachers. Prior to the Revolution, education was rather poor in Russia; and now, in the shortest possible time span, a need has to be met which has been artificially suppressed for far too long. Great possibilities which need to be exploited lie in this land which is almost an entire continent.

— How are the working conditions for a professor?

— I cannot complain, and life over there is quite interesting. Nature itself is also an experience with the infinite steppes in the South, the great rivers and the hilly terrain in the North. The climate is typical continental with short, but warm summers and very cold winters such that the yearly average temperature is almost one degree below freezing. The river which passes the city is therefore only free of ice half of the year. There is low precipitation, only about 384mm a year which falls primarily during the summer since snowfall is normally limited. But winter often begins and ends with severe snowstorms and during the deepest winter one could be prepared for as much as 50 degrees below zero, an immense difference from the 30 degrees one can experience during the hottest summertime.

APPENDIX B

Letter from Heegaard to Engel, 24 July 1936

[Friedrich Engel Nachlass, Universitätsbibliothek Giessen, NE150305]

English translation follows on pp. 264–265

NB. The orthography strictly follows that of the Danish-born Heegaard, who did not yet write the new letter ø for ö, which has become customary in Denmark and Norway since the 1920s. Heegaard also wrote 'Stórmer', whereas the latter signed his name 'Störmer' until his old age, although in print he appeared as 'Størmer'.

POUL HEEGAARD
PROFESSOR DR. PHILOS — 24/7 1936

~~VOLDS TERRASSE 3 JAR~~ — [Giessen 27.7.1936]

Holmenbukta p.å.
pr. Hvalstad — [beantw. (answered) 31. 7. 1936]

Kjære venn

Takk for Deres like mottagne brev. Nu er kongressen jo overstått og efter alles udsagn var den vellykket. For mig går allikevel en linje av mishag gjennem den, fordi De ikke kom. Hvad der egentlig lå bak motstanden, er mig temmelig tåket. De officielle forklaringer er jo nokk så svævende. Blańdt andet var det dog en like så stor formfejl, at jeg, som ikke er medlem av Norsk Matematisk Forening's styre ohne weiter [*sic* for German "ohne weiteres"] forela den for Kongressen. Telegrammet opstod simpelt hen pa den måte, at jeg ved Lie-buste-avslöringen sluttet med å foreslå at et telegram til Dem blev avsendt, hvilket vedtoges med sterk applaus. Journalistene hadde fått avtrykk av min tale, så de var ikke til stede. Derfor hadde avisene ikke noe om dette. Deres svartelegram blev oplest på sluttmötet.

Grækerne hadde sendt en innbydelse til, at kongressen skulde holdes næste gang i Aten, men Zervos og Hazidakis [*sic* for "Hatzidakis"] hadde ikke fått penge til å rejse. Et par unge greske matematikere gjemte sig engstelig bort, men siste dag sendte Stórmer — som hadde fått lyst til å komme til Aten og ikke önskede å fremme Amerikanernes tilbud — bud efter Papaiouannou [*sic* for "Papaioannou"]. Da denne blev spurgt om tilbuddet var alvorlig ment slog han sig for brystet og deklamerte vel og længe. Så hvisket jeg et lite ord i öret på Lefschetz og et öjeblik efter var P. omgitt av amerikanere, som rått og udelikat spurgte om Grekenland hadde penger "Vi har allerede 15000$ å begynne med". Så böiet P. sit hode og erklærte at selvfölgelig böiet Grækenland — omenn med dyp smerte — sig for det rike Amerikas önske. Altså blir Kongressen 1940 i Amerika. 1939 skal vi ha topologkongress [*sic*] i Warschau. Fieldsmedaljen blev [several words crossed out, not legible] gitt til Ahlforss [*sic* for "Ahlfors"] og Douglas. Jeg hadde fått meddelelse fra sekretariatet sålydende: "Max 15m". Jeg blev derfor meget forbauset, da Courant sa, at det var lovet ham 25m. Dette viste sig ved undersökelse av korrespondancen å være riktig. Jeg foreslog nu "triumviratet", at han kunde få den ved Khintchines uteblivelse frigjorte tid til et "stort" foredrag, men dette kunde ikke tilstås. Så slog jeg to ledigblevne foredrag i min seksjon sammen, og Courant holdt et overordentlig interessant foredrag (Plateaus problem) på 3/4 time. Da jeg mærket han var gått videre enn Douglas opstod der en anelse hos mig om, hvorfor foredraget skulde holdes nede. Hvor er der meget intrigevesen ved en kongress. Min kone sa bare "vanity fair". "Unionens" likvidasjonskomité holdt stormfulde möter. Julia vilde, at exekutivkomiteen skulde nedsætte en ny og större Komite, men heldigvis holdt vi på kun at kunne motta forslag fra hele likvidationskomiteen. I siste möte skal Julia ha vært meget uhöflig mot Veblen, men denne tilretteviste ham på en raffineret måte, idet han fingerte ikke å kunne godt nokk fransk og derfor bad Caratheodory om å si Julia, at han ikke kunne ta imot den slags. — Besvim ikke når De ser Tidens Tegn. Jeg er blitt tvunget til å være norsk president ved den 3. internasjonale Kongress for Nationalister og skal åpne mötet i Nobelinstituttet imorgen. Takk for de to separater, dem jeg ennu ikke har fått læst. Jeg håper banken hurtig sender de 400 RM. Jeg ga straks ordre. Hvorledes går det med Deres Kone? Hotellet blev vist avbestilt. Bedste hilsener fra hjem til hjem, Deres hengivne

Poul Heegaard

* * *

Dear friend

Thank you for your letter which I have just received. The congress is now over and everybody has said it was successful. To me, however, a line of discomfort runs through it, because you did not come. What was actually behind the opposition[1] is shrouded in mystery to me. The official explanations are rather vague. And it was also a big formal mistake that I, who am not a member of the board for the Norwegian Mathematical Society, should present it[2] without hesitation to the congress. The telegram simply came into being because at the unveiling of the bust of Lie I ended by suggesting that a telegram be sent to you, which was received with much applause. The journalists had been given a print of my speech, so they were not present.[3] Thus, the papers did not print anything about this. Your telegram in response was read at the final session.

The Greeks had sent an invitation for the next congress to be held in Athens, but Zervos and Hatzidakis had not obtained funding to travel.[4] A couple of young Greek mathematicians hid away nervously, but on the last day Størmer — who had gotten an appetite for going to Athens and did not want to promote the offer of the Americans — called upon Papaioannou. When asked whether the invitation was

---

[1] This probably refers to the opposition of the organising committee to Engel's giving a longer talk on Lie, as discussed in Chapter 3.

[2] It is not quite clear what; probably the telegram that was sent to Engel.

[3] A few journalists must have been present, because there were pictures in the newspapers.

[4] Apparently to Oslo.

meant seriously, he struck his chest and spoke well and at length. Then I whispered a few words in the ear of Lefschetz and within moments, P. was surrounded by Americans who aggressively and indelicately asked whether Greece had the money: "We already have $15,000 to start with". Then P. bent his head and declared that of course Greece would — although with great regret — bend to the will of the rich America. Thus, the 1940 congress will be in America. In 1939 we shall have the congress of topologists in Warsaw. The Fields Medal was given to Ahlfors and Douglas. I had received a notice from the secretariat: "Max 15 minutes". I was therefore very surprised when Courant said that he had been promised 25 minutes. Upon investigation of the correspondence this proved to be right. I then proposed to the "triumvirate" that he could be given the time left by Khinchin's absence for a "great" lecture but this could not be allowed. Then I combined two vacant lecture slots in my section, and Courant gave an extremely interesting lecture (Plateau's problem) for 45 minutes. When I noticed that he had gone further than Douglas, a hunch began to form in me as to why the lecture needed to be suppressed. There is so much intrigue going on at a congress. My wife only said: "vanity fair". The liquidation committee of the "Union" held stormy meetings. Julia wanted that the executive committee should appoint a new and larger committee, but luckily we insisted on only receiving suggestions from the entire liquidation committee. In the last meeting, Julia was allegedly very unkind towards Veblen, who rebuked him in a refined way by pretending not to understand French well enough and therefore asked Carathéodory to tell Julia that we could not accept such. — Do not faint when you see *Tidens Tegn*. I have been forced into becoming the Norwegian president at the 3rd International Congress for Nationalists and I shall open the meeting in the Nobel Institute tomorrow. Thank you for the two offprints, which I have not read yet. I hope the bank will quickly transfer the 400 RM. I gave the order right away. How is your wife? The hotel was cancelled, I believe. Best wishes from home to home, yours truly,

Poul Heegaard

APPENDIX C

Congress Report from *Neue Zürcher Zeitung*, 9 August 1936

[*Neue Zürcher Zeitung*, Sonntag 9. August 1936, Zweite Sonntagsausgabe, No. 1363, Blatt 5; the report is unsigned, but internal evidence indicates that the author was Rudolf Fueter (1880–1950)]

The International Mathematical Congress in Oslo
(13–18 July 1936)

When, on 12 September 1932, the International Mathematician Congress [*sic*] in Zurich decided to hold its next meeting in Oslo, many Swiss mathematicians may have made the decision to travel, if possible, to the beautiful Norwegian capital in 1936. Thus on the evening of 13 July some twenty Swiss and their ladies arrived at the Festival Hall of Oslo University, a respectable number in light of the smallness of our country. The Norwegian State University had invited all congress participants to their lecture hall that evening. From the walls, the incomparably beautiful frescoes of Edward Munch looked down on us, second to none in their modern impact. They embody the severe beauty of Norwegian nature and show the life-giving power of the Nordic sun. But one could scarcely give oneself up to their enjoyment, but had to postpone a more detailed penetration into Munch's masterpieces until later sessions, at which one occasionally turned furtively away from the lectures towards the walls. For now the attendant mathematicians of all countries had to be welcomed, and old friends seen, who had not been spoken to since Zurich. Only the Italians, who did not participate because of the sanctions, and the Russians with their foreign exchange difficulties,[1] unfortunately were missing.

The first session of the congress on the morning of 14 July took place in the same hall. It began punctually at 9 o'clock with the appearance of the king, whose presence proved the esteem that mathematics enjoys in Norway. No wonder! In the nineteenth century one of the greatest mathematicians, the brilliant Henrick Niels Abel [*sic*], and later the notable Sophus Lie, had been Norwegians. Even today, the Oslo mathematical school blossoms, leading via Sylow to Guldberg, Heegard, Störmer and Ore. The first of the latter [namely Guldberg], who in Zurich in 1932 had extended the invitation to Oslo, had unfortunately died last winter; but he was thought of in gratitude.

Herr Störmer warmly greeted the congress participants in the name of Norwegian colleagues, whereupon the Norwegian Minister of Education [*sic*],[2] Halvdan Koht, emphasised the prestige of mathematics and praised its unifying role

[1] This clearly ignores the deeper political reasons.

[2] Koht was in fact foreign minister.

in scientific solidarity. At the request of the author, Herr Störmer was elected congress president by acclamation, whereupon the congress quickly got underway. Pólya (Zurich) was elected as one of the vice presidents. The award of the first golden Fields medals brought something new. Prof. Fields from Toronto (Canada), president of the International Mathematical Congress held there in 1924, had bequeathed a foundation to the Zurich congress. According to this, each successive congress should present golden medals to two young mathematicians who have made the most important scientific achievements during the four years preceding the congress. Prof. Fields had not been able to represent his foundation himself in Zurich because he had died shortly beforehand. The Zurich congress welcomed the foundation with great pleasure, and selected a five-member commission,[3] which submitted its proposals to the congress in Oslo. The first to receive the award was the Finn L. Ahlfors (Helsingfors), who is no stranger in Zurich. Prof. Carathéodory emphasised at the presentation that Ahlfors, as a pupil of Nevanlinna, had, by his geometrical-topological methods, promoted modern function theory in a very decisive way. In addition to him, the American J. Douglas (Cambridge, Mass.) received the medal for his solution of the Plateau Problem.

At 10 o'clock the main lectures began with Herr Störmer, who gave an extract from his outstanding world-renowned work on electron trajectories and the applications to the Northern Lights. The extremely compelling slideshow gave us an inkling of the tremendous work that lies in these important researches. At 11 o'clock the writer had the honour of talking about his latest research on function theory in four dimensions. On all the following days, further lectures were held from 9 to 12 o'clock in all fields of mathematical research; those disciplines that have made significant progress in recent years were favoured in particular. It would take too long to name all the topics and all the speakers in detail.

In the afternoons, the sectional meetings took place; these were held in seven different sections and around 250 communications were delivered. Swiss mathematicians occupied a prominent place here. We refer only to the lectures of Burckhardt, Gut, Pólya, Speiser, Wavre, as well as the historical remarks of Spieß (Basel) and the educational ones of Fehr (Geneva). Various sessions were also presided over by Swiss: Fehr (Geneva), Gut (Zurich), Saxer (Zurich), Speiser (Zurich), Wavre (Geneva).

Our presentation would be incomplete if it stayed only with the scientific events. The other events were a highlight thanks to the great kindness of the Norwegian colleagues and their ladies and to the excellent organisation. This was because they allowed the mathematician the so uncommonly necessary "chatter", without which each Congress would miss its purpose. No other science needs the spoken word and personal contact as much as mathematics. In the shortest possible time, the mathematical conversation can give an account of mutual ideas and investigations, whereas to attempt the same with the printed word is troublesome and time-consuming.

The first event was tea at the Royal Palace, where the King and Queen took part in a charming way, and many congress participants had a casual conversation (14 July). On the next day there followed a glittering banquet, at which following the mayor's speech an American, a German, and a Frenchman expressed admiration and thanks to Norway. A special pleasure on 16 July was the trip on the luxury

[3]See Section 9.22.

FIGURE C.1. The boat trip on the Oslo Fjord (foreground: Carathéordory on the left, Wiener on the right; middle-ground: Veblen towards the left, Fréchet in the centre; background: Weyl on the right) (Source: (**Pólya, 1987**, p. 94))

steamboat of the Norwegian America-Line [Figure C.1]. The Crown Prince and Crown Princess took part in the dinner served on it. The wonderful sunset at 9:15pm will remain unforgettable for every participant. Here, too, their Royal Highnesses conversed casually with the mathematicians.

APPENDIX D

Lietzmann's Unpublished Report on the Congress

[REM 2905, fols. 384–398]

English translation follows on pp. 278–284

Bericht über die deutsche Delegation auf dem Internationalen Mathematiker-Kongress in Oslo 13.–18.7.1936, erstattet vom Führer der Delegation Prof.Dr.W.Lietzmann – Göttingen

Durch Vermittlung der Kongresszentrale war für die Delegation vom Landesfinanzamt statt der für eine vollständige Beteiligung aller der in der Verfügung des Herrn Reichswissenschaftsministers vom 20.Juni 1936 genannten Delegierten angeforderten ca. Rm. 5000.—Devisen nur Rm.3000.—zur Verfügung gestellt worden. Dadurch, dass der Höchstsatz für die einzelnen Teilnehmer auf Rm.185.— herabgesetzt wurde — damit war in der Tat wegen der zahlreichen Einladungen, wenn man nicht in einem der ersten Hotels wohnte, auszukommen — und dass eine ganze Anzahl Herren nicht diesen vollen Satz beanspruchten, war es möglich, nachdem einzelne Herren im letzten Augenblick aus besonderen Gründen zurückgetreten waren, mit folgenden Mitgliedern in Oslo aufzutreten: Behnke, Blaschke, Caratheodory, Geppert, Grunsky, Haenzel, Hamel, Hasse, Hecke, Koebe, Lietzmann, Peschl, Riebesell, E.Schmidt, Seifert, Siegel, Sperner, Ullrich, Vogel, Köthe, Timpe, Reinhardt. Der Umstand, dass die endgültige Festlegung der Summe erst im letzten Augenblick erfolgte und dass die Auszahlung der Devisen nur z.T. in fremden Noten, zum anderen Teil durch Akkreditiv erst in Oslo durch den Delegationsführer erfolgen konnte, stellte eine ausserordentliche Belastung dar.

Dem Führer der Delegation waren ausserdem vom Herrn Minister von deutschen Teilnehmern des Kongresses benannt: Trellfal [*sic* for Threlfall]–Dresden und Rosenthal–Heidelberg. Anschluss an die Delegation erbat ferner Dr.Böhm–Berlin. Von deutschen Mathematikern, die nicht der Delegation angehörten, wurden beobachtet: Prof.Dr.Dehn–Frankfurt/Main, Dr.E.Reissner–Berlin und Dr.Heisig–Leipzig [385] vom Verlag B.G.Teubner–Leipzig. Von "Emigranten" waren, soweit mir bekannt geworden, auf dem Kongress anwesend die Professoren Courant–New-York, Weyl–Princeton, Neugebauer–Kopenhagen, Noether–Tomsk, Fränkel–Jerusalem, schliesslich auch der berüchtigte Herr Gumbel–Lyon. Irgendwelche Störungen oder unliebsame Erörterungen sind von Seiten dieser Herren nicht erfolgt.

Für den Führer der Delegation war sehr wertvoll eine sofortige Besprechung in der Deutschen Gesandtschaft, die namentlich für die Beurteilung der Zeitungen, die in Norwegen gern mit Interviews an Ausländer herantreten, von Wichtigkeit war. Die Mitglieder der Delegation wurden sofort dementsprechend unterrichtet. Auch die Form der geplanten Kranzniederlegung am Denkmal Abels und am Grabe von

Sophus Lie wurde besprochen. Der Gesandte lud mit seiner Familie die deutsche Delegation und einige norwegische Herren an einem Nachmittag zum Tee ein.

Den Kranz am Denkmal von Abel legte der Führer der Delegation zusammen mit dem Vorsitzenden der Deutschen Mathematikervereinigung nieder. Dabei war Polizei anwesend und es war während der ganzen Kongresszeit ein Schutzmann in der Nähe des Denkmals postiert. Der Kranz wurde in einer Tageszeitung im Bild wiedergegeben. Bei der Kranzniederlegung am Grabe von Sophus Lie waren die Mitglieder der Familie Lie zugegen — zwei Töchter sind in Deutschland verheiratet — und erhoben mit uns drei Vertretern der Delegation die Hand zum deutschen Gruss.

Der Führer der Delegation suchte auch den Landesleiter der NSDAP, Pg.Kern, auf und machte am Tage nach dem Kongress mit ihm zusammen einen Ausflug. Es dürfte sich empfehlen, dass die Kongresszentrale in Zukunft den Delegationsführer auf den Besuch beim Vertreter der Partei am Tagungsort ausdrücklich hinweist, da [386] von allem anderen abgesehen für das Verhalten der Delegation wertvolle Anregungen davon zu erwarten sind.

Die Kongressleitung ist in jeder Hinsicht der deutschen Delegation entgegengekommen. Zum ersten Vicepräsidenten, der in der Eröffnungssitzung den Vorsitz beim ersten Vortrag übernahm, wurde E.Schmidt–Berlin ernannt. Der Führer der Delegation wurde anlässlich des Ausfluges im Oslo-Fjord zur Tafel des Kronprinzenpaares hinzugezogen und das Kronprinzenpaar nahm nachher Gelegenheit, sich mit ihm ausführlich zu unterhalten. Auch beim Essen der Stadt Oslo war sein Platz am Kopf der Bürgermeistertafel. Bei einem Frühstück, das die Familie Lie anlässlich der Enthüllung einer Büste von Sophus Lie veranstaltete, gab der Führer der Delegation den Beziehungen Lies zu Deutschland, Koebe–Leipzig insbesondere denen zu Leipzig in kurzer Rede Ausdruck. Auf einem im kleinsten Kreise stattfindenden Abendessen im Hause des Kongress-Präsidenten konnte der Führer der Delegation dem Präsidenten den besonderen Dank der deutschen Delegation aussprechen.

Es waren auf dem Kongress etwa 30 Nationen vertreten. Die Russen, die eine ganze Anzahl von Vorträgen angekündigt hatten, hatten im letzten Augenblick, bis auf eine Ausnahme [apparently F. Noether], abgesagt; aus Sanktionsgründen fehlten auch die Italiener, Frankreich war wirksam (Borel, Cartan, Julia), aber wenig zahlreich vertreten, Amerika sehr stark (Birkhoff, Veblen, Eisenhart, Snyder, D.E.Smith u.a.). Insgesamt zählte man etwa 500 Aktive. Die Beziehungen der Deutschen zu den Vertretern der anderen Nationen waren durchweg entgegenkommend, man kann wohl sagen freundschaftlich. Man war wohl zunächst etwas skeptisch, aber die Art, wie die deutsche [387] Mathematik in den Hauptvorträgen und ebenso in den Sektionsvorträgen vertreten war, erfuhr vielfache Anerkennung von Vertretern anderer Nationen, denen man ein Urteil zuerkennen muss. Eines ist allerdings für viele Ausländer ausserordentlich schwer zu begreifen: unsere Anschauung von der nationalen Eigenart einer Wissenschaft, unbeschadet der internationalen Bedeutung jeder Forschung. Hervorzuheben ist die glänzende Organisation des Kongresses und die dadurch ermöglichte, in jeder Weise vorbildliche Unterstützung der Organisation unserer deutschen Delegation.

Der Kongress hat in diesem Jahre zum ersten Male die Fields-Medaille für besonders hervorragende mathematische Arbeiten verliehen. Die erhielt [*sic* for erhielten] der Finne Ahlfors und der Amerikaner Douglas. Die Begründung gab als Mitglied der vom letzten Kongress eingesetzten Kommission Caratheodory–München. In die neue Kommission, die dem für 1940 auf Einladung der Amerikaner in New-York oder einer kleineren amerikanischen Universitätsstadt in Aussicht genommenen Kongress einen Vorschlag zu machen hat, wählte der Kongress von deutscher Seite Hecke–Hamburg.

Auf dem Osloer Kongress wurde auch die mathematische "Union", das Überbleibsel der Auswirkungen des Versailler Vertrages auf wissenschaftlichem Gebiet, endgültig zu Grabe getragen. In der in Zürich zur Bereinigung dieser Frage eingesetzten Kommission, der seinerzeit nach dem Ausscheiden von Weyl erst Blaschke–Hamburg, jetzt Caratheodory–München angehörte, sprachen sich einige Länder sehr entschieden gegen eine internationale Organisation aus, während z.B. Frankreich zunächst dafür war. Da aber eine allgemeine Beteiligung nicht zu erreichen war — und das lag einerseits an der anrüchigen Entstehung der Union, zum anderen daran, dass sie bisher in fast 20 Jahren nichts geleistet hat — wurde auf Vorschlag [388] der Kommission von dem Kongress einstimmig beschlossen:

"La Commission nommée par le Congrès de Zürich a vivement regretté l'absence de son Président Severi. Elle n'a pu pour diverse raisons arriver à un accord unanime sur la question d'une organization internationale des mathématiciens. Elle souhaite que dans l'avenir la question posée puisse recevoir une solution."

Das Mandat der IMUK [Internationale Mathematische Unterrichtskommission], über deren Arbeiten weiter unten berichtet wird, wurde vom Kongress einstimmig verlängert. Dem Zentralausschuss wurde überlassen, den Gegenstand der nächsten Arbeiten zu bestimmen. Von den Kongress-Sprachen fiel diesmal Italienisch ganz aus. Wenn auch bei den offiziellen Verlautbarungen alle drei Sprachen [apparently German, French, and English] gleichwertig auftraten, ja vielleicht das Französische ein wenig überwog, trat doch die deutsche Sprache stärker, als ursprünglich beabsichtigt, in den Vordergrund. Selbstverständlich waren alle Mitglieder der deutschen Delegation gehalten, auch in der Diskussion deutsch zu sprechen. Aber auch viele Ländervertreter, die sonst Französisch oder Englisch bevorzugten, sprachen deutsch, so z.B. fast alle Berichterstatter der IMUK, ja einige in anderen Sprachen angekündigten Vorträge wurden deutsch gehalten.

Alles in allem darf wohl gesagt werden, dass die Vertretung Deutschlands auf dem Internationalen Mathematiker-Kongress in Oslo dem Ansehen des Reiches und der deutschen mathematischen Forschung entsprochen hat.

[389] Im folgenden gebe ich einen kurzen Bericht über die Verhandlungen, wobei ich Notizen einzelner Mitglieder der Delegation verwerte. Natürlich ist es nicht möglich, über die mehr als 200 Vorträge, die in den Gesamtsitzungen und in den acht, zum Teil noch weiter unterteilten Sektionen gehalten wurden, zu berichten. Nur das für uns Deutsche Wesentlichste ist herausgehoben worden.

1. Allgemeine Vorträge: Interessant ist, dass die drei deutschen Gelehrten, die zu grossen Vorträgen aufgefordert waren, H.Hasse–Göttingen, E.Hecke–Hamburg und C.L.Siegel–Frankfurt/Main, alle Zahlentheoretiker sind. Die führende Stellung Deutschlands in der zahlentheoretischen Forschung ist von Gauss bis auf den heutigen Tag erhalten geblieben. Weitere zahlentheoretische Vorträge wurden noch von Mordell–England und van der Corput–Holland gehalten. Die drei deutschen Vorträge gehörten durch die Fülle neuer und tiefliegender Ergebnisse und durch die glänzende Darstellung zu den grössten Erlebnissen des Kongresses, ein Eindruck,

der uns von zahlreichen ausländischen Gelehrten bestätigt wurde. Die Zahlentheorie steht wegen der Schönheit und Schwierigkeit ihrer Probleme, deren Lösung die souveräne Beherrschung auch grosser Teile der Analysis erfordert, immer wieder im Mittelpunkt des allgemeinen Interesses, obwohl sie vielleicht derjenige Teil der Mathematik ist, der den Anwendungen am fernsten steht.

Demgegenüber fanden die drei von norwegischen und schwedischen Mathematikern, Størmer, Oseen und Bjerknes gehaltenen Vorträge über Gegenstände der angewandten Mathematik keineswegs den grossen Widerhall, wie die Vorträge aus den Gebieten der reinen Mathematik. Zur angewandten Mathematik kann vielleicht noch der Vortrag von Birkhoff–U.S.A. über die Grundlagen der Quantenmechanik gerechnet werden, die nach wie vor nicht nur die Physiker, [390] sondern auch die Mathematiker auf intensivste beschäftigte.

Bei der Enthüllung der Büste von SOPHUS LIE, der Professor in Leipzig war und dessen Werke Prof.Engel–Giessen herausgab, hielt E.Cartan–Frankreich einen viel beachteten, auch die deutschen Arbeiten voll berücksichtigenden Vortrag über die Bedeutung der von Lie geschaffenen Theorie der kontinuierlichen Gruppen für die Geometrie.

L.Ahlfors–Finnland, dessen Leistungen der Kongress, wie oben gesagt, durch Verleihung der Fields-Medaille auszeichnete, sprach über seine Forschungen auf dem Gebiete der Funktionentheorie, S.Banach–Polen trug ebenfalls in deutscher Sprache über die neueste Entwicklung der Theorie der linearen Räume vor.

Oe.Ore–U.S.A. berichtete über seinen Versuch, eine einheitliche algebraische Theorie aufzustellen, der wesentlich auf deutschen Forschungen fusst.

Das Bedürfnis, die allzu grosse Fülle von mathematischen Tatsachen durch übergeordnete Theorien und Prinzipien zu ordnen und zu verringern, wurde auch in anderen Vorträgen (Banach, Cartan, Siegel, Hecke) sehr deutlich; dieser Zug zur Vereinheitlichung und Annäherung verschiedener mathematischer Gebiete kann vielleicht als ein Hauptmerkmal der heutigen Entwicklung der Mathematik angesehen werden.

Zu den weiteren Vorträgen sei nur bemerkt, dass Fueter–Schweiz einen funktionentheoretischen Gegenstand behandelte; geometrische Vorträge hielten Veblen–U.S.A. und Nielsen–Dänemark; analytische Themen behandelten Wiener–U.S.A. und Fréchet–Frankreich; über ein historisches Thema sprach Neugebauer–Dänemark.

[391] 2. Sektions-Vorträge.

In Sektion I (Arithmetik und Algebra) war kein Vortrag eines Mitgliedes unserer Delegation angemeldet. An deutsch-österreichischen Mathematikern trugen Rella–Wien und Hofreiter–Wien vor und ihre Vorträge wurden mit Beifall aufgenommen. Unter den insgesamt 27 Vorträgen war eigentlich nur der etwas längere Vortrag von H.Weyl–Princeton von allgemeinerem wissenschaftlichem Interesse und hervorstechender Leistung. Der gegenseitige persönliche Gedankenaustausch in Sektion I war sehr rege. Es wurde eine grosse Menge neuer Bekanntschaften geknüpft, und die Stimmung war ausgezeichnet. Allgemein dürfte noch interessieren: [This comment is somewhat misplaced because functional analysis actually belonged to Section II] Bei einer gesellschaftlichen Zusammenkunft wurde anlässlich der Anwesenheit des polnischen Mathematikers S.Banach–Lemberg in betonter Form hervorgehoben, wie hoch es anzuerkennen sei, dass in Polen in so kurzer Zeit eine so grosse und leistungsreiche mathematische Schule einer ganz spezifischen Ausprägung entstanden sei.

Die Sitzungen der Sektion II (Analysis) fanden gleichzeitig in drei Unterabteilungen am je vier Tagen statt. Bemerkenswert war zunächst die von Nevanlinna–Finnland geführte skandinavische Mathematikerschule. Nevanlinna selbst legte auf dem Kongress sein neuerschienenes Buch "Eindeutige analytische Funktionen" (Berlin 1936) vor, in dem er die Leitgedanken seines Schaffens zusammenfasst. Sein Schüler, Prof. Ahlfors, stellte in seinem allgemeinen Vortrag einen höchst wichtigen Zusammenhang zwischem dem Picardschen Satz und der Gauss-Bonnetschen Integralformel her, also zwischen Funktionentheorie und Differentialgeometrie. In dieser Richtung lassen sich grosse Fortschritte für die nahe Zukunft erwarten.

[392] Auch die Sektionssitzung, welche ganz der Wertverteilungslehre gewidmet war, machte einen sehr geschlossenen Eindruck. Es sprachen dort u.a. die Herren Speiser–Zürich, Milloux–Bordeaux, Selberg–Oslo, Miss Cartwright–Cambridge. Von seiten der deutschen Abordnung gab Ullrich–Giessen eine Mitteilung über das Umkehrproblem der Wertverteilungslehre, die eine angeregte Aussprache nach sich zog.

In den Sektionssitzungen sprach von den Deutschen ferner Dozent Peschl–Jena; er gab eine Kennzeichnung des Variabilitätsbereichs der Koeffizienten einer schlichten Potenzreihe, indem er die Frage auf eine Untersuchung über partielle Differentialgleichungen zurückführt; damit lassen sich wertvolle Einblicke in die Schwierigkeiten gewinnen, welche dem Beweis der Bieberbachschen Vermutung entgegenstehen — einem seit zwanzig Jahren viel umworbenen Problem.

Ferner sprach Prof. Behnke–Münster in einem einprägsamen und durchsichtigen Vortrag über Funktionen mehrerer komplexer Veränderlichen; dieses Gebiet wird seit einem Jahrzehnt in vorbildlicher, freundschaftlicher Zusammenarbeit zwischen deutschen und französischen Mathematikern bearbeitet.

Von ausländischen Vorträgen muss der von Karamata–Belgrad genannt werden. Er sprach über Umkehrungssätze von Limitierungsverfahren und entwickelte dort eine allgemeine Theorie dieses Gebietes, das ja auch in deutschen Landen, namentlich durch den Kreis um Knopp–Tübingen, viel gepflegt worden ist.

Es mag bei dieser Gelegenheit hervorgehoben werden, wie wichtig es wäre, engere wissenschaftliche Beziehungen zu den südosteuropäischen Ländern zu pflegen. Es fällt auf, dass von französischer Seite — offenbar mit Bedacht — "wissenschaftliche [393] Kulturpropaganda" nach den Balkanstaaten hin gemacht wird. Franzosen halten regelmässig Gastvorträge in jener Gegend und besuchen die Balkanischen Kongresse; von deutscher Seite wurde dem bisher keine Beachtung geschenkt (abgesehen von einer Reise des Herrn Blaschke–Hamburg), und man hält uns — vielleicht — für zu stolz, dorthin zu gehen. Wir haben sehr versucht, gerade mit jüngeren Kräften, die nicht deutschfeindlich eingestellt sind, in Beziehung zu kommen. In dieser Hinsicht scheint Herr Karamata wichtig. Abgesehen davon, dass er seine Arbeiten seit Jahren in deutscher Sprache und meist in deutschen Fachzeitschriften veröffentlicht hat, scheint er uns auch im allgemeinen wohlgesinnt. Vor allem aber ist er weitaus der beste Mann unter allen jüngeren "Südost-Mathematikern". Vielleicht wäre es bei Gelegenheit möglich, ihn zu einer Vortragsreise an einer deutschen Hochschule zu gewinnen.

Ähnlich müssten — wie mir scheint — unsere Verbindungen zu Polen verbessert werden, das wissenschaftlich zur Zeit noch ganz französisch eingestellt scheint.

Freilich ergeben sich da besondere Schwierigkeiten, da offenbar viele polnische Mathematiker Nichtarier sind.

Die Vorträge in Sektion III (Geometrie) waren ziemlich planlos und unsachgemäss zusammengestellt, indem Dinge, die gar keine Berührung miteinander hatten, in die gleiche Sitzung gelegt, zusammengehörige getrennt waren. Sachlich herrschte in den Arbeiten die Topologie weit vor, dann folgte die Differen[394]tialgeometrie, während die synthetische und algebraische Geometrie infolge des Fehlens der Italiener zu kurz kam; auch das Fehlen der Russen bedingte wesentliche Lücken. Von deutscher Seite wurden drei Vorträge gehalten: Blaschke–Hamburg, Integralgeometrie, Haenzel–Karlsruhe, Neue Eigenschaften der linearen Strahlenkongruenz, Geppert–Giessen, Der gemischte Inhalt zweier Bereiche. Im Programm befand sich auch ein gut Teil unverständlicher Exotica.

An den Sitzungen der Sektion IV (Mathematik, Statistik, Versicherungsmathematik, Ökonometrie) war die Deutsche [*sic*] Delegation durch drei Herren (Boehm–Charlottenburg, Riebesell, Timpe) beteiligt. An einem der vier Sitzungstage führte ein Mitglied der Delegation (Riebesell) den Vorsitz. Soweit Sektion IV in Frage steht, hat die Teilnahme am Kongress positive Wirkungen gezeitigt, sowohl im Sinne der deutschen Kulturgeltung als auch im Sinne wertvoller Anregungen für die deutsche Wissenschaft, insbesondere durch Austausch mit den führenden Wissenschaftlern der auf dem Gebiete der Sektion IV stark hervortretenden skandinavischen Länder und des angelsächsischen Kulturkreises. Von den 23 Vorträgen wurden 9 in deutscher Sprache gehalten: 2 Mitglieder der Delegation [Riebesell and Boehm], 2 Österreicher [Lukacz and possibly Alt, whose talk is in English in the proceedings], 1 Schweizer [Arthur Linder], 2 Norweger [Sven Guldberg and Birger Meidell], 1 Schwede [Willy Feller, although also an emigrant], 1 Emigrant [Emil Julius Gumbel, with two talks in German] (im letzteren Falle verhielten sich die deutschen Delegierten passiv). Als Diskussionssprache hielten sich Deutsch und Englisch ziemlich die Wage [*sic* for Waage]. Von den Skandinaviern benutzten 3 Deutsch, 2 Englisch als Vortragssprache [H. Wold and Ragnar Frisch]; bemerkenswert, weil sonst gerade in Norwegen heute das Englische gegenüber Deutsch stark in den Vordergrund tritt. Diskussionen und Verhandlungen vollzogen sich durchweg in liebenswürdigen Formen.

[395] Zusammen mit einem kleineren Kreis der Sektionsteilnehmer waren die Mitglieder der Deutschen Delegation an einem Abend zu Gast bei den Osloer Fachkollegen Ragnar Frisch und Meidell (die übrigens mit Rücksicht auf uns von der Einladung des Emigranten Gumbel Abstand genommen hatten). Wir bekamen ferner Gelegenheit, die ihnen unterstehenden Universitätsinstitute kennen zu lernen. Der mathematischen Wirtschaftsforschung ist in ihnen ein erheblich grösserer Platz eingeräumt, als das bei deutschen Wirtschaftsinstituten der Fall ist. In dem Institut für Ökonomie, das von dem Vertreter der mathematischen Wirtschaftsforschung und einem nichtmathematisch eingestellten Nationalökonomen gemeinsam geleitet wird, sind allein 16 Assistenten tätig, vorwiegend beschäftigt mit der Auswertung der Wirtschaftsstatistik.

Die gegenwärtig gesteigerte wissenschaftliche Aktivität in Deutschland auf dem Gebiet der mathematischen Wirtschaftsforschung und Konjunkturanalysis verfolgen die norwegischen Fachkollegen mit besonderer Sympathie; Mitarbeit an deutschen Fachzeitschriften wurde weiterhin zugesagt.

Über Sektion V (Math. Physik, Astronomie und Geophysik) an deren Arbeiten Deutsche nicht beteiligt waren, kann ich nicht berichten.

Auch über Sektion VI (Mechanik und Ingenieurwissenschaften) kann ich nur wenig sagen: Der Franzose Le Roux gab einen längeren Überblick über sein schon erschienenes Buch: La mécanique invariante, Wavre–Genf machte interessante Bemerkungen über Vieldeutigkeiten Newtonscher Potentiale, dann sprach Hamel–Berlin über Räumliche Strahlen mit konstanter Geschwindigkeit und Reissner–Berlin über erzwungene Schwingungen des Halbraumes.

[396] In Sektion VII (Geschichte der Mathematik) sprach zunächst Spies [*sic* for Spiess] –Schweiz über den Briefwechsel der Bernoullis und Locher–Schweiz über Goethe und die Mathematik. Die Vorträge der Amerikaner Archibald und Gandz sind bereits in der Osiris erschienen. Ein Vortrag von Singh–Indien über magische Quadrate ist vor der Veröffentlichung im Druck nicht nachprüfbar. Heegaard–Norwegen zeigte einen offenbar demotischen Papyrus-Fetzen der Osloer Sammlung, der anscheinend Zahlzeichen enthält, im ganzen aber nicht zu entziffern ist. Im Anschluss daran gab Vogel–München neue Ergebnisse der Forschung auf dem Gebiete der babylonischen Mathematik bekannt. Den Schlussvortrag hielt Jelitai–Ungarn (er erzählte von der Geschichte der Mathematik in seinem Lande).

In der Sektion VIII, die der Pädagogik vorbehalten war, fanden von der Veranstaltung der IMUK, über die gleich im Zusammenhang berichtet werden wird, zwei Vorträge statt. Der eine führte im Film einen Apparat zur Lösung von Differentialgleichungen vor: Fairthorne–Farnborough, The demonstration of qualitative properties of differential equations by means of cinematograph films; der andere Vortrag des Biologen Przibram–Wien [Hans Leo Przibram (1874–1944) was the founder of experimental biology in Austria. As a Jew he was murdered by the Nazis.], wollte das Radizieren mit Hilfe der Logarithmen in der Schule durch ein einfacheres Verfahren ersetzen, bei dem nur die rationalen Rechenoperationen benutzt, ja, wenn nötig, auch auf Division verzichtet werden kann. Der Vortragende führte das an Beispielen aus. Das Verfahren läuft im Wesentlichen darauf hinaus, den Radikanden durch ein Produkt zu ersetzen, deren [*sic* for dessen] erste Faktoren Potenzen sind, die sich leicht radizieren lassen, während der letzte Faktor nahe an 1 liegt und nach dem biologischen [*sic* for binomischen, a question mark added by hand] Lehrsatz angenähert wird. Will man allerdings dieses Verfahren den Schülern wirklich erklären, [397] so dürfte es erheblich mehr Hilfsmittel erfordern, als das bisherige Verfahren.

Die IMUK hatte für ihre Sitzung als Thema gewählt: Die gegenwärtigen Bestrebungen im mathematischen Unterricht der hoheren Schule. Die Verhandlungen fanden im Rahmen der Sektion VIII des Kongresses unter dem Vorsitz des Generalsekretärs der IMUK, Fehr–Genf, bei einer trotz der vielen parallelen Sitzungen anderer Sektionen überraschend zahlreichen Beteiligung statt. Es erstatteten kurze Berichte für Deutschland: Lietzmann (von diesem Bericht liegt ein Durchschlag bei), England: Neville, Österreich: Dinzl, vorgetragen von Kantz, Dänemark: Mollerup, Ungarn: Jelitai, Japan: Kuniyeda, der gleichzeitig einen ausführlichen gedruckten Bericht überreichte, Norwegen: Piene, Polen: Straszewicz, Schweiz: Fehr. Eine Anzahl weiterer Berichte wurde schriftlich vorgelegt oder in Aussicht gestellt. Sie alle werden im Wortlaut im Enseignement Mathematique veröffentlicht werden. Hier sei noch gesagt, dass überraschend viele Änderungen in der Organisation des höheren Schulwesens in den letzten Jahren erfolgt sind. So in Österreich,

Ungarn, Norwegen, Polen. Es dürfte sich vielleicht empfehlen, die sehr interessanten Änderungen, namentlich in Österreich, Ungarn und vielleicht auch in Polen an Ort und Stelle durch Besuch von Schulen und Rücksprache mit den Unterrichtsbehörden näher zu studieren. Die Lage der Mathematik an den höheren Schulen wurde mehrfach als nicht besonders günstig geschildert. Im Anschluss an diese öffentliche Sitzung fand noch eine Tagung der amtlichen Delegierten statt, da der Präsident der IMUK, Hadamard–Paris wegen einer Reise nach China verhindert war, unter dem Vorsitz des unterzeichneten Vizepräsidenten. Es wurde die Fortsetzung der IMUK-Arbeiten in [398] Aussicht genommen und ein entsprechender Beschluss dem Gesamtkongress vorgelegt. Dieser nahm ihn in der Schlussitzung einstimmig an. Ein Thema für die nächsten Arbeiten wurde noch nicht festgelegt. In Aussicht genommen wurde eine ausführliche Behandlung des in den verschiedenen Ländern ausserordentlich verschieden gehandhabten Prüfungswesens.

[added in handwritten letters: Lietzmann]

* * *

Report on the German delegation at the International Mathematical Congress in Oslo, 13–18 July 1936, submitted by the leader [Führer] of the delegation, Prof.Dr.W.Lietzmann–Göttingen

Through the mediation of the Kongresszentrale, ca. Rm. 5000 in foreign currency was requested from the Landesfinanzamt for the full participation of all those delegates named by the Reich Minister of Science on 20 June 1936. However, only Rm. 3000 was made available. Therefore, the maximum rate for each participant had to be reduced to Rm. 185. Due to numerous invitations, if one did not reside in one of the first hotels, that amount proved to be sufficient. Since a number of men did not claim the full amount, it was possible, after individuals had withdrawn at the last moment, for special reasons, for the following to appear as members in Oslo: Behnke, Blaschke, Caratheodory, Geppert, Grunsky, Haenzel, Hamel, Hasse, Hecke, Koebe, Lietzmann, Peschl, Riebesell, E. Schmidt, Seifert, Siegel, Sperner, Ullrich, Vogel, Köthe, Timpe, Reinhardt. The fact that the final determination of the sum was carried out at the last moment, and that the payment of foreign currency could only partly be made in foreign notes, the other part by credit note in Oslo by the delegation leader, represented an extraordinary burden.

Furthermore, the following were also named by the Minister as German participants in the congress: Threlfall–Dresden and Rosenthal–Heidelberg. Dr.Böhm–Berlin also asked for a connection to the delegation. Of German mathematicians in attendance who were not members of the delegation were observed: Prof.Dr.Dehn–Frankfurt/Main, Dr.E.Reissner–Berlin and Dr.Heisig–Leipzig from the publisher B.G.Teubner–Leipzig. The "emigrants" at the congress were, as far as I know, the professors Courant–New York, Weyl–Princeton, Neugebauer–Copenhagen, Noether–Tomsk, Fränkel–Jerusalem, and finally also the infamous Mr.Gumbel–Lyon. There were no disturbances or unwanted discussions caused by these gentlemen.

An immediate meeting at the German Embassy was very valuable for the leader of the delegation, especially important was the assessment of the newspapers, which in Norway like to feature interviews with foreigners. The members of the delegation were immediately informed accordingly. The form of the planned laying of a wreath at the Abel monument and at the grave of Sophus Lie was discussed. The ambassador invited the German delegation and some Norwegian men to tea with his family one afternoon.

The wreath at the Abel monument was laid by the leader of the delegation, together with the Chairman of the Deutsche Mathematikervereinigung. Police were present and a policeman was stationed near the monument throughout the congress. The wreath was reproduced in a daily. Members of the Lie family were present at the laying of a wreath at the grave of Sophus Lie — two daughters are married in Germany — and, along with three representatives of the delegation, they raised their hands in the German salute.

The leader of the delegation also sought out the district leader of the NSDAP, comrade Kern, and went on an excursion with him on the day after the congress. It would be advisable that in future, the Kongresszentrale should explicitly direct delegation leaders to visit the representative of the party at the venue, as, apart from anything else, valuable suggestions on the conduct of the delegation can be expected of them.

The congress leadership accommodated the German delegation in all respects. E.Schmidt–Berlin, who took the chair for the first lecture of the opening session, was named the first vice president. The leader of the delegation was called to the table of the Kronprinzenpaar [the crown prince with his wife] on the occasion of an excursion in the Oslo Fjord, and the Kronprinzenpaar subsequently took the opportunity to speak with him in more detail. Also at the official dinner organised by the City of Oslo, his place was at the head of the mayor's table. At a breakfast that the Lie family hosted on the occasion of the unveiling of a bust of Sophus Lie, the leader of the delegation gave a short speech expressing the ties of Lie to Germany, with Koebe–Leipzig noting those to Leipzig in particular. In a very intimate atmosphere, at a dinner held at the house of the congress president, the leader of the delegation was able to express to the president the special gratitude of the German delegation.

About 30 nations were represented at the congress. The Russians, who had announced a number of lectures, had, with one exception [apparently F. Noether], cancelled at the last moment, the Italians were also absent, for reasons of sanctions; France was represented effectively (Borel, Cartan, Julia), but in small number, America very strongly (Birkhoff, Veblen, Eisenhart, Snyder, D.E.Smith, *et al.*) Overall, there were about 500 active participants. The relations of the Germans with the representatives of other nations were accommodating, one can surely say friendly. People were probably sceptical at first, but the way in which German mathematics was represented in the main lectures and also in the sectional lectures, drew multiple appreciations from the representatives of those nations whose judgement counts. There is one thing that many foreigners have real problems in understanding: our notion of the national peculiarity of science, regardless of the universal validity of its results. The brilliant organisation of the congress ought to be emphasised; this made possible in every way the exemplary support of the organisation for our German delegation.

The Fields Medals for particularly outstanding mathematical work were awarded at the congress this year for the first time. They were received by the Finn Ahlfors and the American Douglas. The reasons were given by a member of the commission established at the last congress, Caratheodory–Munich. In the new commission, for the 1940 congress, which, at the invitation of the Americans, is envisaged for New York or a smaller American university town, the proposal has been made for the congress to elect Hecke–Hamburg from the German side.

Also at the Oslo congress, the mathematical "Union", that remnant of the after-effects for science of the Versailles Treaty, was finally buried. In the commission established in Zurich to address this question, to which, following the departure of Weyl, first Blaschke–Hamburg and now Caratheodory–Munich belonged, some countries spoke out very firmly against an international organisation, while, for example, France was initially in favour. But since a general participation could not be achieved — being partly due to the disreputable origins of the Union, and, on the other hand, the fact that it has so far done nothing in nearly 20 years — the commission of the congress decided unanimously on the following proposal:

"The commission appointed by the congress in Zurich strongly regrets the absence of its chairman Severi. It could not for various reasons come to a unanimous agreement on the question of an international organisation of mathematicians. It hopes that in the future the question can receive a solution."

The mandate of ICMI [International Commission on Mathematical Instruction], whose work is reported upon below, was extended unanimously by the congress. The central committee was left to determine the subject of the next work. Of the congress languages, Italian has this time dropped out completely. Although all three languages [apparently German, French and English] occurred equally in all official statements, French did perhaps predominate a little, yet German did come to the fore more strongly than originally intended. Of course, all members of the German delegation were required to speak German in discussions. But many representatives of other countries, who otherwise preferred French or English, spoke German, so, for example, all those reporting for ICMI, and even some who had announced lectures in other languages, presented in German.

All in all, it may well be said that the representation of Germany at the International Mathematical Congress in Oslo has spoken to the prestige of the Reich and of German mathematical research.

In the following, I give a short report on the proceedings, in which I make use of the notes of the individual members of the delegation. Of course it is not possible to report on the more than 200 lectures that were held in the general sessions and in the eight, in part further subdivided, sections. Only for us are the German essentials singled out.

<u>1. General Lectures:</u> It is interesting that the three German scholars who had been invited to give plenary lectures, H.Hasse–Göttingen, E.Hecke–Hamburg and C.L.Siegel–Frankfurt/Main, are all number theorists. The leading position of Germany in number-theoretic research has been kept from Gauss to the present day. Further talks on number theory were given by Mordell–England and van der Corput–Holland. Due to the wealth of new and deep results and the brilliant presentation, the three German talks were among the most impressive experiences of the congress, an impression that was confirmed to us by numerous foreign scholars. Because of its beauty and the difficulty of its problems, number theory remains at the centre of general interest, although it is perhaps that part of mathematics that is most distant from applications.

By comparison, the three presentations by Norwegian and Swedish mathematicians, Størmer, Oseen and Bjerknes, on topics of applied mathematics met with much less response than those from areas of pure mathematics. In a sense, the talk by Birkhoff–U.S.A. on the foundations of quantum mechanics also belonged to applied mathematics; this topic continues to engage intensely not only physicists but also mathematicians.

At the unveiling of the bust of SOPHUS LIE, who was professor in Leipzig and whose works have been edited by Professor Engel–Giessen, E.Cartan–France delivered a much-noted talk on the importance for geometry of Lie's theory of continuous groups. This presentation also gave full credit to German contributions.

L.Ahlfors–Finland, whose accomplishments the congress, as noted before, celebrated by awarding the Fields Medal, spoke about his research in function theory, S. Banach–Poland, reported, also in German, on the newest developments in the theory of linear spaces.

Oe.Ore–U.S.A. reported on his efforts to create a unifying algebraic theory, an effort which is substantially based on German research.

The need to organise and to reduce the excessive wealth of mathematical facts by resorting to overarching theories and principles manifested itself also in other talks (Banach, Cartan, Siegel, Hecke); this tendency towards unification and convergence of different mathematical domains may be considered a chief characteristic of contemporary developments in mathematics.

With regard to the other lectures, we just mention that Fueter–Switzerland treated a function-theoretic topic; geometric lectures were given by Veblen–U.S.A. and Nielsen–Denmark; analytical themes were treated by Wiener–U.S.A. and Fréchet–France; Neugebauer–Denmark spoke on a historical theme.

2. Sectional Lectures.

In Section I (Arithmetic and Algebra), no lecture was registered by any member of our delegation. The German-Austrian mathematicians Rella–Wien and Hofreiter–Wien presented, and their lectures were met with applause. Among the total of 27 lectures, only the somewhat longer one by H. Weyl–Princeton was of general scientific interest and an outstanding achievement. The mutual personal exchange of views in Section I was very active. It led to a great number of new acquaintances, and the atmosphere was excellent. The following more general information should be of interest: [This comment is somewhat misplaced because functional analysis actually belonged to Section II.] At a social gathering it was stressed with regard to the attending Polish mathematician S. Banach–Lemberg, how admirable it was that Poland had developed such a broad and strong mathematical school of peculiar orientation in a very short period of time.

The meetings of Section II (Analysis) took place at the same time in three subdivisions over four days. Noteworthy first of all was the Scandinavian mathematical school led by Nevanlinna–Finland. Nevanlinna himself laid before the congress his newly published book "Eindeutige analytische Funktionen" (Berlin 1936), in which he summarises the main ideas of his work. His pupil, Prof.Ahlfors, established in his general lecture a highly important relationship between the Picard Theorem and the Gauss-Bonnet Integral Formula, hence between function theory and differential geometry. Significant progress can be expected in this direction in the near future.

The sectional meeting that was devoted entirely to value distribution theory made a very coherent impression. Also there, among others, were Herrn Speiser–Zurich, Milloux–Bordeaux, Selberg–Oslo, Miss Cartwright–Cambridge. On the part of the German delegation, Ulrich [*sic*]–Giessen gave a communication on the inverse problem of value distribution theory which drew a lively debate.

Of the Germans, docent Peschl–Jena also spoke in the sectional meetings; he gave a characterisation of the range of variability of the coefficients of a schlicht power series, by reducing the question to an investigation of partial differential equations; thus, valuable insights were obtained into the difficulties that stand in the way of the proof of the Bieberbach conjecture — for twenty years, a much-courted problem.

Furthermore, Prof. Behnke–Münster spoke in a memorable and transparent lecture on functions of several complex variables; this is a topic that has been cultivated for a decade in an exemplary manner in friendly collaboration between German and French mathematicians.

Of foreign lectures, that of Karamata–Belgrade must be mentioned. He spoke on inversion theorems of limiting processes and developed there a general theory of this area that has also been much cultivated in German lands, in particular by the circle around Kopp–Tübingen.

It might be mentioned in this context that it would be very desirable indeed to intensify scientific relations with the southeast-European countries. It is remarkable that the French — obviously quite deliberately — are making 'scientific cultural propaganda' towards the Balkans. The French give regular lectures in that region and visit congresses in the Balkans. On the German side, we have neglected this so far (except for a trip by Herr Blaschke–Hamburg). Perhaps they consider us too proud to go there. We have tried hard to come into contact particularly with younger mathematicians who are no enemies of Germany. In this regard Herr Karamata seems important. Besides the fact that he has for years published his work in German and mostly in German scientific journals, he also seems generally well disposed towards us. Above all, he is by far the best among the younger 'South-East mathematicians'. Perhaps it would be possible on occasion to invite him for a lecture tour to a German university.

In a similar way — it seems to me — we should improve our contacts with Poland, which is still largely oriented towards France in the scientific respect. However, particular problems may arise, since many Polish mathematicians seem to be non-Aryans.

The lectures in Section III (Geometry) were rather haphazardly and improperly collected, in that things that have no connection to each other were placed together in the same session, and things that are connected were separated. As to content, topology was clearly dominant, then followed differential geometry, while synthetic and algebraic geometry came up short due to the absence of the Italians; also the absence of the Russians caused significant gaps. On the German side, three lectures were given: Blaschke–Hamburg, Integral Geometry, Haenzel–Karlsruhe, New Properties of Linear Ray Congruences, Geppert–Giessen, The Mixed Content of Two Domains. There was also a good deal of incomprehensible exotica in the programme.

The German delegation was involved in the sessions of Section IV (Mathematics, Statistics, Actuarial Mathematics, Econometrics) via three men (Boehm–Charlottenburg, Riebesell, Timpe). On one of the four meeting days a member of the delegation (Riebesell) presided. As far as Section IV is concerned, participation in the congress had positive effects, both in the sense of German cultural influence, and also in terms of valuable stimulation of German science, in particular through exchange with the leading scientists from the Scandinavian countries and Anglo-Saxon cultural circles, which are traditionally strong in the areas of Section IV. Of the 23 lectures, 9 were given in German: 2 members of the delegation [Riebesell and Boehm], 2 Austrians [Lukacz and possibly Alt, whose talk is in English in the proceedings], 1 Swiss [Arthur Linder], 2 Norwegians [Sven Guldberg and Birger Meidell], 1 Swede [Willy Feller, although also an emigrant], 1 emigrant [Emil Julius Gumbel, with two talks in German] (in the latter case, the German delegation reacted passively). As the languages of discussions, German and English were quite balanced. Of the Scandinavians, 3 used German to lecture, and 2 English [H. Wold and Ragnar Frisch]; this is noteworthy because otherwise in Norway English is increasingly becoming dominant against German. Discussions and proceedings took place throughout in an amiable manner.

Together with a small circle of section participants, the members of the German delegation were one evening the guests of the Oslo colleagues Ragnar Frisch and Meidell (who, incidentally had not invited the emigrant Gumbel, out of consideration for us). In addition we had the opportunity to visit their university institutes. Mathematical economics has a much larger space in these institutes than in German economic institutes. In the Institute for Economics, which is headed jointly by the representative of mathematical economics and a non-mathematical political economist, 16 assistants are employed, predominantly busy with the analysis of economical statistical data.

The currently increased scientific activity in Germany in the field of mathematical economy and business cycle analysis is observed, as it turned out, with much sympathy by our Norwegian colleagues; cooperation in German journals was further promised.

I cannot report on Section V (Math. Physics, Astronomy and Geophysics) as no Germans were involved in it.

I can also say little about Section VI (Mechanics and Engineering): The Frenchman Le Roux gave an extended overview of his already published book: La mécanique invariante, Wavre–Geneva made interesting remarks about the multivaluedness of Newtonian potentials, then Hamel–Berlin spoke on spatial radiation with constant speed and Reissner–Berlin on forced oscillations of the half-space.

In Section VII (History of Mathematics), Spies [*sic* for Spiess]–Schweiz spoke first about the exchange of letters of the Bernoullis and Locher–Switzerland spoke about Goethe and mathematics. The lectures of the Americans Archibald and Gandz have already appeared in *Osiris*. A lecture by Singh–India on magic squares has to be checked as to its validity after appearance in print. Heegaard–Norway showed a fragment of an apparently demotic papyrus from the Oslo collection which appears to contain numerals, but is largely indecipherable. Following this, Vogel–Munich announced new results of research in the area of Babylonian mathematics. The final lecture was given by Jelitai–Hungary (he talked about the history of mathematics in his country).

In Section VIII, which was reserved for pedagogy, two lectures took place within the ICMI event (to be reported on below). One showed in a film an apparatus for solving differential equations: Fairthorne–Farnborough, The demonstration of qualitative properties of differential equations by means of cinematograph films; the other lecture, by the biologist Przibram–Vienna [Hans Leo Przibram (1874–1944) was the founder of experimental biology in Austria. As a Jew he was murdered by the Nazis.] wanted to replace root extraction by means of logarithms, as taught in school, by a simpler method that uses only the rational arithmetic operations, indeed, if necessary, division can also be dispensed with. The lecturer led from examples. The process is in essence to replace the radicand by a product whose first factors are powers, from which the square root is easily extracted, while the last factor is close to 1 and is approximated by the binomial theorem. However, if one were truly to explain this process to students, it would likely require considerably more resources than the traditional method.

ICMI had chosen the following theme for its meeting: The current endeavours in mathematical teaching in high school. Proceedings were held within the context of Section VIII of the congress, chaired by the Secretary General of ICMI, Fehr–Geneva, and were surprisingly well attended in light of the many parallel sessions of other sections. Short reports were given for Germany: Lietzmann (a copy of this report is enclosed), England: Neville, Austria: Dinzl, presented by Kantz, Denmark: Mollerup, Hungary: Jelitai, Japan: Kuniyeda, who at the same time distributed a detailed printed report, Norway: Piene, Poland: Straszewicz, Switzerland: Fehr. A number of further reports was presented in writing or were promised. All texts of them will be published in *L'enseignement mathématique.* It should be said here that surprisingly many changes in the organisation of higher education have been made in recent years. This is the case in Austria, Hungary, Norway, Poland. It may perhaps be recommended that the very interesting changes, notably in Austria, Hungary and perhaps also in Poland, be studied more closely through visits to schools and by consultation with teaching authorities. The situation of mathematics in secondary schools was repeatedly described as not very auspicious. Following this open session, a meeting of official delegates took place, which, the President of ICMI, Hadamard–Paris, being prevented from attending because of a trip to China, was chaired by the undersigned vice-president. A continuation of the work of ICMI was proposed and a corresponding decision was submitted to the whole congress. This was accepted unanimously in the closing session. A theme for further work has not yet been determined. A detailed treatment of the extraordinarily different handling of the examination systems in different countries was considered.

[Lietzmann]

APPENDIX E

Report by E. G. Kern, the Nazi functionary in Oslo, to Nazi headquarters in Berlin, 15 August 1936

[DKZ 3]

Quote from a report apparently written by Kern. The original contains some small linguistic and syntactic slips that we have corrected in the translation.

Im Vordergrund stand bei unseren Kongressteilnehmern die Frage der Emigranten. Ausserhalb der Delegation waren anwesend die in Deutschland noch tätigen oder ansässigen Professoren Dehn, Rosenthal, Landau, sämtlich Juden. Davon sind die beiden erstgenannten, wenn ich recht verstanden habe, noch im Dienst. Ausserdem waren anwesend die früheren deutschen Professoren Neugebauer (Kopenhagen), Gumbel (Kohlrübengumbel, jetzt in Frankreich), Weil (USA) und Courant (USA). Alle haben sich nicht in einer Art bemerkbar gemacht, die uns aufgefallen wäre. Was hinter dem Rücken unserer Delegierten geschehen ist, lässt sich natürlich nicht überblicken. Von den Russen und Italienern sind aus politischen Gründen vermutlich keine Delegationen geschickt worden, trotzdem die Russen Vorträge zugesagt hatten. Die Franzosen Cartan und Julia haben sich in verständnisvoller Weise in ihren Vorträgen geäussert. Ein holländischer Vertreter lehnte in seinem Vortrage die Auffassung von volksmässig gebundener Wissenschaft ab. In seiner Rede erwiderte Prof. Dr. Lietzmann in einer Art und Weise, die den Holländer veranlasste, sich nachher mit ihm freundschaftlich über unsere Auffassung auseinanderzusetzen. Die deutsche Delegation ist überall gebührend berücksichtigt worden. Vizepräsident des Kongresses war der Deutsche Professor Erhard Schmidt. Von den Deutschen sind drei Vorträge gehalten worden, die als die besten Leistungen anerkannt worden sind. Als beste Leistung des Kongresses soll der Vortrag von Professor Siegel aus Frankfurt/Main bewertet worden sein.

Eine Reihe der deutschen Delegierten konnte aus Devisenrücksichten nicht reisen. Ich habe festgestellt, dass verhältnismässig grosse Beträge für Jugendwanderungen in Norwegen ausgegeben worden sind, wofür auch Devisenbewilligung erteilt wurde. Da sich diese Wanderungen, die sich auf die zu erwartende politische und volkskundliche Schulung stützen, meistens in Hochgebirgs- und von Ausländern überlaufenen Touristengebieten bewegen, in denen weder von politischer noch von volkskundlicher Schulung überhaupt die Rede sein kann, würde ich es für besser und zweckmässiger halten, dass man in erster Linie solche deutschen wissenschaftlichen Delegationen, deren Wert bei der richtigen Auswahl in Deutschland für unsere Sache unübersehbar ist, mit den nötigen Devisen bedenkt. Ich bitte besonders diesen Punkt bei der entsprechenden Stelle ernstlich zur Sprache zu bringen. Ich habe bereits früher vorgeschlagen, dass die HJ [Hitlerjugend] eine bis zwei Fahrten, die dann wirkliche Schulungsfahrten sind, im Laufe des Jahres vornimmt. Für solche Zwecke stelle ich selbstverständlich gerne meine Wohnung, das Parteiheim und vor allem mein Landhaus zur Verfügung. Die augenblickliche Art der Durchführung dieser Fahrten halte ich für eine unnötige Belastung unseres Devisenkontos.

* * *

Most important for the members of our delegation was the question of emigrants. Outside the delegation were present the professors Dehn, Rosenthal,[1] Landau,[2] all Jews. Of those, if I understood it correctly, the first two are still in their positions. In addition the former German professors Neugebauer (Copenhagen), Gumbel (turnip-Gumbel now in France),[3] Weil (USA) [*sic* for Weyl] and Courant (USA). All of them did not draw any attention which we would have noticed. What happened behind the backs of our delegates, we cannot know, of course. The Russians and Italians did not send delegations, probably for political reasons, although the Russians had promised presentations. The Frenchmen Cartan and Julia have shown consideration [apparently for the Germans] in their talks. In his talk a Dutch representative refused the view of science being based on national peculiarities [volksmässig gebunden]. In his reply Prof. Dr. Lietzmann managed to be so circumspect that the Dutchman afterwards discussed with him our view in a friendly manner. The German delegation has been respected everywhere appropriately. Vice-president of the congress was the German professor Erhard Schmidt [in fact he was only one of twelve vice-presidents]. The Germans gave three presentations [obviously meaning plenary talks], which have been acknowledged as the best. The very best presentation of the Congress was allegedly a talk by Professor Siegel from Frankfurt/Main.

A number of German delegates could not travel due to foreign currency considerations. I have found that relatively large sums have been spent on youth travel in Norway, which also obtained foreign currency permissions. These walking tours, which are motivated by anticipated political and ethnological [volkskundliche] instruction, mostly take place in higher mountains or in regions which are flooded by tourists, where there is no political or ethnological instruction to be expected at all. When it comes to foreign currency allocations I would therefore find it better and more expedient to prioritise German scientific delegations, whose importance for our cause — provided a right selection is taken in Germany — cannot be doubted. I ask you to stress this point seriously with the responsible agency. Earlier I had suggested that the HJ [Hitler Youth] restrict its walking tours to one or two per year, however then with real instructional intent. For such a purpose I would of course like to offer my apartment, the party home and, above all, my cottage. The current way of carrying out these travels I deem a useless burden on our foreign currency account.

[1]Both Dehn and Rosenthal had already been dismissed from their positions in Germany at the time: see (**Siegmund-Schultze, 2009**).

[2]The claim for Landau is obviously wrong and typical of the superficial character of Kern's report, which is probably based on oral communication. See Section 9.21.3 on Landau.

[3]On Gumbel, see Chapter 6. The name "turnip-Gumbel" derives from Gumbel's alleged proposal to erect a turnip statue in memory of the First World War, because the turnip represented the hunger winter of 1916–17. This provocation by Gumbel enraged the German nationalists during the Weimar Republic.

APPENDIX F

Newspaper Profile of E. G. Kern, the Nazi functionary in Oslo, 4 December 1937 (with Størmer's Annotation, and Commentary, and Ketil Kern's Memoirs)

[Ukens Portrett, *Dagbladet*, 4 December 1937, pp. 7, 14; also: Størmer Archive, Box 11, Folder nov–des 1937]

Commentary based on the memoirs of Kern's grandson Ketil Kern follows on pp. 292–293

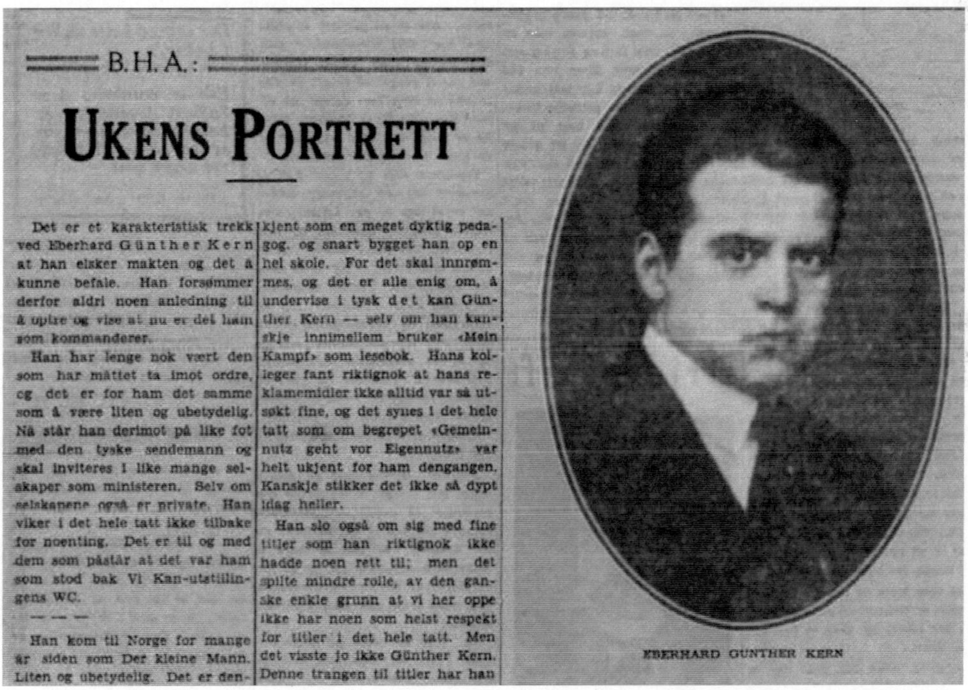

B.H.A.:

UKENS PORTRETT

Det er et karakteristisk trekk ved Eberhard Günther Kern at han elsker makten og det å kunne befale. Han forsømmer derfor aldri noen anledning til å uptre og vise at nu er det ham som kommanderer.

Han har lenge nok vært den som har måttet ta imot ordre, og det er for ham det samme som å være liten og ubetydelig. Nå står han derimot på like fot med den tyske sendemann og skal inviteres i like mange selskaper som ministeren. Selv om selskapene også er private. Han viker i det hele tatt ikke tilbake for noenting. Det er til og med dem som påstår at det var ham som stod bak Vi Kan-utstillingens WC.

— — —

Han kom til Norge for mange år siden som Der kleine Mann. Liten og ubetydelig. Det er denkjent som en meget dyktig pedagog, og snart bygget han op en hel skole. For det skal innrømmes, og det er alle enig om, å undervise i tysk det kan Günther Kern — selv om han kanskje innimellem bruker «Mein Kampf» som lesebok. Hans kolleger fant riktignok at hans reklamemidler ikke alltid var så utsøkt fine, og det synes i det hele tatt som om begrepet «Gemeinnutz geht vor Eigennutz» var helt ukjent for ham dengangen. Kanskje stikker det ikke så dypt idag heller.

Han slo også om sig med fine titler som han riktignok ikke hadde noen rett til; men det spilte mindre rolle, av den ganske enkle grunn at vi her oppe ikke har noen som helst respekt for titler i det hele tatt. Men det visste jo ikke Günther Kern. Denne trangen til titler har han

EBERHARD GUNTHER KERN

FIGURE F.1. *Dagbladet*, 4 December 1937, p. 7 (extract from profile of Kern)

Dagbladet *was a Norwegian daily with a distribution across the whole country, founded in 1869 and usually under the influence of the traditional party Venstre ("left"), which was and is today politically liberal but not necessarily leftist. After*

1933, Dagbladet *gained a reputation for very sharp criticism of National Socialism in Germany. The editor-in-chief during these years was Einar Skavlan. The article included here is a very polemical profile of the NSDAP leader in Oslo, Eberhard Günther Kern, to whom, as we saw in Chapter* 3, *the German ICM delegation had to report. The original article features a large photograph of Kern on p. 7, which underscores the sarcastic character of the format "portrait of the week"; see Figure* F.1.

According to Wikipedia,[1] *Kern lived 1899–1974. He was deported from Norway to Germany in 1947, after the Norwegians had not found enough material to put him on trial. Wikipedia also says that Kern was sacked by the Nazis from his post as party representative in Oslo in 1937, the same year that this article appeared. The reason was allegedly that he did not activate enough Germans in Norway for political work. This seems to be corroborated by parts of this article. Further commentary on the article follows the English translation below, based on the 2013 memoirs of Kern's grandson Ketil Kern.*

A clipping of this article may be found in the Størmer Papers; a handwritten note appears on Størmer's copy, just under the article title: "Naar skitt kommer til ære!" (see Figure F.2*). This is the first half of a Norwegian proverb which ends "... vet det ikke hvordan det skal være", and may be rendered into English as "When shit comes to honour it does not know how to behave". Indeed, the entire article is written by the anonymous journalist B.H.A. in the same scathing style.*

As an indication of the atmosphere in which this article appeared, we note that at the bottom right-hand corner of p. 7 of the newspaper, one finds an advertisement for a company "Jotun" which sells paint, and is still thriving today (see Figure F.3, *p. 292). The text of the advert, "Hele verden ruster undtatt det som er malt med ARCANOL" ("The entire world arms for war (is rusting), except that part of it which is painted with ARCANOL"), bases its pun on the ambiguity of the Norwegian verb "ruste". We reproduce here facsimiles of the two details from the article (Figures* F.2 *and* F.3*).*

The article on Kern allows us to draw interesting conclusions about the problematic relationship between Norwegians and the German authorities at this time, and provides some reflection of the diminishing role of the German language in Norwegian surroundings.

[1] `https://no.wikipedia.org/wiki/Eberhard_G\unhbox\voidb@x\bgroup\accent127u\penalty\@M\hskip\z@skip\egroupnther_Kern` (accessed 31 May 2019)

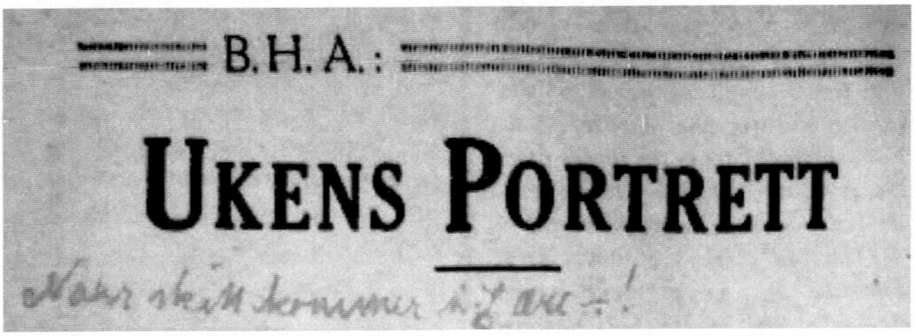

FIGURE F.2. Størmer's handwritten comment on the newspaper profile of Kern, *Dagbladet*, 4 December 1937, p. 7 (Source: National Library, Oslo, unprocessed 115 Carl Størmer Archives)

B.H.A.: Portrait of the week[2]

It is a characteristic trait of Eberhard Günther Kern that he loves power and to give orders. Therefore he never misses an opportunity to show up and to demonstrate that it is now he who is in command.

Long enough it has been he who had to obey orders, and this to him is the same as being small and unimportant. But now he is on the same level as the German ambassador[3] and is [supposed] to be invited to as many parties as the minister. Even if the parties are private. He does not flinch from anything. There are those who claim that it was he who was behind the WC of the 'We can' exhibition.[4]

— — —

He came to Norway many years ago as Der kleine Mann [German for 'The little man'].[5] Small and unimportant. From this perspective one has to look at his appearance now. He had nothing to say and nothing to do, and now he is sitting as the representative of the big state party and, in addition, perceives himself bigger than he is in reality.

But he used to be an aggressive young man and knew already then how to put himself into the limelight at appropriate, and particularly at inappropriate, times. When he started accepting students in German he became known as a very able pedagogue and he quickly built an entire school. And it should be admitted,[6] and

[2]We have not been able to decipher the acronym B.H.A.

[3]Heinrich Sahm, who is mentioned (though not by name) in Lietzmann's congress report: see, for example, p. 279.

[4]The writer alludes to the exhibit about modern industrial design that took place in Oslo in 1938 with political overtones, and which was apparently already being planned in December 1937 when this article appeared. It is not clear to which incident the remark alludes, but the very word "WC" indicates an embarrassing episode. One might also suspect that this is only a hyperbolic, jocular remark and that there was no real involvement of Kern in that episode.

[5]This is probably meant both in a transferred sense and in a concrete physical sense (**Kern, 2013**, p. 41).

[6]This "admission" appears to be ambiguous, particularly in connection with Hitler's *Mein Kampf.* Kern was probably an able linguist, but the fact that — at least officially (see the commentary at the end of the appendix) — he as a German translated Hitler's infamous book into Norwegian (1941) may also point to the lack of native Norwegian translators who wanted to compromise themselves with the job. Also the term "to teach German" has the political connotation of Germans pushing around other nationals and the episode about Kern's association

all agree on it, this, to teach German, Günther Kern knows how to do — even though he may occasionally use 'Mein Kampf' as textbook. His colleagues found, however, that his methods of advertisement were not always very decent, and it seems as though the notion 'Gemeinnutz geht vor Eigennutz'[7] was totally unknown to him then. Maybe the notion is not so deeply rooted in him today either.

He also showed off with nice titles to which he actually had no right; but this was of little importance simply because we up here [in Norway] have no respect for titles anyway. But this Günther Kern did not know. He still has this affinity for titles, by the way. Therefore he fancies to call himself 'bearer of authority' ['der Hoheitsträger'].

After a few years he became German teacher at the Broadcast[8] and thus received a considerable push ahead. He became known in broad circles and his school received many applications — the able teacher that he was.

But then the educational plan at the Broadcast was changed; there was no space anymore for three languages in the programme and it was decided to alternate between German, English, and French. The first to go was German. And thus Günther Kern too.[9] But this was the last straw. A storm of criticism rose in German newspapers against Norway and the Norwegian radio. It was called heaping scorn on Germany and the Third Reich. Exactly the same arguments which we hear on similar occasions. Who was behind this press campaign in Germany has never been found out, but it was in any case criticism which would not exactly contribute to a good understanding between the two countries.

Günther Kern was in many respects an energetic man. He was present everywhere, and the fact that he was a foreigner and head of the local branch of the NSDAP did not matter. He even succeeded in becoming translator for the Royal Norwegian Department of Foreign Affairs.[10]

— — —

Günther Kern shows the same aggressiveness as a party leader. Although he was not always as convinced a Nazi as he is today. It was not long before Hitler definitely seized power in Germany that Günther Kern took part in a debate in Oslo where Nazism was discussed. On that occasion he stood in strong opposition to Hitler and NS. But he quickly saw which way the wind blew, and he did not waste time to re-saddle.

And so he became party leader. He was present at the Eastern train station [østbanestasjon in Oslo] and gave the Heil Hitler salute when an official visit came from Germany, and he popped up here and there as a representative for the Third Reich. He was not always welcomed with enthusiasm. He has no talent at all for being liked or being popular.

with the Norway radio (see below) is quite revealing about the gradual falling behind of the German language in comparison to English and French.

[7] "Common benefits come before private benefits": this is an old German legal phrase which had already been conspicuously misused by the Nazis in the NSDAP programme of 1920. It is clear that the writer here alludes to the Nazi demagogy.

[8] This refers to the Norwegian Broadcasting Service, *Norsk rikskringkasting* (NRK), which was founded in 1933 and still exists under the same name today.

[9] Ketil Kern dates this to 1935 (**Kern, 2013**, p. 138).

[10] One probably has to read this as a criticism of naive Norwegian politicians.

He worked hard to gain influence within the Norwegian-German-Austrian Association [Norsk-tysk-østerriksk forening] but was bluntly refused. It was only professor Klaus Hansen[11] who did not turn his back on him. But it was only through Günther Kern that his association could exist and receive guest speakers from Germany. And exactly through this association the professor has consolidated his position.[12]

— — —

— The other association has never shown any interest in what I have been able to offer them, Günther Kern said in an interview with T.T.[13] *As a consequence I have never insisted [on my offers].*

But to the contrary! Not only did he insist, he forced himself upon them. When the Norwegian-German-Austrian Association once arranged a dinner for one of its German speakers, the organisers were forced to invite Günther Kern as well. On the whole so much was done in order to avoid scandal. It was only Günther Kern who was not nervous about such things. After all, he had the entire party behind him. If the leadership of the Association would not comply, at least the speaker would have to. And the German ambassador. When therefore a reception for a German guest was organised on another occasion the simple result was that the Association had to celebrate without an honorary guest. And without the minister. But also without Günther Kern! And with this demonstration the Association won respect.[14]

It did not go better for Günther Kern when he tried to gather into his hands all power when it came to the German organisations in Oslo. He called a conference and a meeting and proposed plans according to which he should become the supreme leader for the (German) colony associations in Norway. But he was turned down, practically unanimously.

As a compensation, he rules supremely over the party meetings at Majorstuehuset.[15] [14] He does not have there terribly many members, there are even rumours that several Germans have become Norwegian citizens in order to not to come under his command. But there are of course problems connected to such open resistance. It happens that one wishes to revisit one's old country every once in a while, and then one never knows what Günther Kern may have to report.

— — —

[11]Klaus Hansen (1895–1971), a prominent Norwegian pharmacologist, whose work on blood alcohol levels led to the passing in Norway in 1936 of the world's first law on legal levels for driving. He was the key person of the Nazi-friendly Norwegian-German Association between 1934 and 1944. From 1940 he collaborated with the German occupiers. In 1948, he was deprived of his professorship, and went on to work as a general practitioner and dermatologist. See: `https://nbl.snl.no/Klaus_Hansen` (accessed 31 May 2019).

[12]This remark is curious. It would be surprising if Hansen's political engagement for the Germans had helped him at this point of time in his position in Norway, where he was apparently influential as a pharmacologist.

[13]Probably the newspaper *Tidens Tegn.*

[14]This somewhat unclear description seems to say the following: the Norwegian-German-Austrian Association refrained from inviting Kern on the next occasion after he had created embarrassment at the previous dinner. Kern therefore forced the next German speaker not to take part either, which in turn made it impossible for the Association to invite the Norwegian minister. On top of that the Association apparently went on with the reception without any honorary guests at all, using this as a public demonstration of its independence.

[15]According to Ketil Kern (**Kern, 2013**, p. 124), this is the house where Kern had his language school and where the family lived. It is located in the Oslo quarter Majorstua.

FIGURE F.3. Advertisement for the paint company "Jotun" appearing next to the newspaper profile of Kern (*Dagbladet*, 4 December 1937, p. 7)

Unfortunately, the Kern type is not rare. The little man who suddenly starts climbing up the ladder and who climbs so quickly as to lose his balance. It is not even the first time that we have come to know such men in Oslo. Some years ago we were saddled with press attaché Bogs.[16] He had the same ability to turn up in places he should have stayed away from. And he had the same ability to create turmoil.

But Bogs was under the command of the Foreign Office in Berlin. And there they quickly understood that he was not suited as successor to Karl Behrens[17] in Oslo. Instead he was sent to South America. Maybe one finds another sphere of activity also for Günther Kern, a place where his abilities are more appropriate than as Ortsgruppenführer and Hoheitsträger in Norway.

Commentary on the anonymous 1937 article on Günther Kern, based on the 2013 memoirs of his grandson Ketil

Wikipedia's article on Günther Kern mentions Kern's grandson Ketil Kern as a source. Ketil Kern is a photographer in Oslo and the only grandchild of Kern's still living in Norway under the original name Kern. Ketil Kern published personal memoirs in 2013 whose Norwegian title translates as *What do you have to do with Hitler, Dad?* (**Kern, 2013**). The cause for publishing his memoirs was that Ketil Kern had been informed by a journalist in 2002 that the Norwegian publisher Nexus had finished the republication of Hitler's *Mein Kampf* using the Norwegian translation which had been made in 1941 by Kern's grandfather Günther Kern. Ketil Kern had been ignorant of this particular fact although he knew in general terms about his grandfather's Nazi past. He went to court and stopped and had destroyed the largely unsold edition. The incident triggered his desire to engage with his grandfather's life in Norway. Kern studied the file kept in the Norwegian State Archives (Riksarkiv) in Oslo, which deals with the preparation of the unsuccessful Norwegian trial against his grandfather. Ketil Kern's 2013

[16]We have no information on this Bogs.

[17]Again, we have no information on this Behrens.

book largely confirms the impression given by the article in *Dagbladet* in 1937. Family letters seem to show that Günther Kern and his Norwegian wife Sølvi, who met whilst studying German and philosophy in Hamburg before 1922, were both early supporters of German National Socialism (**Kern, 2013**, pp. 100–122). Sølvi found this ideology a broader and more satisfactory world view than the (in her opinion) narrow Norwegian nationalism which had been on the rise since Norwegian independence in 1905. The couple came to Norway in 1922 and the remark (see p. 290 above) in the article regarding Kern's changing his mind with Hitler's rise to power has therefore to be considered with caution. Ketil Kern discovered that Günther Kern had not sent his four children to a Norwegian school because he considered the schools "too communist" (**Kern, 2013**, p. 197). He preferred private teaching for his children, but this was often wanting and thus a liability for the future career of Ketil's father.

The real punch line of Kern's 2013 book, however, follows on its p. 216: the discovery that it was not his grandfather Günther, but his grandmother Sølvi who had translated Hitler's book into Norwegian. For Ketil this explained also the very special Norwegian dialect "samnorsk" used in the translation, which would probably have been beyond the reach even of the very able Norwegian speaker Günther Kern.[18] This discovery and the apparent exploitation of his grandmother by her husband confirmed the traditional gender roles; at the same time Sølvi — by hiding behind her husband — evaded the accusation of having translated the infamous book from the language of the occupiers. Sølvi and her daughter Krimhildt were killed in late 1944 when their ship was sunk near Göteborg by a torpedo (of unknown origin), as they travelled from Germany to Norway (**Kern, 2013**, p. 45).

After his deportation in 1947, Günther Kern founded another language school in the German state of Schleswig Holstein. Ketil's occasional visits confirmed that Günther Kern was not given to self-critical inspection of his role in Norway; for example, the latter condemned the Allied "Vandals" for their destruction of a church in Lübeck during the war (**Kern, 2013**, p. 47).

[18]Ironically, this combination of the (even today separate) main Norwegian dialects *bokmål* and *nynorsk* was at the time supported above all by Social Democrats such as the historian and foreign minister Halvdan Koht.

APPENDIX G

Transcription of Two Recommendations Made by the Subcommittee on Conferences to the Organizing Committee of the Planned 1940 ICM

[MAA Headquarters, Austin, TX, round letter, 15 April 1938, 5pp., pp. 2–3, M. H. Stone, chairman]

(A) RECOMMENDATION. That in connection with the 1940 International Congress of Mathematicians, conferences be organized on the following four topics: (1) algebra; (2) mathematical logic; (3) theory of measure and integration, probabilities, and allied topics; (4) topology.

Both algebra and topology are well-defined fields in which the steady progress of recent years continues unchecked. It is clear that conferences in these fields would be of great value to specialists in organizing surveys and discussions of recent advances. The strengthening of the bonds between these fields, especially algebra, and other branches of mathematics provides a sound basis for general interest on the part of non-specialists. The Organizing Committee has already approved the organization of a conference on topology by a vote taken at its informal meeting of September 7, 1937. In this connection it should be noted that the tentative invitation of Polish mathematicians for a topological conference to be held in Warsaw in 1939 has been permitted to lapse.

The intense specialized interest in symbolic logic and the remarkable fundamental progress which has lately been made virtually guarantee the success of a conference in this field. There is some reason to hope that discussion and debate in such a conference may contribute to deeper understanding, inasmuch as the bitter feeling which once divided various schools of thought has been replaced by more conciliatory attitudes. The cooperation of the Association for Symbolic Logic in organizing the proposed conference seems assured. The interest of non-specialists in such a conference would naturally depend upon its scope. To enhance its general appeal it might concern itself not only with symbolic logic in the strict sense but also with the metamathematical relations between logical theory and such branches of mathematics as algebra, topology, and set-theory.

A conference on the theory of measure and integration, probabilities, and allied topics can perform a very valuable service in correlating and organizing a body of knowledge which has been growing in a peculiarly scattered and unobtrusive way in recent years. For example, while outstanding contributions to ergodic theory and harmonic analysis are well known, the tendencies in the theory of measure and integration are less generally appreciated. A conference of the kind suggested would interest a large number of mathematicians because of its numerous and varied contacts with other branches of mathematics, ranging from abstract functional analysis to statistics. It would, for instance, be appropriate to include in the program of such a conference one or two addresses on the relatively elaborate modern statistical techniques.

With regard to the topics of statistics and mathematical economics, your Subcommittee entertains doubts which it does not feel competent to resolve. Both fields are branches of applied mathematics which derive their interest from the immediacy and the practical nature of the problems they are designed to meet, rather than from their intrinsic mathematical content or from their progress toward mastery of those problems. Recognizing that conferences on these fields might not only interest specialists in them but also serve a desirable and useful end in strengthening the relations between pure mathematics and these branches of applied mathematics, your Subcommittee therefore submits the following

(B) RECOMMENDATION. That the Organizing Committee approve conferences on statistics and mathematical economics, either as separate programs or as a single joint program, provided that the appropriate scientific associations are sufficiently interested to cooperate financially and otherwise in organizing them.

APPENDIX H

Transcription of an Attachment to the Round Letter of the Organizing Committee of the Planned 1940 ICM, Listing the Mathematicians Invited

[MAA Headquarters, Austin, TX, 17 May 1939, signed Graustein and Stone, the copy from which we quote was addressed to W. D. Cairns]

INTERNATIONAL CONGRESS OF MATHEMATICIANS
Cambridge, Massachusetts
September 4–12, 1940

Speakers Invited for Plenary Sessions

Albert, A. A.
Alexandroff, Paul
Artin, Emile
Cramér, Harald
Denjoy, Arnaud
Dirac, Paul Adrien Maurice
Gödel, Kurt
Hodge, William Vallance Douglas
Hopf, Heinz
Khintchine, A.
Lefschetz, Solomon
Moore, R. L.
Morse, Marston
von Neumann, John
Pearson, Egon Sharpe
Rademacher, H. A.
Riesz, Marcel
Stone, M. H.
Synge, J. L.
Tarski, Alfred
Taylor, Geoffrey Ingram
Tinbergen, Jan
Tonelli, Leonida
Vinôgradoff, Ivan M.
van der Waerden, B. L.
Weyl, Hermann

Additional Speakers Invited for Sessions of Conferences

Conference on Algebra

Aitken, Alexander Craig
Birkhoff, Garrett
Brauer, Richard
Chevalley, Claude
Deuring, M.

Conference on Theory of Measure and Integration, Probabilities, and Allied Topics

Banach, Stefan
Cantelli, Francesco Paolo
Hedlund, G. A.
Hostinsky, Bohuslav
Kolmogoroff, A.

Hall, Philip
Jacobson, Nathan
Krull, Wolfgang
MacDuffee, C. C.
MacLane, Saunders
Ore, Oystein
Schilling, O. F. G.
Zariski, Oscar

Lévy, Paul
von Mises, Richard
Ridder, J.
Riesz, Friedrich
Steinhaus, Hugo
Ulam, S. M.
Young, Laurence Chisholm
Zygmund, Antoni

Conference on Mathematical Logic

Bernays, Paul
Carnap, Rudolf
Church, Alonzo
Chwistek, Leon
Gentzen, Gerhard
Heyting, Arend
Kalmár, László
Quine, W. V.
Scholz, Heinrich
Skolem, Thoralf Albert

Conference on Topology

Alexander, J. W.
Borsuk, Karl
Čech, Eduard
Hurewicz, Witold
Kline, J. R.
Kuratowski, Kazimierz
Pontrjagin, Lev Semiovich
Stoïlow, Siméon
Whitney, Hassler

Bibliography

Archival sources

AAMS: Archives of the American Mathematical Society, Providence, RI.

Archivio Centrale dellos Stato, Roma, Presidenza del Consiglio dei Ministri, Gabinetto (1934–1936), b. (folder) 2022, f. (file) 14.3.6375 "Oslo. Congresso Internazionale dei Matematici".

Bessel-Hagen, Nachlass E. Bessel-Hagen, Bonn University Archives, George David Birkhoff Papers, no. 328.

Birkhoff Papers, Harvard University Archives, George David Birkhoff Papers. HUG 4213.

Brevsamling 386: Letters to Koht, Halvdan, Manuscript Collection, National Library, Oslo.

Brevsamling 698B-8: Letters to Ore, Øystein, Manuscript Collection, National Library, Oslo.

Brevsamling 719: Letters to Brun, Viggo, Manuscript Collection, National Library, Oslo.

DKZ 1: Hoover Institution, Stanford University California, Deutsche Kongresszentrale (DKZ), Box 362, f. DKZ, Berlin Richtlinien für die Leiter deutscher Abordnungen zu Kongressen im Ausland, 17 pages, type-written.

DKZ 2: Hoover Institution, DKZ, Box 362, f. DKZ, Berlin, Die Deutsche Kongress-Zentrale (Jahresbericht 1936), 8 pages, type-written.

DKZ 3: Hoover Institution, DKZ, box 203.

EN: Friedrich Engel Nachlass, Universitätsbibliothek Giessen.

Freiburg University Archives, E4 37, DMV (Schriftführer Müller) 1936–1942.

Göttingen University Archives.

Hasse Papers: code Ms. H. Hasse in Handschriftenabteilung, Niedersächsische Staats- und Universitätsbibliothek, Göttingen.

Hecke Papers: Papers of Erich Hecke, University Archives Hamburg.

IML: Institut Mittag Leffler, correspondence between, among others, Mittag-Leffler and Størmer.

KUD: Kirke- og undervisningsdepartementet, 1. skolekontor D, serie En, Nordisk og internasjonalt samarbeid, nr. 880, Internasjonale Kongresser, hyllenummer 3A03432, Riksarkivet, Oslo.

MAA Archives, Mathematical Association of America, Austin, TX.

Mordell Papers, St John's College Library, Cambridge.

Norges Bank: Norges Bank, Statistisk avdeling, serie D, nr. 83, Møter o.l. Matematikerkongress, 1935, hyllenummer 3619566. Rigsarkivet, Oslo.

RCP: Richard Courant Papers, New York City, New York University Archives.

REM: files of Reichs- und Preussische Minister für Wissenschaft, Erziehung und Volksbildung, Bundesarchiv Berlin, nos. 2905 and 3190.

R. G. D. Richardson Papers, Brown University Archives, Providence, RI.

RVM: Richard von Mises Papers, Harvard University Archives.

Marshall Harvey Stone Papers at AAMS.

Størmer Archive, Ubehandlet 115, Carl Størmer Brevarkiv m.m., Boxes 10 and 11, Manuscript Collection, National Library, Oslo.

Veblen Papers, Library of Congress, Oswald Veblen Papers, Washington, DC.

van der Waerden, Personal file B. L. van der Waerden PA 70, University Archives Leipzig.

WLP: Walther Lietzmann Papers, Göttingen, Municipal Archives, Depositum 89.

WLR 1936: Unpublished Report by W. Lietzmann to REM, Bericht über die deutsche Delegation auf dem Internationalen Mathematiker-Kongress in Oslo, 13.-18.7.1936 (15 pages type-written), in REM 2905, fols. 384–398.

WP: E. B. Wilson Papers, Harvard University Archives, HUG 4878.

Newspaper articles

Aftenposten, 27 October 1920: "Den internationale matematikerkongres i Strassbourg", p. 4.

Aftenposten, 13 July 1936: "Abel møter vi stadig", no. 346, pp. 1+5.

Aftenposten, 14 July 1936a: "Utenriksministeren hylder matematikken som åndens stormagt", no. 348, p. 2.

Aftenposten, 14 July 1936b: "Universitetet ønsker matematikerne velkommen", no. 348, p. 3.

Aftenposten, 15 July 1936a: "Tyskerne bekranser Abel-støtten", no. 349, p. 3.

Aftenposten, 15 July 1936b: "Matematikerkongressen — hyldest til Sophus Lie", no. 350, p. 1.

Aftenposten, 18 July 1936: "Matematikerne i den rumenske legasjon", no. 356, p. 7.

Arbeiderbladet, 14 July 1936: "700 berømte matematikere", no. 161, p. 5.

Arbeiderbladet, 16 July 1936:[1] "1500 tyske videnskapsmenn jaget av nazistene", no. 163, p. 3.

Dagbladet, 15 July 1936: "Sophus Lies minne hedres", no. 162, p. 3.

Morgenbladet, 15 July 1936: "Den ledende videnskap i åndens ekspansjon", no. 194, p. 5.

Morgenbladet, 16 July 1936: "Sophus Lie får sin byste på Blindern universitet", no. 195, p. 5.

Politiken, 23 June 1936: "Brogede Skæbner hjem med Amerika-baaden", pp. 11–12.

Politiken, 26 July 1936: "Et let tilgængeligt og billigt Land hvorfra man kommer hjem med Sundhed", p. 7.

Tidens Tegn, 14 July 1936a: "500 studenter ved Jerusalems universitet", no. 162, p. 2.

Tidens Tegn, 14 July 1936b: "Matematikerkongressen hylder geniet Sophus Lie", no. 162, p. 6.

Tidens Tegn, 18 July 1936: "Seks døgns jernbanereise for aa snakke 15 minutter i Oslo", no. 164, p. 1.

Congress proceedings[2]

ICM (1904). *Verhandlungen des dritten Internationalen Mathematiker-Kongresses in Heidelberg vom 8. bis 13. August 1904* (A. Krazer, ed.), Teubner, Leipzig (published 1905).

ICM (1908). *Atti del IV Congresso Internazionale dei Matematici (Roma, 6–11 Aprile 1908)* (G. Castelnuovo, ed.), 3 vols., Tipografia della R. Accademia dei Lincei, Roma (published 1909).

ICM (1912). *Proceedings of the Fifth International Congress of Mathematicians (Cambridge, 22–28 August 1912)* (E. W. Hobson and A. E. H. Love, eds.), 2 vols., Cambridge University Press (published 1913).

ICM (1920). *Comptes rendus du Congrès international des mathématiciens (Strasbourg, 22–30 Septembre 1920)* (Henri Villat, ed.), Imprimerie et Librairie Édouard Privat, Toulouse (published 1921).

ICM (1924). *Proceedings of the International Mathematical Congress held in Toronto, August 11–16, 1924* (J. C. Fields, ed.), 2 vols., University of Toronto Press (published 1928).

ICAM (1924). *Proceedings of the First International Congress for Applied Mechanics (Delft, 1924)* (C. B. Biezeno and J. M. Burgers, eds.), Technische Boekhandel en Drukkerij J. Waltman Jr., Delft (published 1925).

ICM (1928). *Atti del congresso internazionale dei matematici, Bologna, 3–10 Set. 1928 (VI)*, 6 vols., Nicola Zanichelli, Bologna (published 1929).

SCM (1929). *Den syvende skandinaviske matematikerkongress i Oslo 19–22 August 1929.* A. W. Brøggers Boktrykkeri A/S, Oslo (published 1930).

---

[1]We note a curious issue with dating here: the second author has found the same issue of *Arbeiderbladet* with a different p. 13, and an issue from 17 July with the correct article on Gumbel, also on p. 13. However, a copy of the newspaper article that is in the possession of Henrik Kragh Sørensen has the Gumbel article on p. 13 for 16 July.

[2]Listed chronologically.

ICM (1932). *Verhandlungen des Internationalen Mathematiker-Kongresses, Zürich 1932* (Walter Saxer, ed.), 2 vols., Orell Füssli, Zürich.

ICM (1936). *Comptes rendus du congrès international des mathématiciens, Oslo 1936*, 2 vols, A. W. Brøggers Boktrykkeri A/S, Oslo (published 1937).

ICAM (1938). *Proceedings of the Fifth International Congress for Applied Mechanics held at Harvard University and the Massachusetts Institute of Technology (Cambridge, Massachusetts, September 12–16, 1938)* (J. P. den Hartog and H. Peters, eds.), Wiley/Chapman Hall (published 1939).

ICM (1950). *Proceedings of the International Congress of Mathematicians, Cambridge, Massachusetts, U.S.A., August 30–September 6, 1950* (Lawrence M. Graves, Einar Hille, Paul A. Smith, Oscar Zariski, eds.), 2 vols., AMS (published 1952).

ICM (1962). *Proceedings of the International Congress of Mathematicians, Stockholm, 15–22 August 1962*, Institut Mittag-Leffler, Djursholm, Sweden (published 1963).

ICM (1978). *Proceedings of the International Congress of Mathematicians, Helsinki 1978* (Olli Lehto, ed.), 2 vols., Academia Scientarum Fennica, Helsinki (published 1980).

ICM (1986). *Proceedings of the International Congress of Mathematicians, Berkeley, California, August 3–11, 1986* (Andrew M. Gleason, ed.), 2 vols., AMS (published 1987).

ICM (2014). *Proceedings of the International Congress of Mathematicians: August 13–21, 2014, Seoul, Korea* (Sun Young Jang, Young Rock Kim, Dae-Woong Lee and Ikkwon Yie, eds.), 4 vols., Kyung Moon Sa, Seoul.

1936 plenary lectures

Ahlfors, Lars V. (1936/7). *Geometrie der Riemannschen Flächen*, in (**ICM, 1936**), vol. I, pp. 239–248. [JFM 63.0300.05; Zbl 0018.26202]

Banach, Stefan (1936/7). *Die Theorie der Operationen und ihre Bedeutung für die Analysis*, in (**ICM, 1936**), vol. I, pp. 261–268. [JFM 63.0349.01; Zbl 0018.36402]

Birkhoff, G. D. (1936/7). *The foundation of quantum mechanics*, in (**ICM, 1936**), vol. I, pp. 207–225. [JFM 63.1385.03; Zbl 0018.33602]

Bjerknes, V. (1936/7). *New lines in hydrodynamics*, in (**ICM, 1936**), vol. I, pp. 186–188. [JFM 63.0764.02]

Cartan, Élie (1936/7). *Le rôle de la théorie des groupes de Lie dans l'évolution de la géométrie moderne*, in (**ICM, 1936**), vol. I, pp. 92–103. [JFM 63.1155.04; Zbl 0018.27202]

van der Corput, Jan G. (1936/7). *Diophantische Approximationen*, in (**ICM, 1936**), vol. I, pp. 249–260. [JFM 63.0918.03; Zbl 0018.29501]

Fréchet, Maurice (1936/7). *Mélanges mathématiques*, in (**ICM, 1936**), vol. I, pp. 269–283. [JFM 63.1004.01]

Fueter, Rud (1936/7). *Die Theorie der regulären Funktionen einer Quaternionenvariablen*, in (**ICM, 1936**), vol. I, pp. 75-91. [JFM 63.0310.04; Zbl 0018.31601]

Hasse, H. (1936/7). *Über die Riemannsche Vermutung in Funktionenkörpern*, in (**ICM, 1936**), vol. I, pp. 189–206. [JFM 63.0881.01; Zbl 0018.34302]

Hecke, E. (1936/7). *Neuere Fortschritte in der Theorie der elliptischen Modulfunktionen*, in (**ICM, 1936**), vol. I, pp. 140–156. [JFM 63.0339.02; Zbl 0018.35703]

Mordell, Louis Joel (1936/7). *Minkowski's theorems and hypotheses on linear forms*, in (**ICM, 1936**), vol. I, pp. 226–238. [JFM 63.0921.04; Zbl 0018.29502]

Neugebauer, O. (1936/7). *Über griechische Mathematik und ihr Verhältnis zur vorgriechischen*, in (**ICM, 1936**), vol. I, pp. 157–170. [JFM 63.0799.01; Zbl 0018.33902]

Nielsen, J. (1936/7). *Einige Methoden und Ergebnisse aus der Topologie der Flächenabbildungen*, in (**ICM, 1936**), vol. I, pp. 128–139. [JFM 63.0553.02; Zbl 0018.42503]

Ore, Ø. (1936/7). *On the decomposition theorems of algebra*, in (**ICM, 1936**), vol. I, pp. 297–307. [JFM 63.0868.03; Zbl 0018.29002]

Oseen, C. W. (1936/7). *Probleme der geometrischen Optik*, in (**ICM, 1936**), vol. I, pp. 171–185. [JFM 63.0717.04; Zbl 0019.04703]

Siegel, C. L. (1936/7). *Analytische Theorie der quadratischen Formen*, in (**ICM, 1936**), vol. I, pp. 104–110. [JFM 63.0895.04; Zbl 0018.24404]

Størmer, C. (1936/7). *Programme for the quantitative discussion of electron orbits in the field of a magnetic dipole, with applications to cosmic rays and kindred phenomena*, in (**ICM, 1936**), vol. I, pp. 61–74. [JFM 63.0736.01]

Veblen, O. (1936/7). *Spinors and projective geometry*, in (**ICM, 1936**), vol. I, pp. 111–127. [JFM 63.0693.02; Zbl 0018.32604]

Wiener, Norbert (1936/7). *Gap theorems*, in (**ICM, 1936**), vol. I, pp. 284–296. [JFM 63.0974.01; Zbl 0019.03402]

Other published sources

Ahlfors, Lars V. (1935a). *Über eine Methode in der Theorie der meromorphen Funktionen*, Commentat. Phys.-Math., Soc. Sci. Fennica, Helsinki **8**, no. 10, 1–14.

Ahlfors, Lars (1935b). *Zur Theorie der Überlagerungsflächen*, Acta Math. **65**, no. 1, 157–194, DOI 10.1007/BF02420945. MR1555403

Ahlfors, Lars V. (1936/7). *Geometrie der Riemannschen Flächen*, in (**ICM, 1936**), vol. I, pp. 239–248. [JFM 63.0300.05; Zbl 0018.26202]

Ahlfors, Lars (1937). *Über die Anwendung differentialgeometrischer Methoden zur Untersuchung von Überlagerungsflächen*, Acta Soc. Sci. Fennica. Nova series A, Opera phys.-math. **2**, no. 6, 1–17.

Ahlfors, Lars V. (1963). *Teichmüller spaces*, Proc. Internat. Congr. Mathematicians (Stockholm, 1962), Inst. Mittag-Leffler, Djursholm, pp. 3–9. MR0176055

Ahlfors, Lars V. (1980). *Quasiconformal mappings, Teichmüller spaces, and Kleinian groups*, in (**ICM, 1978**), vol. I, pp. 71–84. MR562598; reprinted in (**Atiyah and Iagolnitzer, 1997**), pp. 10–23

Ahlfors, L. V. (1989). *The joy of function theory*, in (**Duren, 1988/1989**), vol. 3, pp. 443–447. MR1025357

Ahlfors, Lars V. (1997a). *Autobiography of Lars V. Ahlfors [reprinted from MR0688648 (84k:01066a)]*, in (**Atiyah and Iagolnitzer, 1997**), pp. 2–7. MR1622938

Ahlfors, Lars V. (1997b). *Commentary on: "Zur Theorie der Überlagerungsflächen" (1935) [reprinted from MR0688648 (84k:01066a)]*, in (**Atiyah and Iagolnitzer, 1997**), pp. 8–9, DOI 10.1142/9789812385215_0002. MR1622939

Albers, Donald J. (1998). *An interview with Lars V. Ahlfors*, College Math. J. **29**, no. 2, 82–92, DOI 10.2307/2687700. MR1615498

Albers, Donald J., G. L. Alexanderson, and Constance Reid (1987). *International mathematical congresses: An illustrated history, 1893–1986*, Springer-Verlag, New York. MR874336

Aleksandrov, P. S. (1935). Obituary of Emmy Noether, in A. Dick, *Emmy Noether 1882–1935*, Birkhäuser, Boston, Basel, Stuttgart, 1981, pp. 153–179.

Aleksandrov, P. S. 1936. *First International Topological Congress in Moscow*, Uspekhi mat. nauk, no. 1, 260–262 (in Russian).

Aleksandrov, P. S. 1979. Pages from an autobiography, *Uspekhi mat. nauk* **34**(6), 219–249; *ibid.* **35**(3), 241–278 (in Russian); English transl.: *Russian Math. Surveys* **34**(6) (1979), 267–302; *ibid.* **35**(3), 315–358. MR562847

Aleksandrov, P. S. and H. Heinz (1935). *Topologie*, Springer, Berlin.

Anderson, John D., Jr. (2006). *Research in supersonic flight and the breaking of the sound barrier*, Chapter 3 in Pamela Mack (ed.), From engineering science to big science: the NACA and NASA Collier Trophy Research Project winners, University Press of the Pacific, pp. 59–90.

Anon. (1929a). *Notes*, Bull. Amer. Math. Soc. **35**, no. 4, 582–586.

Anon. (1929b). *Notes*, Bull. Amer. Math. Soc. **35**, no. 5, 740–746.

Anon. (1931). *Soviet mathematicians, support your journal!*, Mat. sbornik **38**, nos. 3–4, 1 (in Russian).

Anon. (1935). *VII. Philosophie et Histoire des Mathématiques*, Enseign. math. **34**, 111.

Anon. (1936). *Congrès international des mathématiciens*, Enseign. math. **34** (1935), 377–379.[3]

Anon. (1946). *Aleksandr Yakovlevich Khinchin (on the occasion of his fiftieth birthday)*, Uspekhi mat. nauk **1**, no. 2(12), 192–193 (in Russian).

Anon. (1949a). *Relations scientifiques entre les mathématiciens polonais et étrangers*, Ann. Soc. polon. math. **22**, 304–308.

[3]Although dated 1935, this cannot have been printed before 1936, as it mentions Guldberg's death in February 1936, and his subsequent replacement by Størmer as head of the Oslo ICM organising committee. Moreover, the page-overlap with (**Fehr, 1936**) indicates that the volume number here must be correct.

Anon. (1949b). *The list of mathematical papers by Prof. M. Fujiwara*, Tôhoku Math. J. **49**, 133–138; supplement: *Tôhoku Math. J. (2)* **4**, no. 3, (1952), 316–317.

Anon. (1955). *Aleksandr Yakovlevich Khinchin (on his sixtieth birthday)*, Uspekhi mat. nauk **10**, no. 3(65), 197–212 (in Russian).

Anon. (1962). *Otto Neugebauer. Bibliography*, Osiris **14**, 4–12.

Anon. (1970). Oystein [*sic*] Ore (1899–1968), *J. Comb. Theory* **8**, i–iii. MR250841

Archibald, R. C. (1938a). *A semicentennial history of the American Mathematical Society, 1888–1938*, AMS.

Archibald, Raymond Clare (ed.) (1938b). *Semicentennial addresses of the American Mathematical Society*, AMS.

Aronszajn, N. (1935). *Sur les décompositions des fonctions analytiques uniformes et sur leurs applications*, Acta Math. **65**, no. 1, 1–156, DOI 10.1007/BF02420944. MR1555402

Atiyah, Michael and Daniel Iagolnitzer (eds.) (1997). *Fields Medallists' lectures*, World Scientific Series in 20th Century Mathematics, vol. 5, World Scientific Publishing Co., Inc., River Edge, NJ; Singapore University Press, Singapore. MR1622945

Aubert, K. E. (1970). *Øystein Ore and his mathematical work* (Norwegian, with English summary), Nordisk Mat. Tidskr. **18**, 121–126. MR274254

Babbitt, Donald and Judith Goodstein (2009). *Guido Castelnuovo and Francesco Severi: two personalities, two letters*, Notices Amer. Math. Soc. **56**, no. 7, 800–808. MR2546822

Babbitt, Donald and Judith Goodstein (2011). *Federigo Enriques's quest to prove the "completeness theorem"*, Notices Amer. Math. Soc. **58**, no. 2, 240–249. MR2768117

Baer, George W. (1976). *Test case: Italy, Ethiopia, and the League of Nations*, Hoover Institution Press, Stanford.

Bakke, Bjørn Lien (2016). *Norge og konflikten mellom Italia og Etiopia 1935–1936: interesser og påvirkningskraft*, Masteroppgave i Historie, Trondheim.

Banach, Stefan (1932). *Théorie des opérations linéaires*, Monografie Matematyczne 1, Warszawa.

Banach, Stefan (1936/7). *Die Theorie der Operationen und ihre Bedeutung für die Analysis*, in (**ICM, 1936**), vol. I, pp. 261–268. [JFM 63.0349.01; Zbl 0018.36402]

Barany, Michael Jeremy (2016). *Distributions in postwar mathematics*, ProQuest LLC, Ann Arbor, MI. Thesis (Ph.D.)–Princeton University. MR3579528

Barany, Michael Jeremy (2018). *The Fields Medal should return to its roots*, Nature **553** (18 Jan), 271–273.

Barany, Michael J., Anne-Sandrine Paumier, and Jesper Lützen (2017). *From Nancy to Copenhagen to the world: the internationalization of Laurent Schwartz and his theory of distributions*, Historia Math. **44**, no. 4, 367–394, DOI 10.1016/j.hm.2017.04.002. MR3716327

Barrow-Green, June (1994). *International Congresses of Mathematicians from Zurich 1897 to Cambridge 1912*, Math. Intelligencer **16**, no. 2, 38–41, DOI 10.1007/BF03024282. MR1270839

Barrow-Green, June (2011). *An American goes to Europe: three letters from Oswald Veblen to George Birkhoff in 1913/1914*, Math. Intelligencer **33**, no. 4, 37–47, DOI 10.1007/s00283-011-9248-x. MR2865297

Battimelli, Giovanni (1996). *"With no official connection": Tullio Levi-Civita and the International Congresses of Applied Mechanics* (Italian, with English summary), Riv. Stor. Sci. (2) **4**, no. 1, 51–80 (1997). MR1611838

Battimelli, Giovanni (2016). *About the early International Congresses of Applied Mechanics*, in Peter Eberhard and Stephen Juhasz (eds.), IUTAM: a short history, 2nd ed., Springer, pp. 21–30.

Beaulieu, Liliane (1989). *Bourbaki: une histoire du groupe de mathématiciens français et de ses travaux (1934–1944)*, PhD thesis, Université de Montréal.

Behnke, Heinrich (1978). *Semesterberichte: Ein Leben an deutschen Universitäten im Wandel der Zeit*, Vandenhoeck & Ruprecht, Göttingen. MR514204

Behnke, Heinrich and P. Thullen (1934). *Theorie der Funktionen mehrerer komplexer Veränderlichen*, Ergebnisse der Mathematik und ihrer Grenzgebiete, Band 3, no. 3, Springer, Berlin.

Bertrand, Louis and Hans Keller (1936). *L'Internationale, ennemie des nations*, A. Nauck, Zurich.

Birkeland, B. (1993a). *Norske matematikere: Litt om deres liv og virke*, Gyldendal Norsk Forlag, Oslo.

Birkeland, Bent (1993b). *75 years of the Norwegian Mathematical Society* (Norwegian), Normat **41**, no. 3, 125–127. MR1248342

Birkhoff, Garrett (1935). *Combinatorial relations in projective geometries*, Ann. of Math. (2) **36**, no. 3, 743–748, DOI 10.2307/1968656. MR1503250

Birkhoff, Garrett (1976a). *The rise of modern algebra to 1936*, in (**Tarwater *et al.* 1976**), pp. 41–63.

Birkhoff, Garrett (1976b). *The rise of modern algebra, 1936–1950*, in (**Tarwater *et al.* 1976**), pp. 65–85.

Birkhoff, Garrett (1980). *Computer developments 1935–1955, as seen from Cambridge, U.S.A.*, in N. Metropolis, J. Howlett and G.-C. Rota (eds.), A history of computing in the twentieth century. A collection of essays, Academic Press, pp. 21–30.

Birkhoff, Garrett (1989). *Mathematics at Harvard, 1836–1944*, in (**Duren, 1988/1989**), vol. 2, pp. 3–58. MR1003118

Birkhoff, Garrett and Erwin Kreyszig (1984). *The establishment of functional analysis*, Historia Math. **11**, no. 3, 258–321, DOI 10.1016/0315-0860(84)90036-3. MR765342

Birkhoff, G. D. (1927). *A mathematical critique of some physical theories*, Bull. Amer. Math. Soc. **33**, no. 2, 165–181, DOI 10.1090/S0002-9904-1927-04346-4. MR1561343

Birkhoff, G. D. (1931). *Proof of the ergodic theorem*, Proc. Nat. Acad. Sci. USA **17**, no. 12, 656–660.

Birkhoff, G. D. (1936/7). *The foundation of quantum mechanics*, in (**ICM, 1936**), vol. I, pp. 207–225. [JFM 63.1385.03; Zbl 0018.33602]

Birkhoff, G. D. (1942). *What is the ergodic theorem?*, Amer. Math. Monthly **49**, 222–226, DOI 10.2307/2303229. MR6619

Bjerknes, V. (1936/7). *New lines in hydrodynamics*, in (**ICM, 1936**), vol. I, pp. 186–188. [JFM 63.0764.02]

Bjerknes, V. (1937). *Application of line integral theorems to the hydrodynamics of terrestrial and cosmic vortices*, Astrophys. Norv. **2**, no. 6, 263–339.

Blom, Gunnar (1987). *Harald Cramér 1893–1985*, Ann. Statist. **15**, no. 4, 1335–1350, DOI 10.1214/aos/1176350596. MR913560

Bohr, H. (1932). *Fastperiodische Funktionen*, Ergebnisse der Mathematik und ihrer Grenzgebiete, Band 1, no. 5, Springer, Berlin.

Bolondi, Giorgio, Angelo Guerraggio, and Pietro Nastasi (2011). *The way we were: the protagonists of the "Italian spring" in the first decades of the twentieth century*, Mathematical lives, Springer, Berlin, pp. 11–23, DOI 10.1007/978-3-642-13606-1_2. MR2743927

Bonnesen, T. and W. Fenchel (1934). *Theorie der konvexen Körper*, Ergebnisse der Mathematik und ihrer Grenzgebiete, Band 3, no. 1, Springer, Berlin.

Boobbyer, Philip (2012). *The Stalin era*, Routledge.

Born, Max and Albert Einstein (1971). *The Born–Einstein letters: correspondence between Albert Einstein and Max and Hedwig Born from 1916 to 1955 with commentaries by Max Born*, translated by Irene Born, Macmillan.

Bott, Raoul, Clifford Earle, Dennis Hejhal, James Jenkins, Troels Jorgensen, Albert Marden, and Robert Osserman (1998). *Lars Valerian Ahlfors (1907–1996)*, Notices Amer. Math. Soc. **45**, no. 2, 248–255. Edited and with an introduction by Steven G. Krantz. MR1601820

Bottazzini, Umberto (1989). *I matematici nell'Italia del Risorgimento: osservazioni sull'emergere di una comunità scientifica*, in Vittorio Ancarani (ed.), La scienza accademica nell'Italia post-unitaria: discipline scientifiche e ricerca universitaria, Franco Angeli, Milan, pp. 37–52.

Bottazzini, Umberto (1994). *Va' pensiero: Immagini della matematica nell'Italia dell'ottocento. [Images of mathematics in nineteenth-century Italy]* (Italian, with Italian summary), Società Editrice Il Mulino, Bologna. MR1789608

Bottazzini, Umberto and Pietro Nastasi (2001). *La matematica*, in (**Simili and Paolini, 2001**), vol. II, pp. 141–164.

Bottazzini, Umberto and Pietro Nastasi (2013). *La patria ci vuole eroi: Matematici e vita politica nell'Italia del Risorgimento. [Mathematicians and political life in the Italian Risorgimento]*, Zanichelli Editore S.p.A., Bologna. MR3379930

Brack-Bernsen, Lis (2016). *Otto Neugebauer's visits to Copenhagen and his connection to Denmark*, in (**Jones *et al.*, 2016**), pp. 107–126, DOI 10.1007/978-3-319-25865-2_3. MR3496895

Brauer, R., E. Noether, and H. Hasse (1932). *Beweis eines Hauptsatzes in der Theorie der Algebren*, J. Reine Angew. Math. **167**, 399–404, DOI 10.1515/crll.1932.167.399. MR1581351

Brauer, Richard and Hermann Weyl (1935). *Spinors in n Dimensions*, Amer. J. Math. **57**, no. 2, 425–449, DOI 10.2307/2371218. MR1507084

Brechenmacher, Frédéric, Guillaume Jouve, Laurent Mazliak and Rossana Tazzioli (eds.) (2016). *Images of Italian mathematics in France: the Latin sisters, from Risorgimento to fascism*, Trends in the History of Science, Birkhäuser.

Brelot, M. (1966). *Norbert Wiener and potential theory*, Bull. Amer. Math. Soc. **72**, no. 1, pt. 2, 39–41, DOI 10.1090/S0002-9904-1966-11457-X. MR183888

Brigaglia, Aldo (2002). *The first international mathematical community: the Circolo matematico di Palerno*, in (**Parshall and Rice, 2002**), pp. 179–200. MR1907176

Brigaglia, Aldo and Simonetta Di Sieno (2009). *L'opera politica di Luigi Cremona attraverso la sua corrispondenza. Prima parte. Gli anni dell'entusiasmo e della creatività*, Mat. Soc. Cult. Riv. Unione Mat. Ital. (1) **2**, no. 3, 358–388; *Seconda parte. Il crollo delle speranze e il lavoro organizzativo*, *ibid.* **3**, no. 2, 137–179+311. MR2656428

Brigaglia, Aldo, Simonetta Di Sieno, Angelo Guerraggio, and Pietro Nastasi (eds.) (1998). *La matematica italiana dopo l'Unità: Gli anni tra le due guerre mondiali. [The years between the two World Wars]*, La Matematica Italiana dopo l'Unità [Italian Mathematics after Unity], vol. 1, Marcos y Marcos, Milan. MR1682026

Brooks, Robert and Mikhail Sodin (eds.) (2000). *Lectures in memory of Lars Ahlfors*, Israel Mathematical Conference Proceedings, vol. 14, The Gelbart Research Institute for Mathematical Sciences and The Emmy Noether Research Institute of Mathematics, Bar-Ilan University.

Bru, Bernard (2003). *Souvenirs de Bologne*, Journal de la Société Française de Statistique **144**, nos. 1–2, 135–226.

de Bruijn, N. G. (1977). *Johannes G. van der Corput (1890–1975): a biographical note*, Acta Arith. **32**, no. 3, 207–208, DOI 10.4064/aa-32-3-207-208. MR0424486

Brun, Viggo (1958). *Carl Störmer in memoriam*, Acta Math. **100**, i-vii, DOI 10.1007/BF02559599. MR0098007

Bryan, G. H. (1908). *The International Mathematical Congress at Rome*, Nature **77**(2008), 582–584.

Buchstaber, Victor M. (2001). *Interview with Sergey P. Novikov*, EMS Newsletter, no. 42 (December), 17–20.

Burckhardt, Johann Jakob (1980). *Die Mathematik an der Universität Zürich, 1916–1950 unter den Professoren R. Fueter, A. Speiser, P. Finsler. [Under Professors R. Fueter, A. Speiser and P. Finsler]*, Beihefte zur Zeitschrift *Elemente der Mathematik* [Supplements to the Journal *Elemente der Mathematik*], vol. 16, Birkhäuser Verlag, Basel-Boston, Mass. MR632941

Burckhardt, Johann Jakob (2008). *Fueter, Karl Rudolf*, in Dictionary of scientific biography (C. C. Gillispie, ed.), Charles Scribner's Sons, Detroit, vol. 5, p. 206.

Cagiano de Azevedo, Paola and Elvira Gerardi (eds.) (2005). *Reale Accademia d'Italia. Inventario dell'archivio*, Pubblicazioni degli Archivi di Stato: Strumenti, CLXVII, Ministero per i Beni Culturali e Ambientali — Dipartimento per i Beni Archivistici e Librari — Direzione Generale per gli Archivi, Roma.

Capristo, Annalisa (2015). *Volterra, fascism, and France*, Sci. Context **28**, no. 4, 637–674.

Capristo, Annalisa (2016). *French mathematicians at the Bologna Congress (1928). Between participation and boycott*, in (**Brechenmacher et al. 2016**), pp. 289–309. MR3776679

Carathéodory, C. (1936/7). *Bericht über die Verleihung der Fieldsmedaillen*, in (**ICM, 1936**), vol. I, pp. 308–314.

Carruccio, Ettore (2008). *Bianchi, Luigi*, in Dictionary of scientific biography (C. C. Gillispie, ed.), Charles Scribner's Sons, Detroit, vol. 2, p. 121.

Cartan, É. (1935). *La méthode du repère mobile, la théorie des groupes continus, et les espaces généralisés*, Actualités Scientifiques et Industrielles no. 194, Hermann, Paris.

Cartan, Élie (1936/7). *Le rôle de la théorie des groupes de Lie dans l'évolution de la géométrie moderne*, in (**ICM, 1936**), vol. I, pp. 92–103. [JFM 63.1155.04; Zbl 0018.27202]

Cartan, É. (1937). *La théorie des groupes finis et continus et la géométrie différentielle traitées par la méthode du repère mobile*, Gauthier-Villars, Paris.

Cartan, Élie (1945). *Les systèmes différentiels extérieurs et leurs applications géométriques*, Actualités Sci. Ind., no. 994, Hermann et Cie., Paris. MR0016174

Cassels, J. W. S. (1973). *Louis Joel Mordell 1888–1972*, Biogr. Mem. Fellows Roy. Soc. **19**, 493–520.

Cassels, J. W. S. (1974). *L. J. Mordell*, Bull. London Math. Soc. **6**, 69–96. (1 plate), DOI 10.1112/blms/6.1.69. MR0384417

Cassels, J. W. S. (1986). *Mordell's finite basis theorem revisited*, Math. Proc. Cambridge Philos. Soc. **100**, no. 1, 31–41, DOI 10.1017/S0305004100065841. MR838651

Cassels, J. W. S. (2008). *Mordell, Louis Joel*, in Dictionary of scientific biography (C. C. Gillispie, ed.), Charles Scribner's Sons, Detroit, vol. 18, pp. 653–655.

Catellier, Rémi and Laurent Mazliak (2012). *The emergence of French probabilistic statistics. Borel and the Institut Henri Poincaré around the 1920s*, Rev. Histoire Math. **18**, no. 2, 271–335. MR3059066

Chandler, Bruce and Wilhelm Magnus (1982). *The history of combinatorial group theory: A case study in the history of ideas*, Studies in the History of Mathematics and Physical Sciences, vol. 9, Springer-Verlag, New York. MR680777

Chapman, S. (1958). *Fredrik Carl Mülertz Störmer*, Biogr. Mem. Fellows Roy. Soc. **4** 257–279.

Chern, S. (1978). *A summary of my scientific life and works*, in Selected papers, Springer, vol. 1, pp. xxi–xxxi.

Chern, Shiing-Shen and Claude Chevalley (1952). *Obituary: Elie Cartan and his mathematical work*, Bull. Amer. Math. Soc. **58**, 217–250, DOI 10.1090/S0002-9904-1952-09588-4. MR46972

Cioli, Monica (2016). *Arte e scienza internazionale. Il "modernismo" fascista negli anni Venti*, Scienza & Politica, no. 5, Dipartimento di Scienze politiche e sociali, Università di Bologna.

Cogliati, Alberto (2018). *Writing small omegas: Élie Cartan's contributions to the theory of continuous groups, 1894–1926*, Studies in the History of Mathematical Enquiry, Academic Press, London. MR3729308

Comoglio, Mauro (2015). *A mathematician goes to war: the brief life of Eugenio Elia Levi*, Lett. Mat. **3**, no. 1-2, 43–52, DOI 10.1007/s40329-015-0082-4. MR3348277

van der Corput, J. G. (1919). *Over roosterpunten in het platte vlak*, PhD thesis, Leiden.

van der Corput, Jan G. (1936/7). *Diophantische Approximationen*, in (**ICM, 1936**), vol. I, pp. 249–260. [JFM 63.0918.03; Zbl 0018.29501]

van der Corput, J.-G. and J.-F. Koksma (1930). *Sur l'ordre de grandeur de la fonction $\zeta(s)$ de Riemann dans la bande critique*, Ann. Fac. Sci. Toulouse Sci. Math. Sci. Phys. (3) **22**, 1–39. MR1508407

Corry, Leo (2004). *Modern algebra and the rise of mathematical structures*, 2nd ed., Birkhäuser Verlag, Basel. MR2033171

Corry, Leo and Norbert Schappacher (2010). *Zionist internationalism through number theory: Edmund Landau at the opening of the Hebrew University in 1925*, Sci. Context **23**, no. 4, 427–471, DOI 10.1017/S0269889710000177. MR2745962

Cramér, Harald (1962). *A. I. Khinchin's work in mathematical probability*, Ann. Math. Statist. **33**, 1227–1237, DOI 10.1214/aoms/1177704355. MR139512

Crone, Christian (1923). *Matematisk Forening gennem 50 aar*, Matematisk Forening (Jul. Gjellerups Forlag), København.

Cross Roads (1931). *Science at the cross roads (Papers presented to the International Congress of the History of Science and Technology held in London from June 29th to July 3rd 1931 by the delegates of the USSR)*, Kniga (England); 2nd ed., Frank Cass & Co., London, 1971.

Curbera, Guillermo P. (2009). *Mathematicians of the world, unite!: The International Congress of Mathematicians—a human endeavor; With a foreword by Lennart Carleson*, A K Peters, Ltd., Wellesley, MA. MR2499757

van Dalen, Dirk and L. E. J. Brouwer (2011). *The selected correspondence of L. E. J. Brouwer*, Sources and Studies in the History of Mathematics and Physical Sciences, Springer-Verlag London, Ltd., London. Edited and with an introduction by van Dalen. MR2868096

Dauben, Joseph W. (1980). *Mathematicians and World War I: the international diplomacy of G. H. Hardy and Gösta Mittag-Leffler as reflected in their personal correspondence*, Historia Math. **7**, no. 3, 261–288, DOI 10.1016/0315-0860(80)90026-9. MR586421

Davenport, H. (1964). *L. J. Mordell*, Acta Arith. **9**, 3–22, DOI 10.4064/aa-9-1-3-12. MR0172777

Davis, Philip J. (1994). *Otto Neugebauer: reminiscences and appreciation*, Amer. Math. Monthly **101**, no. 2, 129–131, DOI 10.2307/2324359. MR1259824

Dawson, John W., Jr. (2002). *Max Dehn, Kurt Gödel, and the trans-Siberian escape route*, Int. Math. Nachr. **56**, no. 189, 1–13; reprinted in Notices Amer. Math. Soc. **49**, no. 9, 1068–1075. MR1927456

Deavours, C. A. (1973). *The quaternion calculus*, Amer. Math. Monthly **80**, 995–1008, DOI 10.2307/2318774. MR325986

Decaens, Simon (2018). *Une histoire de la théorie des treillis au sein de l'American Mathematical Society entre 1933 et 1948*, PhD thesis, Université Paris Diderot.

Demidov, S. S. and B. V. Levshin (eds.) (1999/2016). *The case of Academician Nikolai Nikolaevich Luzin*, Russian Christian Humanitarian Institute, St Petersburg; English transl. by Roger Cooke: History of Mathematics, vol. 43, AMS, 2016.

Deuring, Max (1935). *Algebren*, Ergebnisse der Mathematik und ihrer Grenzgebiete, Band 4, no. 1, Springer, Berlin.

Deuring, Max (1985). *Carl Ludwig Siegel, December 12, 1896–April 4, 1981*, Acta Arith. **45**, no. 2, 93–113, DOI 10.4064/aa-45-2-93-107. MR797256

Dickson, L. E. (1947). *Obituary: Hans Frederik Blichfeldt. 1873–1945*, Bull. Amer. Math. Soc. **53**, 882–883, DOI 10.1090/S0002-9904-1947-08874-1. MR21508

Dieudonné, J. (1972). *The historical development of algebraic geometry*, Amer. Math. Monthly **79**, 827–866, DOI 10.2307/2317664. MR308117

Dieudonné, Jean (1978). *Abrégé d'histoire des mathematiques 1700–1900*, 2 vols., Hermann, Paris; German transl.: *Geschichte der Mathematik 1700–1900. Ein Abriß*, DVW, Berlin, 1985.

Dieudonné, Jean (1981). *History of functional analysis*, North-Holland Mathematics Studies, vol. 49, North-Holland Publishing Co., Amsterdam-New York. Notas de Matemática [Mathematical Notes], 77. MR605488

Dieudonné, Jean (1984). *Une brève histoire de la topologie*, in (**Pier, 1994a**), pp. 35–155.

Dieudonné, Jean (1985a). *Introductory remarks on algebra, topology and analysis*, Hist. Math. **2**, 537–548.

Dieudonné, Jean (1985b). *History of algebraic geometry: an outline of the history and development of algebraic geometry*, CRC Press.

Dieudonné, Jean (1986). *The beginnings of Italian algebraic geometry*, Symposia mathematica, Vol. XXVII (Cortona, 1983), Sympos. Math., XXVII, Academic Press, London, pp. 245–263. MR863152

Dieudonné, Jean (1989). *A history of algebraic and differential topology. 1900–1960*, Birkhäuser Boston, Inc., Boston, MA. MR995842

Dieudonné, Jean (2008a). *Cartan, Élie*, in Dictionary of scientific biography (C. C. Gillispie, ed.), Charles Scribner's Sons, Detroit, vol. 3, pp. 95–96.

Dieudonné, Jean (2008b). *Siegel, Carl Ludwig*, in Dictionary of scientific biography (C. C. Gillispie, ed.), Charles Scribner's Sons, Detroit, vol. 18, pp. 826–832.

Doel, Ronald E., Dieter Hoffmann, and Nikolai Krementsov (2005). *National states and international science: a comparative history of international science congresses in Hitler's Germany, Stalin's Russia, and Cold War United States*, Osiris, **20**, 49–76.

Doob, J. L. (1961). *Appreciation of Khinchin*, Proc. 4th Berkeley Sympos. Math. Statist. and Prob., Vol. II, Univ. California Press, Berkeley, Calif., pp. 17–20. MR0131342

Doob, Joseph L. (1994). *The development of rigor in mathematical probability (1900–1950)*, in (**Pier, 1994a**), pp. 157–170, DOI 10.1007/978-3-0348-9114-1_3. MR1298633

Douglas, Jesse (1931). *Solution of the problem of Plateau*, Trans. Amer. Math. Soc. **33**, no. 1, 263–321, DOI 10.2307/1989472. MR1501590

Douglas, Jesse (1938). *The problem of Plateau*, Scripta Math. **5**, 159–164.

Duda, Roman (2014). *Pearls from a lost city: The Lvov school of mathematics*, History of Mathematics, vol. 40, American Mathematical Society, Providence, RI. Translated from the 2007 Polish original by Daniel Davies. MR3222779

Dugac, Pierre, Beno Eckmann, Jean Mawhin, and Jean-Paul Pier (1994). *Guidelines 1900–1950*, in (**Pier, 1994a**), pp. 1–34, DOI 10.1007/978-3-0348-9114-1_1. MR1298631

Dunnington, G. Waldo (1936a). *Oslo under the Integral Sign*, Natl. Math. Mag. **11**, no. 2, 85–94. MR1569359

Dunnington, G. Waldo (1936b). *Vigeland's Monument to Abel in Oslo*, Natl. Math. Mag. **11**, no. 3, 145–146. MR1569372

Dunnington, G. Waldo (1955). *Carl Friedrich Gauss: titan of science. A study of his life and work*, Exposition Press, New York. MR0072814

Duparc, H. J. A. and J. Korevaar (1982). *Johannes Gualtherus van der Corput (4 September 1890–13 September 1975)* (Dutch), Nieuw Arch. Wisk. (3) **30**, no. 1, 1–39. MR657083

Durand, Antonin (2017). *Mathematicians and politics: new research scenarios?*, Lett. Mat. **4**, no. 3-4, 161–165, DOI 10.1007/s40329-016-0149-x. Translated from the Italian. MR3619598

Duren, P. L. (ed.) (1988/1989). *A century of mathematics in America*, 3 vols., Amer. Math. Soc., Providence, RI.

Eckert, Michael (2017). *Ludwig Prandtl — Strömungsforscher und Wissenschaftsmanager. Ein unverstellter Blick auf sein Leben*, Springer, Berlin.

Edwards, Harold M. (2008). *Hasse, Helmut*, in Dictionary of scientific biography (C. C. Gillispie, ed.), Charles Scribner's Sons, Detroit, vol. 17, pp. 385–387.

Egeland, Alv and William J. Burke (2013). *Carl Størmer: auroral pioneer*, Springer.

Ehinger, Berndt and Andrzej Grzybowski (2011). *Allvar Gullstrand (1862–1930) — the gentleman with the lamp*, Acta Ophthal. **89**, 701–708.

Eilenberg, Samuel (1936). *Sur les espaces multicohérents I*, Fund. Math. **27**, 153–190.

Ellison, W. and F. Ellison (1978/1985). *Théorie des nombres*, Chapter V in (**Dieudonné, 1978**), vol. I, pp. 165–334; in German transl.: pp. 171–358.

Epple, Moritz, Andreas Karachalios, and Volker R. Remmert (2005). *Aerodynamics and mathematics in National Socialist Germany and Fascist Italy: a comparison of research institutes*, Osiris (2) **20**, 131–158, DOI 10.1086/649416. MR2166390

Eremenko, Alexandre (2000). *Ahlfors' contribution to the theory of meromorphic functions*, in (**Brooks and Sodin, 2000**), pp. 41–63. MR1786560

Esakov, V. D. (ed.) (2000). *The Academy of Sciences in resolutions of the Politburo 1922–1991*, vol. 1: 1922–1952, ROSSPEN, Moscow (in Russian).

Evgrafov, M. A., N. M. Korobov, Yu. V. Linnik, I. I. Pyatetskii-Shapiro and N. I. Fel'dman (1969). *Aleksandr Osipovich Gel'fond (obituary)*, Uspekhi mat. nauk **24**, no. 3(147), 219–220 (in Russian); English transl.: Russian Math. Surveys **24**, no. 3, 177–178.

Fairbairn, Ben (2017). *Louis Joel Mordell's time in London*, BSHM Bull. **32**, no. 2, 160–169, DOI 10.1080/17498430.2016.1233692. MR3645975

Faltings, G. (1983). *Endlichkeitssätze für abelsche Varietäten über Zahlkörpern*, Invent. Math. **73**, no. 3, 349–366, DOI 10.1007/BF01388432. MR718935

Fehr, H. (ed.) (1932). *Union Internationale Mathématique: Troisième Assemblée Générale tenue à Zurich le 11 septembre 1932*, Enseign. Math. **31**, 276–278.

Fehr, H. (ed.) (1936). *Le 10e Congrès international des mathématiciens. Oslo, 13–18 juillet 1936*, Enseign. Math. **35**, 373–385.

Fenchel, Werner (1960a). *Jakob Nielsen in memorian*, Acta Math. **103**, vii-xix, DOI 10.1007/BF02546355. MR0111668

Fenchel, Werner (1960b). *Jakob Nielsen in memoriam* (Danish, with English summary), Nordisk Mat. Tidskr. **8**, 5–10, 63. MR122687

Ferrari, Carlo (1996). *Recalling the Vth Volta Congress: high speeds in aviation*, Annual review of fluid mechanics, Vol. 28, Annual Reviews, Palo Alto, CA, pp. 1–9. MR1371160

Fichera, Gaetano (1994). *Vito Volterra and the birth of functional analysis*, in (**Pier, 1994a**), pp. 171–183. MR1298634

Fischer, Gerd, Friedrich Hirzebruch, Winfried Scharlau, and Willi Törnig (eds.) (1990). *Ein Jahrhundert Mathematik 1890–1990*, Dokumente zur Geschichte der Mathematik [Documents on the History of Mathematics], vol. 6, Friedr. Vieweg & Sohn, Braunschweig; Deutsche Mathematiker Vereinigung, Freiburg. Festschrift zum Jubiläum der DMV. [Festschrift on the anniversary of the DMV]. MR1085961

Forster, Walter (2008). *Schauder, Juliusz Pawel*, in Dictionary of scientific biography (C. C. Gillispie, ed.), Charles Scribner's Sons, Detroit, vol. 18, pp. 782–783.

Fox, Robert (2016). *Science without frontiers: cosmopolitanism and national interests in the world of learning, 1870–1940*, Oregon State University Press.

Fraenkel, Abraham A. (1967/2016). *Lebenskreise: Aus den Erinnerungen eines jüdischen Mathematikers,* Deutsche Verlags-Anstalt, Stuttgart*; Recollections of a Jewish mathematician in Germany*, Birkhäuser/Springer, [Cham]. With a foreword by Menachem Magidor; With a foreword to the 1967 German edition by Y. Bar-Hillel; Edited and with a chapter "Afterword: 1933–1965" by Jiska Cohen-Mansfield; Translated from the German by Allison Brown. MR3561164

Frank, Philipp (1932). *Das Kausalgesetz und seine Grenzen*, Springer, Wien; English transl.: *The law of causality and its limits*, Kluwer, Dordrecht, 1997.

Frank, Philipp and Richard von Mises (1925–1927). *Die Differential- und Integralgleichungen der Mechanik und Physik*, 2 vols., Vieweg, Braunschweig.

Fréchet, Maurice (1906). *Sur quelques points du calcul fonctionnel*, PhD thesis, École Normale Supérieure Paris.

Fréchet, Maurice (1930). *Sur certaines décompositions de la fonction complexe uniforme la plus générale*, Acta Math. **54**, no. 1, 37–79, DOI 10.1007/BF02547517. MR1555302

Fréchet, Maurice (1936/7). *Mélanges mathématiques*, in (**ICM, 1936**), vol. I, pp. 269–283. [JFM 63.1004.01]

Freudenthal, Hans (2008). *Wiener, Norbert*, in Dictionary of scientific biography (C. C. Gillispie, ed.), Charles Scribner's Sons, Detroit, vol. 14, pp. 344–347.

Friedman, R. M. (1989). *Appropriating the weather: Vilhelm Bjerknes and the construction of a modern meteorology*, Cornell University Press, Ithaca and London.

Fueter, R. (1934). *Die Funktionentheorie der Differentialgleichungen* $\Theta u = 0$ *und* $\Theta\Theta u = 0$ *mit vier reellen Variablen*, Comment. Math. Helv. **7**, no. 1, 307–330, DOI 10.1007/BF01292723. MR1509515

Fueter, R. (1935). *Über die analytische Darstellung der regulären Funktionen einer Quaternionenvariablen*, Comment. Math. Helv. **8**, no. 1, 371–378, DOI 10.1007/BF01199562. MR1509533

Fueter, R. (1936). *Die Singularitäten der eindeutigen regulären Funktionen einer Quaternionenvariablen*, Comment. Math. Helv. **9**, no. 1, 320–334, DOI 10.1007/BF01258196. MR1509562

Fueter, R. (1936/7). *Die Theorie der regulären Funktionen einer Quaternionenvariablen*, in (**ICM, 1936**), vol. I, pp. 75-91. [JFM 63.0310.04; Zbl 0018.31601]

Fueter, Rudolf (1950). *Nachgelassene Aufzeichnungen*, Elemente der Math. **5**, 99–104. MR0037794

Fujiwara, Matsusaburô (1941). *A brief sketch of the Wazan, the mathematics of the old Japanese school*, J. Sendai Cultural Soc. **1941**, 64–84. MR4007

Gehring, Frederick (2005). *Lars Valerian Ahlfors 1907–1996*, Biogr. Mem. Nat. Acad. Sci. USA **87**, 3–28.

Gehring, Frederick, Irwin Kra, and Robert Osserman (1998). *The mathematics of Lars Valerian Ahlfors*, Notices Amer. Math. Soc. **45**, no. 2, 233–242. Edited by Steven G. Krantz. MR1601816

Gel'fond, A. (1929a). *Sur les nombres transcendants*, C. R. Acad. sci. Paris **189**, 1224–1226.

Gel'fond, A. (1929b). *Sur les propriétés arithmétiques des fonctions entières*, Tôhoku Math. J. **30**, 280–285.

Gel'fond, A. (1934). *Sur le septième problème de Hilbert*, Izvest. Akad. Nauk SSSR, no. 4, 623–634.

Gel'fond, A. O. (1952). *Transcendental and algebraic numbers*, GITTL, Moscow-Leningrad (Russian); translated from the first Russian edition by Leo F. Boron, Dover Publications, Inc., New York, 1960. MR0111736

Gel'fond, A. O. (1969). *On Hilbert's seventh problem*, in P. S. Aleksandov (ed.), Hilbert's problems, Nauka, Moscow, pp. 121–127 (in Russian).

Gel'fond, A. O. (1977). *Some impressions of a scientific excursion to Germany in 1930*, Istor.-mat. issled., no. 22, 246–251 (in Russian).

Gerovitch, Slava (2002). *From Newspeak to Cyberspeak: a history of Soviet cybernetics*, MIT Press.

Ghizzetti, Aldo (1986). *Mauro Picone e l'Istituto Nazionale per le Applicazioni del Calcolo*, in G. Montalenti, L. Amerio, G. Acquaro and E. Baiada (eds.), Convegno celebrativo del centenario della nascita di Mauro Picone e Leonida Tonelli (6–9 maggio 1985), Atti dei Convegni Lincei, vol. 77, Accademia Nazionale dei Lincei, Roma, pp. 111–116.

Giacardi, Livia (2009). *The Italian contribution to the International Commission on Mathematical Instruction from its founding to the 1950s, in K. Bjarnadóttir, F. Furinghetti and G. Schubring* (eds.), "Dig where you stand". Proceedings of the conference "On-going research in the History of Mathematics Education", University of Iceland, Reykjavik, pp. 47–64.

Gnedenko, B. V. (1960). *Aleksandr Yakovlevich Khinchin*, Teoriya veroyat. primen. **5**, no. 1, 3–6 (in Russian); English transl.: *Theory Prob. Appl.* **5**, no. 1, 1960, 1–4.

Gnedenko, B. V. (1961). *Alexander Iacovlevich Khinchin*, Proc. 4th Berkeley Sympos. Math. Statist. and Prob., Vol. II, Univ. California Press, Berkeley, Calif., pp. 1–15. (1 plate). MR0131956

Gnedenko, B. V. and A. N. Kolmogorov (1960). *Aleksandr Jakovlevič Hinčin: Obituary*, Uspekhi mat. Nauk **15**, no, 4(94), 97–110; English transl., Russian Math. Surveys **15**, no. 4, 93–106, DOI 10.1070/RM1960v015n04ABEH004097. MR0126363

Gnedenko, B. V. and A. I. Markushevich (1968). *A. Ya. Khinchin: a biographical sketch, Appendix 1* in B. V. Gnedenko (ed.), The teaching of mathematics: essays by A. Ya. Khinchin, English Universities Press, pp. 102–111.

Godement, R. (1948/1971). *Les méthodes modernes et l'avenir des mathématiques concrètes*, in (**Le Lionnais, 1948/1971**), vol. I, pp. 337–342.

Gold, E. (1951). *Obituary: Vilhelm Friman Koren Bjerknes, 1862–1951*, Obit. Notices Roy. Soc. London **7**, 303–317. MR0045633

Goodstein, Judith R. (2007). *The Volterra chronicles: The life and times of an extraordinary mathematician 1860–1940*, History of Mathematics, vol. 31, American Mathematical Society, Providence, RI; London Mathematical Society, London. MR2287463

Goodstein, Judith and Donald Babbitt (2012). *A fresh look at Francesco Severi*, Notices Amer. Math. Soc. **59**, no. 8, 1064–1075, DOI 10.1090/noti881. MR2985808

Gordin, Michael D. (2015). *Scientific Babel: the language of science from the fall of Latin to the rise of English*, Profile Books.

Graham, Loren and Jean-Michel Kantor (2009). *Naming infinity: A true story of religious mysticism and mathematical creativity*, The Belknap Press of Harvard University Press, Cambridge, MA. MR2526973

Grattan-Guinness, Ivor (1986). *The Società Italiana, 1782–1815: a survey of its mathematics and mechanics*, Symposia mathematica, Vol. XXVII (Cortona, 1983), Sympos. Math., XXVII, Academic Press, London, pp. 147–168. MR863147

Gray, Jeremy (1999). *Otto Neugebauer (b. 1899)*, EMS Newsletter, no. 34 (December), 23–24.

Gray, Jeremy J. (2000). *The Hilbert challenge*, Oxford University Press, Oxford. MR1828558

Gray, Jeremy and Mario Micallef (2008). *About the cover: the work of Jesse Douglas on minimal surfaces*, Bull. Amer. Math. Soc. (N.S.) **45**, no. 2, 293–302, DOI 10.1090/S0273-0979-08-01192-0. MR2383307

Gray, Jeremy and Mario Micallef (forthcoming). *Jesse Douglas, minimal surfaces, and the first Fields Medal*, forthcoming.

Greenaway, Frank (1996). *Science international: a history of the International Council of Scientific Unions*, Cambridge University Press.

Guerraggio, Angelo and Pietro Nastasi (2005a). *Italian mathematics between the two World Wars*, Science Networks. Historical Studies, vol. 29, Birkhäuser Verlag, Basel. MR2188015

Guerraggio, Angelo and Pietro Nastasi (2005b). *Matematica in camicia nera: il regime e gli scienziati*, Matematica e dintorni, Bruno Mondadori, Milano.

Guerraggio, Angelo and Pietro Nastasi (2008). *Roma 1908: il Congresso internazionale dei matematici*, Bollati Boringhieri, Torino.

Guerraggio, Angelo and Giovanni Paoloni (2008). *Vito Volterra*, Muzzio editore, Montereggio; translated from the 2008 Italian original by Kim Williams, Springer, Heidelberg, 2012. MR3025507

Guerraggio, Angelo, Frédéric Jaëck, and Laurent Mazliak (2016). *Lines on the horizon: Hadamard and Fréchet, readers of Volterra*, in (**Brechenmacher *et al.* 2016**), pp. 127–179. MR3776675

Gumbel, E. J. (ed.) (1938). *Freie Wissenschaft. Ein Sammelbuch aus der deutschen Emigration*, Sebastian Brant, Strasbourg.

Gumbel, E. J. (1958). *Statistics of extremes*, Columbia University Press, New York. MR0096342

Hadamard, J. (1928/9). *Le développement et le rôle scientifique du Calcul fonctionnel*, in (**ICM, 1928**), vol. I, pp. 143–161.

Hahn, E. (1925). *Note sur l'application aux turbomachines des théories modernes de l'hydrodynamique*, in (**ICAM, 1924**), pp. 427–431 + Tables XLIII–XLIV.

Hansen, Vagn Lundsgaard (1993). *Jakob Nielsen (1890–1959)*, Math. Intelligencer **15**, no. 4, 44–53, DOI 10.1007/BF03024323. MR1240668

Hansen, Vagn Lundsgaard (1999). *Jakob Nielsen and his contributions to topology*, in (**James, 1999a**), pp. 979–989, DOI 10.1016/B978-044482375-5/50038-9. MR1721132

Hardy, G. H. and H. Heilbronn (1938). *Edmund Landau*, J. London Math. Soc. **13**, no. 4, 302–310, DOI 10.1112/jlms/s1-13.4.302. MR1574982

Hasse, Helmut (1924). *Symmetrische Matrizen im Körper der rationalen Zahlen*, J. Reine Angew. Math. **153**, 12–43, DOI 10.1515/crll.1924.153.12. MR1581019

Hasse, Helmut (1935). *Zur Theorie der abstrakten elliptischen Funktionenkörper*, Nachr. Ges. Wiss. Göttingen. Math.-Phys. Kl. Fachgruppe 1, Math. Neue Folge, 119–129.

Hasse, Helmut (1936). *Kurt Hensel zum 75. Geburtstag*, Forschungen und Fortschritte **12**, no. 35/36.

Hasse, H. (1936/7). *Über die Riemannsche Vermutung in Funktionenkörpern*, in (**ICM, 1936**), vol. I, pp. 189–206. [JFM 63.0881.01; Zbl 0018.34302]

Hawkins, Thomas (2000). *Emergence of the theory of Lie groups: An essay in the history of mathematics 1869–1926*, in Sources and Studies in the History of Mathematics and Physical Sciences, Springer-Verlag, New York. MR1771134

Hecke, E. (1936). *Über die Bestimmung Dirichletscher Reihen durch ihre Funktionalgleichung*, Math. Ann. **112**, no. 1, 664–699, DOI 10.1007/BF01565437. MR1513069

Hecke, E. (1936/7). *Neuere Fortschritte in der Theorie der elliptischen Modulfunktionen*, in (**ICM, 1936**), vol. I, pp. 140–156. [JFM 63.0339.02; Zbl 0018.35703]

Hecke, E. (1937). *Über Modulfunktionen und die Dirichletschen Reihen mit Eulerscher Produktentwicklung. I*, Math. Ann. **114**, no. 1, 1–28, DOI 10.1007/BF01594160. MR1513122

van Heijenoort, Jan (1978). *With Trotsky in exile: from Prinkipo to Coyoacán*, Harvard University Press, Cambridge, MA.

Heyting, A. (1934). *Mathematische Grundlagenforschung. Intuitionismus. Beweistheorie*, Ergebnisse der Mathematik und ihrer Grenzgebiete, Band 3, no. 4, Springer, Berlin.

Hlawka, Edmund (1985). *Carl Ludwig Siegel (31/12/1896–4/4/1981)*, J. Number Theory **20**, no. 3, 373–404, DOI 10.1016/0022-314X(85)90028-9. MR797184

Hodge, W. V. D. (1953). *Obituary: Elie Cartan*, J. London Math. Soc. **28**, 115–119, DOI 10.1112/jlms/s1-28.1.115. MR0051759

Hollcroft, T. R. (1940). *The February meeting in New York*, Bull. Amer. Math. Soc. **46**, no. 5, 368–373, DOI 10.1090/S0002-9904-1940-07186-1. MR1564110

Hollings, Christopher (2014). *Mathematics across the Iron Curtain: A history of the algebraic theory of semigroups*, History of Mathematics, vol. 41, American Mathematical Society, Providence, RI. MR3222721

Hollings, Christopher D. (2016). *Scientific communication across the Iron Curtain*, SpringerBriefs in History of Science and Technology, Springer, Cham. MR3494795

Holst, Elling, Carl Størmer, and L. Sylow (eds.) (1902). *Festskrift ved Hundredeaarsjubilæet for Niels Henrik Abels Fødsel*, Jacob Dybwad, Kristiania.

Holst, Per A. (1996). *Svein Rosseland and the Oslo Analyzer*, IEEE Ann. Hist. Comput. **18**, no. 4, 16–26.

Honda, Kin-ya (1975/76). *Teiji Takagi: a biography—on the 100th anniversary of his birth*, Comment. Math. Univ. St. Paul. **24**, no. 2, 141–167. MR0389505

Hopf, Eberhard (1937). *Ergodentheorie*, Springer.

Høyrup, Jens (2017). *What is "geometric algebra", and what has it been in historiography?* AIMS Math. **2**, no. 1, 128–160.

Ingham, A. E. (1938). *Review: Über einige neuere Fortschritte der additiven Zahlentheorie by E. Landau*, Math. Gaz. **22**, no. 249, 190–192.

Israel, Giorgio (2000). *Italian mathematics, Fascism and the politics of race* (Italian), Mathematics and culture, 2000 (Italian) (Venice, 1999), Collana Mat. Cult., Springer Italia, Milan, pp. 21–48. MR2000626

Jackson, Allyn (1999). *Interview with Henri Cartan [b. 1904]*, Notices Amer. Math. Soc. **46**, no. 7, 782–788. MR1697836

Jacobs, Konrad (1986). *Entwicklungslinien in der Mathematik im 20. Jahrhundert*, Math. Semesterber. **33**, no. 1, 26–54. MR845934

James, I. M. (ed.) (1999a). *History of topology*, Elsevier.

James, I. M. (1999b). *Topologists at conferences*, in (**James, 1999a**), pp. 837–848, DOI 10.1016/B978-044482375-5/50030-4. MR1721124

Jammer, M. (1966). *The conceptual development of quantum mechanics*, McGraw-Hill Book Company.

Jammer, M. (1974). *The philosophy of quantum mechanics. The interpretations of quantum mechanics in historical perspective*, John Wiley & Sons.

Javillier, Maurice (1951). *Notice nécrologique sur Élie Cartan (1869–1951)*, C. R. Acad. Sci. Paris **232**, 1735–1791. MR41769

Jerison, David and Daniel Stroock (1995). *Norbert Wiener*, Notices Amer. Math. Soc. **42**, no. 4, 430–438. MR1319275

Jessen, B. (1993). *Mathematiker unter den deutschsprachigen Emigranten: Otto Neugebauer, Werner und Käte Fenchel, Herbert Busemann, Willy Feller*, in W. Dähnhardt and B. S. Nielsen (eds.), Exil in Dänemark. Deutschsprachige Wissenschaftler, Künstler und Schriftsteller im dänischen Exil nach 1933, Westholsteinische Verlagsanstalt Boysen & Co., Heide, pp. 127–133.

Johansen, Nils Voje (2006). *Einstein's visit to Norway*, in (**Siegmund-Schultze and Sørensen, 2006**), pp. 65–82.

Jones, Alexander, Christine Proust, and John M. Steele (eds.) (2016). *A mathematician's journeys: Otto Neugebauer and modern transformations of ancient science*, Archimedes: New Studies in the History and Philosophy of Science and Technology, vol. 45, Springer, Cham. Including revised and expanded papers from the conference held at New York University, New York, 2010. MR3410193

Joravsky, David (1970). *The Lysenko affair*, University of Chicago Press.

Katětov, Miroslav (2008). *Banach, Stefan*, in Dictionary of scientific biography (C. C. Gillispie, ed.), Charles Scribner's Sons, Detroit, vol. 1, pp. 427–428.

Kern, Ketil (2013). *Hva har du med Hitler å gjøre, pappa?*, Juritzen forlag, Oslo.

Khinchin, A. [Khinchine, A. Ya.] (1924). *Über einen Satz der Wahrscheinlichkeitsrechnung*, Fund. Math. **6**, 9–20.

Khinchin, A. [Khinchine, A. Ya.] (1933). *Asymptotische Gesetze der Wahrscheinlichkeitsrechnung*, Ergebnisse der Mathematik und ihrer Grenzgebiete, Band 2, no. 4, Springer, Berlin.

Klein, Felix (1894). *The Evanston colloquium: lectures on mathematics, delivered from Aug. 28 to Sept. 9,1893, before members of the Congress of Mathematics held in connection with the World's Fair in Chicago, at Northwestern University, Evanston, Ill.*, reported by Alexander Ziwet, Macmillan, New York.

Klein [Klein-Barmen], Fritz (1929). *Einige distributive Systeme in Mathematik und Logik*, Jahresber. Deutsch. Math.-Verein. **38**, 35–40.

Klein, Fritz (1931). *Zur Theorie der abstrakten Verknüpfungen*, Math. Ann. **105**, no. 1, 308–323, DOI 10.1007/BF01455821. MR1512717

Klingen, H. (1983). *Das Werk C. L. Siegels in der Funktionentheorie*, Jahresber. Deutsch. Math.-Verein. **85**, no. 4, 158–173. MR722518

Knopp, Konrad (1951). *Edmund Landau*, Jahresber. Deutsch. Math. Verein. **54**, no. Abt. 1, 55–62. MR0039654

Koksma, J. F. (1936). *Diophantische Approximationen*, Ergebnisse der Mathematik und ihrer Grenzgebiete, Band 4, no. 4, Springer, Berlin.

Kol'man [Colman], E. (1931). *The present crisis in mathematical sciences and general outlines for their reconstruction*, in (**Cross Roads, 1931**), pp. 215–229.

Kolmogoroff, A. (1933). *Grundbegriffe der Wahrscheinlichkeitsrechung*, Ergebnisse der Mathematik und ihrer Grenzgebiete, Band 2, no. 3, Springer, Berlin; English transl. by Nathan Morrison: *Foundations of the theory of probability*, Chelsea, New York, 1950; 2nd ed., 1956.

Korevaar, J. (1975). *Levensbericht van Johannes Gualtherus van der Corput (4 september 1890–13 september 1975)*, in Jaarboek. Koninklijke Nederlandse Akademie van Wetenschappen, pp. 198–203; expanded English transl.: (**Korevaar, 2015**).

Korevaar, Jaap (2015). *Johannes Gualtherus van der Corput (4 September 1890–13 September 1975)*, Indag. Math. (N.S.) **26**, no. 5, 715–722, DOI 10.1016/j.indag.2015.08.008. MR3425872

Krull, W. (1935). *Idealtheorie*, Ergebnisse der Mathematik und ihrer Grenzgebiete, Band 4, no. 3, Springer, Berlin.

Kubach, Fritz (1936). *Studenten in Front!* Deutsche Math. **1**, no. 1, pp. 5–8.

Kubota, Tadahiko (1949). *Obituary note: Matsusaburô Fujiwara (1881–1946)*, Tohoku Math. J. (2) **1**, 1–2, DOI 10.2748/tmj/1178245760. MR34325

Landau, Edmund (1903). *Neuer Beweis des Primzahlsatzes und Beweis des Primidealsatzes*, Math. Ann. **56**, no. 4, 645–670, DOI 10.1007/BF01444310. MR1511191

Landau, Edmund (1909). *Handbuch der Lehre von der Verteilung der Primzahlen*, 2 vols., Teubner, Leipzig.

Landau, Edmund (1912). *Gelöste und ungelöste Probleme aus der Theorie der Primzahlverteilung und der Riemannschen Zetafunktion*, Jahresber. Deutsch. Math.-Verein. **21**, 208–228.

Landau, Edmund (1927). *Vorlesungen über Zahlentheorie*, 3 vols., S. Hirzel, Leipzig.

Landau, Edmund (1937). *Über einige neuere Fortschritte der additiven Zahlentheorie*, Cambridge University Press.

Lang, Serge (1995). *Mordell's review, Siegel's letter to Mordell, Diophantine geometry, and 20th century mathematics*, Gaz. Math., no. 63, 17–36; also published in Notices Amer. Math. Soc. **42**, no. 3, 339–350. MR1316133

Lange-Nielsen, F. (1931). *Alf Victor Emanuel Guldberg (1866)*, in Norsk biografisk leksikon (E. Bull *et al.*, eds.), 19 vols, Aschehoug, Oslo, vol. 5, pp. 68–69.

Lehmer, D. H. (1988). *A half century of reviewing*, in (**Duren, 1988/1989**), vol. 1, pp. 265–266. MR1003174

Lehto, Olli (1998a). *Mathematics without borders: A history of the International Mathematical Union*, Springer-Verlag, New York. MR1488698

Lehto, Olli (1998b). *On the life and work of Lars Ahlfors*, Math. Intelligencer **20**, no. 3, 4–8, DOI 10.1007/BF03024796. MR1646577

Lehto, Olli (2013). *Tieteen huipulla—Lars Ahlforsin elämä*, Bidrag till Kännedom av Finlands Natur och Folk [Contributions to the Knowledge of Finnish Nature and People], vol. 192, Societas Scientiarum Fennica, Helsinki; English transl. by William Hellberg, *Lars Ahlfors: at the summit of mathematics*, Amer. Math. Soc., Providence, RI, 2015. MR3222281

Le Lionnais, F. (ed.) (1948/1971). *Les Grands Courants de la Pensée Mathématique*, Cahiers du Sud; English transl. in 2 vols., Dover, New York. MR0026982

Lemaitre, G. and M. S. Vallarta (1936). *On the allowed cone of cosmic radiation*, Phys. Rev. **50**, no. 6, 493–504.

Leopoldt, H.-W. (1982). *Zum wissenschaftlichen Werk Helmut Hasses*, Mitt. Math. Ges. Hamburg **11**, no. 1, 9–23. MR689140

Leray, Jean (1980). *My friend Julius Schauder*, in Numerical solution of highly nonlinear problems (Sympos. Fixed Point Algorithms and Complementarity Problems, Univ. Southampton, Southampton, 1979), North-Holland, Amsterdam-New York, pp. 427–439. MR559136

Levin, B. V., N. I. Fel′dman, and A. B. Šidlovski (1971). *Alexander O. Gelfond*, Acta Arith. **17**, 315–336, DOI 10.4064/aa-17-4-315-336. MR0281571

Levinson, N. (1966). *Wiener's life*, Bull. Amer. Math. Soc. **72**, no. 1, pt. 2, 1–32, DOI 10.1090/S0002-9904-1966-11450-7. MR183625

Lietzmann, W. (1936). *Der Internationale Mathematikerkongreß in Oslo vom 13. bis 18. August 1936*, Z. math. naturwiss. Unterricht **67**, 300–304.

Lietzmann, W. (1937). *Die gegenwärtigen Bestrebungen im mathematischen Unterricht der höheren Schulen Deutschlands*, Z. math. naturwiss. Unterricht **68**, 19–22.

Lietzmann, W. (1960). *Aus meinen Lebenserinnerungen* (K. Fladt, ed.), Vandenhoeck & Ruprecht, Göttingen.

Linguerri, Sandra (2015). *Vito Volterra at the front: from the Office of Inventions to the Italian National Research Council*, Lett. Mat. **3**, no. 1-2, 53–61, DOI 10.1007/s40329-015-0081-5. Translated from the Italian by Kim Williams. MR3348278

Linnik, Yu. V. and A. I. Markushevich (1956). *Aleksandr Osipovich Gel′fond (on the fiftieth anniversary of his birth)*, Uspekhi mat. Nauk **11**, no. 5(71), 239–248 (in Russian).

Lorentz, G. G. (2002). *Mathematics and politics in the Soviet Union from 1928 to 1953*, J. Approx. Theory **116**, no. 2, 169–223, DOI 10.1006/jath.2002.3670. MR1911079

Lotsch, Fritz (1932). *Internationale Kongresse und Konferenzen im Jahre 1931*, Weltwirtschaftliches Archiv **35**, 617–652.

Lützen, Jesper (1982). *The prehistory of the theory of distributions*, Studies in the History of Mathematics and Physical Sciences, vol. 7, Springer-Verlag, New York-Berlin. MR667854

Mac Lane, Saunders (1964). *Oswald Veblen*, Biogr. Mem. Nat. Acad. Sci. USA **37**, 325–341.

Mac Lane, Saunders (2008). *Veblen, Oswald*, in Dictionary of scientific biography (C. C. Gillispie, ed.), Charles Scribner's Sons, Detroit, vol. 13, pp. 599–600.

Maiocchi, Roberto (1999). *Scienza italiana e razzismo fascista*, La nuova Italia, Scandicci.

Mandelbrojt, S. (1966). *Les taubériens généraux de Norbert Wiener*, Bull. Amer. Math. Soc. **72**, no. 1, pt. 2, 48–51, DOI 10.1090/S0002-9904-1966-11461-1. MR184008

Mandrekar, V. (1995). *Mathematical work of Norbert Wiener*, Notices Amer. Math. Soc. **42**, no. 6, 664–669. MR1330407

Masani, P. R. (1989). *Norbert Wiener: a survey of a fragment of his life and work*, in (**Duren, 1988/1989**), vol. 3, pp. 299–341, DOI 10.1016/0377-8401(89)90151-X. MR1025351

Masani, P. R. (1990). *Norbert Wiener: 1894–1964*, Vita Mathematica, vol. 5, Birkhäuser Verlag, Basel. MR1032520

Mazliak, Laurent and Rossana Tazzioli (2009). *Mathematicians at war: Volterra and his French colleagues in World War I*, Archimedes: New Studies in the History and Philosophy of Science and Technology, vol. 22, Springer, New York. With a preface by Ivor Grattan-Guinness. MR2583733

Mazliak, Laurent and Rossana Tazzioli (2015a). *"To each according to his trade": Italian mathematicians at war*, Lett. Mat. **3**, no. 1-2, 3–15, DOI 10.1007/s40329-015-0076-2. MR3348274

Mazliak, Laurent and Rossana Tazzioli (2015b). *From the Italian-Austrian front: selected writings by Italian mathematicians at war*, Lett. Mat. **3**, no. 1-2, 73–84, DOI 10.1007/s40329-015-0078-0. Translated from the Italian by Kim Williams. MR3348280

McMillan, Brockway (1989). *Norbert Wiener and chaos*, in (**Duren, 1988/1989**), vol. 2, pp. 479–491. MR1003151

Meagher, Laura R. and Ursula Martin (2017). *Slightly dirty maths: the richly textured mechanisms of impact*, Res. Eval. **26**, no. 1, 15–27.

Mehrtens, Herbert (1979). *Die Entstehung der Verbandstheorie*, Arbor Scientiarum: Beiträge Wissenschaftsgeschichte, Reihe A: Abhandlungen [Arbor Scientiarum: Contributions to the History of Science, Series A: Proceedings], VI, Gerstenberg Verlag, Hildesheim. MR552144

Mehrtens, H. (1989). *The Gleichschaltung of mathematical societies in Nazi Germany*, Math. Intelligencer **11**, no. 3, 48–60, DOI 10.1007/BF03025192. Translated from the German by Victoria M. Kingsbury. MR1007038

Meidell, B. (1936). *Professor dr. Alf Guldberg*, Norsk mat. tidsskr., 33–39.

Menger, K. (1928). *Bemerkungen zu Grundlagenfragen IV: Axiomatik der endlichen Mengen und der elementargeometrischen Verknüpfungsbeziehungen*, Jahresber. Deutsch. Math.-Verein. **37**, 309–325.

Menger, Karl (1936). *New foundations of projective and affine geometry*, Ann. of Math. (2) **37**, no. 2, 456–482, DOI 10.2307/1968458. In collaboration with Franz Alt and Otto Schreiber. MR1503293

Monastyrsky, Michael (1997). *Modern mathematics in the light of the Fields medals*, A K Peters, Ltd., Wellesley, MA. Translated from the 1991 Russian original by Roger Cooke and revised by the author; With a foreword by Freeman Dyson. MR1427488

Monna, A. F. (1994). *Oswald Veblen*, Math. Intell. **16**, no. 1, (1994), 50–51.

Montgomery, Deane (1963). *Oswald Veblen*, Bull. Amer. Math. Soc. **69**, 26–36; reprinted in (**Duren, 1988/1989**), vol. 1, pp. 118–129, DOI 10.1090/S0002-9904-1963-10838-1. MR148526

Mordell, L. J. (1922). *On the rational solutions of the indeterminate equations of the 3rd and 4th degrees*, Proc. Camb. Phil. Soc. **21**, 179–192.

Mordell, Louis Joel (1936/7). *Minkowski's theorems and hypotheses on linear forms*, in (**ICM, 1936**), vol. I, pp. 226–238. [JFM 63.0921.04; Zbl 0018.29502]

Mordell, L. J. (1971). *Reminiscences of an octogenarian mathematician*, Amer. Math. Monthly **78**, 952–961, DOI 10.2307/2317799. MR292632

Morse, Marston (1936). *The international Congress in Oslo*, Bull. Amer. Math. Soc. **42**, no. 11, 777–781, DOI 10.1090/S0002-9904-1936-06421-9. MR1563422

Morse, Marston (1946). *George David Birkhoff and his mathematical work*, Bull. Amer. Math. Soc. **52**, 357–391, DOI 10.1090/S0002-9904-1946-08553-5. MR16341

Morse, Marston (2008). *George David Birkhoff*, in Dictionary of scientific biography (C. C. Gillispie, ed.), Charles Scribner's Sons, Detroit, vol. 2, pp. 143–146.

Munkholm, Ellen S. and Hans J. Munkholm (1998). *Poul Heegaard (1871–1948), a Danish-Norwegian topologist* (Danish, with English summary), Normat **46**, no. 4, 145–169, 188. MR1682372

Munkholm, Ellen S. and Hans J. Munkholm (1999). *Poul Heegaard*, in (**James, 1999a**), pp. 925–946, DOI 10.1016/B978-044482375-5/50035-3. MR1721129

Nastasi, Pietro (1998). *Italian mathematics from the "Manifesto of Fascist Intellectuals" to the Race Laws* (Italian), Boll. Unione Mat. Ital. Sez. A Mat. Soc. Cult. (8) **1**, no. 3, 317–345 (1999). MR1719421

Nastasi, Pietro (2001). *La matematica*, in (**Simili and Paolini, 2001**), vol. I, pp. 243–280.

Nastasi, Pietro (2015). *A mathematician in the "Great War": Mauro Picone*, Lett. Mat. **3**, no. 1-2, 17–26, DOI 10.1007/s40329-015-0080-6. MR3348275

Nastasi, Pietro and Rossana Tazzioli (2005). *Toward a scientific and personal biography of Tullio Levi-Civita (1873–1941)*, Historia Math. **32**, no. 2, 203–236, DOI 10.1016/j.hm.2004.03.003. MR2137320

Nastasi, Pietro and Rossana Tazzioli (2013). *Italian mathematicians and scientific internationalism (1914–1924)* (Italian, with English and Italian summaries), Mat. Soc. Cult. Riv. Unione Mat. Ital. (I) **6**, no. 3, 355–405, 501. MR3243740

Nastasi, Pietro and Rossana Tazzioli (2014). *Italian mathematicians and the First World War: intellectual debates and institutional innovations*, in David Aubin and Catherine Goldstein (eds.), The war of guns and mathematics, Hist. Math., vol. 42, Amer. Math. Soc., Providence, RI, pp. 181–227. MR3308770

Neuenschwander, E. (1986). *Der Aufschwung der italienischen Mathematik zur Zeit der politischen Einigung Italiens und seine Auswirkungen auf Deutschland*, Symposia mathematica, Vol. XXVII (Cortona, 1983), Sympos. Math., XXVII, Academic Press, London, pp. 213–237. MR863150

Neugebauer, Otto (1926). *Die Grundlagen der ägyptischen Bruchrechnung*, dissertation, Göttingen; published by Springer the same year.

Neugebauer, Otto (1934a). *Vorlesungen über Geschichte der antiken mathematischen Wissenschaften*, Erster Band: Vorgiechische Mathematik, Die Grundlehren der Mathematischen Wissenschaften, vol. 43, Springer; 2nd ed., 1969.

Neugebauer, O. (1934b). *Babylonian mathematics*, Scripta Math. **2**, no. 4, 312–315.

Neugebauer, O. (1936/7). *Über griechische Mathematik und ihr Verhältnis zur vorgriechischen*, in (**ICM, 1936**), vol. I, pp. 157–170. [JFM 63.0799.01; Zbl 0018.33902]

Neugebauer, O. (1945). *The history of ancient astronomy: Problems and methods*, J. Near East. Stud. **4**, no. 1, 1–38; reproduced in Publ. Astr. Soc. Pacific **58**, 17–43, 104–142. MR0014950

von Neumann, J. (1932a). *Proof of the quasi-ergodic hypothesis*, Proc. Nat. Acad. Sci. USA **18**, no. 1, 70–82.

von Neumann, J. (1932b). *Mathematische Grundlagen der Quantenmechanik*, Springer, Berlin.

Nevanlinna, Rolf (1925). *Zur Theorie der Meromorphen Funktionen*, Acta Math. **46**, no. 1-2, 1–99, DOI 10.1007/BF02543858. MR1555200

Nielsen, Jakob (1927–1932). Untersuchungen zur Topologie der geschlossenen zweiseitigen Flächen , Acta Math. **50** (1927), no. 1, 189–358; II, *ibid.* **53** (1929), 1–76; III, *ibid.* **58** (1932), 87–167. DOI 10.1007/BF02421324. MR1555256

Nielsen, J. (1936/7). *Einige Methoden und Ergebnisse aus der Topologie der Flächenabbildungen*, in (**ICM, 1936**), vol. I, pp. 128–139. [JFM 63.0553.02; Zbl 0018.42503]

Nielsen, Jakob (1986). *Jakob Nielsen: collected mathematical papers. Vol. 1*, Contemporary Mathematicians, Birkhäuser Boston, Inc., Boston, MA. Edited and with a preface by Vagn Lundsgaard Hansen. MR865335

Ore, Øystein (1935). *On the foundation of abstract algebra. I*, Ann. of Math. (2) **36**, no. 2, 406–437, DOI 10.2307/1968580. MR1503232

Ore, Ø. (1936/7). *On the decomposition theorems of algebra*, in (**ICM, 1936**), vol. I, pp. 297–307. [JFM 63.0868.03; Zbl 0018.29002]

Ore, Øystein (1954/7). *Niels Henrik Abel. Et geni og hans samtid*, Gyldendal Norsk Forlag, Oslo; English transl.: *Niels Henrik Abel: mathematician extraordinary*, University of Minnesota Press, Minneapolis, 1957. MR0062674

Oseen, C. W. (1935). *Allvar Gullstrand: minnesteckning* [No. 107], K. Svenska Vetenskapsakad. Levnadsteckn. över Kungl. Svenska vetenskapsakad. led., pp. 1–35.

Oseen, C. W. (1936/7). *Probleme der geometrischen Optik*, in (**ICM, 1936**), vol. I, pp. 171–185. [JFM 63.0717.04; Zbl 0019.04703]

Osserman, Robert (1999). *From Schwarz to Pick to Ahlfors and beyond*, Notices Amer. Math. Soc. **46**, no. 8, 868–873. MR1704258

Palazzo, E. (1938). *Una proposizione di geometria elementare e le sue applicazioni*, in Atti del primo Congresso dell'Unione Matematica Italiana: tenuto in Firenze nei giorni 1-2-3 Aprile 1937-XV, Zanichelli, Bologna.

Palazzo, E. (1961). *"Le développement nouveau de la géométrie élémentaire" par Louis Bertrand*, Arch. int. hist. sci. **13**, 223–236.

Pancaldi, Giuliano (1993). *Vito Volterra: cosmopolitan ideas and nationality in the Italian scientific community between the Belle époque and the First World War*, Minerva **31**, 21–37.

Parshall, Karen Hunger and Adrian C. Rice (eds.) (2002). *Mathematics unbound: the evolution of an international mathematical research community, 1800–1945. Papers from the International Symposium held at the University of Virginia, Charlottesville, VA, May 27–29, 1999*, History of Mathematics, vol. 23, AMS/LMS.

Parshall, Karen Hunger and David E. Rowe (1994). *The emergence of the American mathematical research community, 1876–1900: J. J. Sylvester, Felix Klein, and E. H. Moore*, History of Mathematics, vol. 8, American Mathematical Society, Providence, RI; London Mathematical Society, London. MR1290994

Patterson, S. J. (1990). *Erich Hecke und die Rolle der L-Reihen in der Zahlentheorie*, in (**Fischer *et al.*, 1990**), pp. 629–655, DOI 10.1007/978-3-322-80265-1_14. MR1085975

Perron, Oskar (1944/48). *Obituary: Erich Hecke*, Jber. Bayer. Akad. Wiss. München **1944/48**, 274–276 (1948). MR0034329

Petersson, Hans (1949). *Das wissenschaftliche Werk von E. Hecke*, Abh. Math. Sem. Univ. Hamburg **16**, no. nos. 1-2, 7–31, DOI 10.1007/BF02941083. MR0034328

Piene, Kay (1937). *Die gegenwärtigen Entwicklungstendenzen im mathematischen Unterricht in Norwegen*, Enseign. math. **36**, 374–376.

Pier, J.-P. (ed.) (1994a). *Development of mathematics 1900–1950*, Birkhäuser.

Pier, Jean-Paul (1994b). *Intégration et mesure 1900–1950*, in (**Pier, 1994a**), pp. 517–564, DOI 10.1007/978-3-0348-9114-1_12. MR1298642

Pier, Jean-Paul (ed.) (2000). *Development of mathematics 1950–2000*, Birkhäuser Verlag, Basel. MR1796836

Pihl, Mogens (2008a). *Bjerknes, Carl Anton*, in Dictionary of scientific biography (C. C. Gillispie, ed.), Charles Scribner's Sons, Detroit, vol. 2, pp. 166–167.

Pihl, Mogens (2008b). *Bjerknes, Vilhelm Frimann Koren*, in Dictionary of scientific biography (C. C. Gillispie, ed.), Charles Scribner's Sons, Detroit, vol. 2, pp. 167–169.

Pingree, David (1990). *In memoriam: Otto Neugebauer (1899–1990)*, Arch. Internat. Hist. Sci. **40**, no. 124, 82–84. MR1137300

Pitcher, Everett (1988). *A history of the second fifty years, American Mathematical Society, 1939–1988*, American Mathematical Society centennial publications, vol. I, American Mathematical Society, Providence, RI. MR1002190

von Plato, Jan (1994). *Creating modern probability: Its mathematics, physics and philosophy in historical perspective*, Cambridge Studies in Probability, Induction, and Decision Theory, Cambridge University Press, Cambridge. MR1265718

Pólya, George (1987). *The Pólya picture album: encounters of a mathematician*, Birkhäuser Boston, Inc., Boston, MA. Edited and with an introduction and a biography by G. L. Alexanderson. MR898436

Pyatetskii-Shapiro, I. I. and A. B. Shidlovskii (1967). *Aleksandr Osipovich Gel'fond (on his sixtieth birthday)*, Uspekhi mat. Nauk **22**, no. 3(135), 247–256 (in Russian); English transl.: *Russian Math. Surveys* **22**, no. 3, (1967), 234–242.

Pyenson, Lewis (1999). *Neugebauer, Otto Eduard (26 May 1899–19 February 1990)*, in American National Biography, Oxford University Press, vol. 16, pp. 302–303.

Radó, Tibor (1933). *On the problem of Plateau*, Ergebnisse der Mathematik und ihrer Grenzgebiete, Band 2, no. 2, Springer, Berlin.

Ramskov, Kurt (1995). *Matematikeren Harald Bohr*, Licentiatafhandling, Institut for de eksakte videnskabers historie, Aarhus Universitet.

Ramskov, Kurt (2000). *The Danish Mathematical Society through 125 years*, Historia Math. **27**, no. 3, 223–242, DOI 10.1006/hmat.2000.2283. MR1784430

Rassias, Th. (ed.) (1992). *The problem of Plateau: a tribute to Jesse Douglas & Tibor Radó*, World Scientific.

Reid, Constance. 1970. *Hilbert*, Springer-Verlag, Berlin, Heidelberg and New York.

Reid, Constance (1976). *Courant in Göttingen and New York: The story of an improbable mathematician*, Springer-Verlag, New York-Heidelberg. MR0532647

Reidemeister, Kurt (1932). *Knotentheorie*, Ergebnisse der Mathematik und ihrer Grenzgebiete, Band 1, no. 1, Springer, Berlin.

Reingold, Nathan (1981). *Refugee mathematicians in the United States of America, 1933–1941: reception and reaction*, Ann. of Sci. **38**, no. 3, 313–338, DOI 10.1080/00033798100200251. MR612372

Remmert, Volker (2012). *The German Mathematical Association during the Third Reich: professional policy within the web of National Socialist ideology*, in Dieter Hoffmann and Mark Walker (eds.), The German Physical Society in the Third Reich: physicists between autonomy and accommodation, Cambridge University Press, pp. 246–279.

Richardson, R. G. D. (1932). *International Congress of Mathematicians, Zurich, 1932*, Bull. Amer. Math. Soc. **38**, no. 11, 769–774, DOI 10.1090/S0002-9904-1932-05491-X. MR1562505

Riehm, Elaine McKinnon and Frances Hoffman (2011). *Turbulent times in mathematics: The life of J. C. Fields and the history of the Fields Medal*, American Mathematical Society, Providence, RI; Fields Institute for Research in Mathematical Sciences, Toronto, ON. MR2850575

Ritter, Jim (2016). *Otto Neugebauer and ancient Egypt*, in (**Jones *et al.*, 2016**), pp. 127–163, DOI 10.1007/978-3-319-25865-2_4. MR3496896

Roghi, Gino (2005). *Materiale per una storia dell'Istituto Nazionale di Alta Matematica dal 1939 to 2003*, Boll. Unione Mat. Ital. Sez. A: Mat. soc. cult. **8**, no. 3-2, pp. X+301.

Rogora, Enrico (2016). *Guido Castelnuovo e la matematica a Roma tra Risorgimento e Belle Époque*, Rend. Mat. (7) **37**, 219–233. MR3622302

Rohatgi, Vijay K. (1988). *Obituary: Eugene Lukacs*, J. Appl. Probab. **25**, no. 3, 642–646. MR954514

Root, William L. (1966). *Contributions of Norbert Wiener to communication theory*, Bull. Amer. Math. Soc. **72**, no. 1, pt. 2, 126–134, DOI 10.1090/S0002-9904-1966-11472-6. MR183626

Roquette, P. (1989). *Über die algebraisch-zahlentheoretischen Arbeiten von Max Deuring*, Jahresber. Deutsch. Math.-Verein. **91**, no. 3, 109–125. MR1009320

Roquette, Peter (2013). *Contributions to the history of number theory in the 20th century*, Heritage of European Mathematics, European Mathematical Society (EMS), Zürich. MR3026650

Rota, Gian-Carlo (1997). *The many lives of lattice theory*, Notices Amer. Math. Soc. **44**, no. 11, 1440–1445. MR1488572

Roth, L. (1963). *Francesco Severi*, J. London Math. Soc. **38**, 282–307, DOI 10.1112/jlms/s1-38.1.282. MR0157887

Rowe, David E. (2003). *On projecting the future and assessing the past—the 1946 Princeton Bicentennial Conference*, Math. Intelligencer **25**, no. 4, 8–15, DOI 10.1007/BF02984857. MR2012559

Rowe, David E. (2016). *From Graz to Göttingen: Neugebauer's early intellectual journey*, in (**Jones *et al.*, 2016**), pp. 1–59, DOI 10.1007/978-3-319-25865-2_1. MR3496893

Royden, Halsey (1989). *A history of mathematics at Stanford*, in (**Duren, 1988/1989**), vol. 2, pp. 237–277. MR1003132

Sachs, J. and G. J. Toomer (1978/79). *Otto Neugebauer, bibliography, 1925–1979*, Centaurus **22**, no. 4, 258–280, DOI 10.1111/j.1600-0498.1979.tb00592.x. MR536372

Salvadori, Mario G. (1939). *The activity of the Italian National Institute for the Application of Calculus*, in (**ICAM, 1938**), pp. 260–264.

Schappacher, Norbert (1987). *Das Mathematische Institut der Universität Göttingen 1929–1950*, in H. Becker, H.-J. Dahms, and C. Wegeler (eds.), Die Universität Göttingen unter dem Nationalsozialismus. Das verdrängte Kapitel ihrer 250 jährigen Geschichte, Saur, München, pp. 345–373.

Schappacher, Norbert (2006). *Seventy years ago: the Bourbaki congress at El Escorial and other mathematical (non) events of 1936*, in F. Chamizo and A. Quirós (eds.), Madrid Intelligencer 2006, Springer, pp. 8–15.

Schaumberger, Norman (2008). *Douglas, Jesse*, in Dictionary of scientific biography (C. C. Gillispie, ed.), Charles Scribner's Sons, Detroit, vol. 4, pp. 173–174.

Schlote, Karl-Heinz (1991). *Fritz Noether—Opfer zweier Diktaturen. Tod und Rehabilitierung*, NTM Schr. Geschichte Natur. Tech. Medizin **28**, no. 1, 33–41. MR1152727

Schneider, Th. (1983). *Das Werk C. L. Siegels in der Zahlentheorie*, Jahresber. Deutsch. Math.-Verein. **85**, no. 4, 147–157. MR722517

Schoeneberg, Bruno (1974). *Elliptic modular functions: an introduction*, Die Grundlehren der mathematischen Wissenschaften, vol. 203, Springer-Verlag, New York-Heidelberg. Translated from the German by J. R. Smart and E. A. Schwandt. MR0412107

Schoeneberg, B. (1989). *Erich Hecke 1887–1947*, Jahresber. Deutsch. Math.-Verein. **91**, no. 4, 168–190. MR1027063

Schoenberg, B. (2008a). *Hecke, Erich*, in Dictionary of scientific biography (C. C. Gillispie, ed.), Charles Scribner's Sons, Detroit, vol. 6, pp. 214–215.

Schoenberg, B. (2008b). *Landau, Edmund*, in Dictionary of scientific biography (C. C. Gillispie, ed.), Charles Scribner's Sons, Detroit, vol. 7, pp. 615–616.

Schouten, J. A. and J. Haantjes (1937). *On the theory of the geometric object*, Proc. London Math. Soc. (2) **42**, no. 5, 356–376, DOI 10.1112/plms/s2-42.1.356. MR1577037

Schroeder-Gudehus, Brigitte (1973). *Challenge to transnational loyalties: international scientific organizations after the First World War*, Sci. Stud. **3**, 93–118.

Schwarz, Wolfgang (1994). *Some remarks on the history of the prime number theorem from 1896 to 1960*, in (**Pier, 1994a**), pp. 565–616. MR1298643

Schwermer, Joachim (2007). *Minkowski, Hensel, and Hasse: on the beginnings of the local-global principle*, in Jeremy J. Gray and Karen Hunger Parshall (eds.), Episodes in the history of modern algebra (1800–1950), Hist. Math., vol. 32, Amer. Math. Soc., Providence, RI, pp. 153–177. MR2353495

Scirè, Pietro (1942). *Algebra per la scuola media. Con millecento esercizi e note storiche*, G. Barbera, Firenze.

Scirè, Pietro (1946). *Geometria elementare per i lincei. Con note storiche esercizi problemi*, Edizioni Pantea, Palermo.

Segal, S. L. (1980). *Helmut Hasse in 1934*, Historia Math. **7**, no. 1, 46–56, DOI 10.1016/0315-0860(80)90063-4. MR559835

Segal, Sanford L. (2003). *Mathematicians under the Nazis*, Princeton University Press, Princeton, NJ. MR1991149

Segre, Beniamino (2008). *Severi, Francesco*, in Dictionary of scientific biography (C. C. Gillispie, ed.), Charles Scribner's Sons, Detroit, vol. 12, pp. 330–332.

Senechal, Marjorie (1998). *Oberwolfach, 1944–1964*, Math. Intelligencer **20**, no. 4, 17–24, DOI 10.1007/BF03025224. MR1646693

Shafer, Glenn and Vladimir Vovk (2006). *The sources of Kolmogorov's Grundbegriffe*, Statist. Sci. **21**, no. 1, 70–98, DOI 10.1214/088342305000000467. MR2275967

Siegel, C. L. (1935–1937). *Über die analytische Theorie der quadratischen Formen*, Ann. of Math. (2) **36** (1935), no. 3, 527–606; II, *ibid.* **37** (1936), no. 1, 230–263; III, *ibid.* **38** (1937), no. 1, 212–291. DOI 10.2307/1968694. MR1503276

Siegel, C. L. (1936/7). *Analytische Theorie der quadratischen Formen*, in (**ICM, 1936**), vol. I, pp. 104–110. [JFM 63.0895.04; Zbl 0018.24404]

Siegel, C. L. (1937). *Formes quadratiques et modules des courbes algébriques*, Bull. sci. math. (2) **61**, 331–352.

Siegel, C. L. (1965). *Lectures on advanced analytic number theory*, Notes by S. Raghavan. Tata Institute of Fundamental Research Lectures on Mathematics, No. 23, Tata Institute of Fundamental Research, Bombay. MR0262150

Siegel, C. L. (1978/79). *On the history of the Frankfurt Mathematics Seminar*, Math. Intelligencer **1**, no. 4, 223–230, DOI 10.1007/BF03028242. Address given on June 13, 1964, in the Mathematics Seminar of the University of Frankfurt on the occasion of the 50th anniversary of the Johann-Wolfgang-Goethe-University Frankfurt. MR547750

Siegmund-Schultze, Reinhard (1993). *Mathematische Berichterstattung in Hitlerdeutschland: Der Niedergang des Jahrbuchs über die Fortschritte der Mathematik*, Studien zur Wissenschafts-, Sozial- und Bildungsgeschichte der Mathematik [Studies on the Scientific, Social and Educational History of Mathematics], vol. 9, Vandenhoeck & Ruprecht, Göttingen. MR1253317

Siegmund-Schultze, Reinhard (1994). *"Scientific control" in mathematical reviewing and German–U.S.-American relations between the two world wars*, Historia Math. **21**, no. 3, 306–329, DOI 10.1006/hmat.1994.1027. MR1295722

Siegmund-Schultze, Reinhard (2001). *Rockefeller and the internationalization of mathematics between the two world wars*, Science Networks. Historical Studies, vol. 25, Birkhäuser Verlag, Basel. Documents and studies for the social history of mathematics in the 20th century. MR1826906

Siegmund-Schultze, Reinhard (2002). *The effects of Nazi rule on the international participation of German mathematicians: an overview and two case studies*, in (**Parshall and Rice, 2002**), pp. 335–357. MR1907182

Siegmund-Schultze, Reinhard (2003). *The origins of functional analysis*, in Niels Jahnke (ed.), A history of analysis, Hist. Math., vol. 24, Amer. Math. Soc., Providence, RI, pp. 385–407. MR1998255

Siegmund-Schultze, Reinhard (2004). *Mathematicians forced to philosophize: an introduction to Khinchin's paper on von Mises' theory of probability*, Sci. Context **17**, no. 3, 373–390, DOI 10.1017/S0269889704000171. MR2106023

Siegmund-Schultze, Reinhard (2005). *Maurice Fréchet à Strasbourg*, in Elisabeth Crawford and Josiane Olff-Nathan (eds.), La Science sous influence: L'université de Strasbourg enjeu des conflits franco-allemands 1872–1945, La Nuée bleue, pp. 185–196.

Siegmund-Schultze, Reinhard (2007). *Philipp Frank, Richard von Mises, and the Frank-Mises*, Phys. Perspect. **9**, no. 1, 26–57, DOI 10.1007/s00016-006-0288-0. MR2308126

Siegmund-Schultze, Reinhard (2009). *Mathematicians fleeing from Nazi Germany: Individual fates and global impact*, Princeton University Press, Princeton, NJ. MR2522825

Siegmund-Schultze, Reinhard (2011). *Opposition to the boycott of German mathematics in the early 1920s: letters by Edmund Landau (1877–1938) and Edwin Bidwell Wilson (1879–1964)*, Rev. Histoire Math. **17**, no. 1, 139–165. MR2849161

Siegmund-Schultze, Reinhard (2012). *Landau and Schur—documents of a friendship until death in an age of inhumanity [translation of MR2895882]*, Eur. Math. Soc. Newsl. **84**, 31–36. MR2934663

Siegmund-Schultze, Reinhard (2014). *One hundred years after the Great War (1914–2014): a century of breakdowns, resumptions and fundamental changes in international mathematical communication*, in (**ICM, 2014**), vol. 4, pp. 1231–1253. MR3751174

Siegmund-Schultze, Reinhard (2016a). *"Mathematics knows no races": a political speech that David Hilbert planned for the ICM in Bologna in 1928*, Math. Intelligencer **38**, no. 1, 56–66, DOI 10.1007/s00283-015-9559-4. MR3466891

Siegmund-Schultze, Reinhard (2016b). *"Not in possession of any Weltanschauung": Otto Neugebauer's flight from Nazi Germany and his search for objectivity in mathematics, in reviewing, and in history*, in (**Jones *et al.*, 2016**), pp. 61–106, DOI 10.1007/978-3-319-25865-2_2. MR3496894

Siegmund-Schultze, Reinhard and H. Kragh Sørensen (eds.) (2006). *Perspectives on Scandinavian science in the early twentieth century*, The Norwegian Academy of Science and Letters, I. Mat.-Nat. Klasse. Skrifter og avhandlinger 1. Novus forlag, Oslo.

Sigmund, Karl (2004). *Failing phoenix: Tauber, Helly, and Viennese life insurance*, Math. Intelligencer **26**, no. 2, 21–33, DOI 10.1007/BF02985648. MR2067894

Sigmund, Karl, Josef Teichmann, and Walter Schachermayer (2011). *Franz Alt 1910–2011*, Int. Math. Nachr., no. 218, 1–10.

Simili, Raffaella and Giovanni Paolini (eds.) (2001). *Per una storia del Consiglio nazionale delle ricerche*, 2 vols., GLF editori Laterza, Roma.

Slagstad, Rune (1998). *De nasjonale strateger*, Pax Forlag, Oslo.

Smith, David Eugene (1935). *Review: Vorlesungen über Geschichte der antiken mathematischen Wissenschaften*. Erster Band, Vorgriechische Mathematik. O. Neugebauer. Berlin, J. Springer, 1934, Scripta Math. **3**, 162–164.

Sodin, Mikhail (2000). *Lars Ahlfors' thesis*, in (**Brooks and Sodin, 2000**), pp. 113–134. MR1786563

Somsen, Geert J. (2008). *A history of universalism: conceptions of the internationality of science from the enlightenment to the Cold War*, Minerva **46**, no. 3, 361–379.

Sørensen, Henrik Kragh (2006a). *The Scandinavian Congresses of Mathematics until the end of World War I* (Danish, with English and Danish summaries), Normat **54**, no. 1, 1–17, 47. MR2278614

Sørensen, Henrik Kragh (2006b). *Niels Henrik Abel's political and professional legacy in Norway*, in (**Siegmund-Schultze and Sørensen, 2006**), pp. 197–219.

Sørensen, Henrik Kragh (2012). *Reading Mittag-Leffler's biography of Abel as an act of mathematical self-fashioning*, in Benjamin Wardhaugh (ed.), The history of the history of mathematics, Peter Lang, Oxford, pp. 115–144. MR2975642

Sørensen, Henrik Kragh (2014). *Confluences of agendas: emigrant mathematicians in transit in Denmark, 1933–1945*, Historia Math. **41**, no. 2, 157–187, DOI 10.1016/j.hm.2013.10.002. MR3191521

Speiser, A. (1950). *Ansprache anläßlich der Abdankungsfeier am 11. August 1950 in Zürich*, Elem. Math. **5**, 98–99.

Spjeldnaes, Nils (2008). *Størmer, Fredrik Carl Mülertz*, in Dictionary of scientific biography (C. C. Gillispie, ed.), Charles Scribner's Sons, Detroit, vol. 13, pp. 82–83.

Stadler, F. (2001). *The Vienna circle. Studies in the origins, development, and influence of logical empiricism*, Springer, New York.

Steinhaus, H. (1963). *Stefan Banach*, Studia Math. (Ser. Specjalna) Zeszyt **1**, 7–15. MR0175746

Stone, Marshall H. (1932). *Linear transformations in Hilbert space and their applications to analysis*, AMS, New York.

Størmer, C. (1936/7). *Programme for the quantitative discussion of electron orbits in the field of a magnetic dipole, with applications to cosmic rays and kindred phenomena*, in (**ICM, 1936**), vol. I, pp. 61–74. [JFM 63.0736.01]

Struik, Dirk J. (1962). *A concise history of mathematics*, G. Bell and sons, London.

Struik, Dirk J. (2008). *Fano, Gino*, in Dictionary of scientific biography (C. C. Gillispie, ed.), Charles Scribner's Sons, Detroit, vol. 4, pp. 522–523.

Stubhaug, Arild (2002). *The mathematician Sophus Lie: It was the audacity of my thinking*, Springer-Verlag, Berlin. Translated from the 2000 Norwegian original by Richard H. Daly. MR1876528

Stubhaug, Arild (2010). *Gösta Mittag-Leffler: A man of conviction*, Springer-Verlag, Berlin. Translated from the 2007 Norwegian original by Tiina Nunnally. MR2743189

Sudbery, A. (1979). *Quaternionic analysis*, Math. Proc. Cambridge Philos. Soc. **85**, no. 2, 199–224, DOI 10.1017/S0305004100055638. MR516081

Swerdlow, Noel M. (1993a). *Otto E. Neugebauer (26 May 1899–19 February 1990)*, Proc. Amer. Philos. Soc. **137**, no. 1, 137–165. MR3363428

Swerdlow, N. M. (1993b). *Otto E. Neugebauer (26 May 1899–19 February 1990)*, J. Hist. Astronom. **24**, no. 4, 289–299, DOI 10.1177/002182869302400405. MR1252079

Tarwater, J. D., J. T. White, and J. D. Miller (eds.) (1976). *Men and institutions in American Mathematics*, Graduate Studies, Texas Tech University, Lubbock.

Taylor, Angus E. (1982–1987). *A study of Maurice Fréchet: I. His early work on point set theory and the theory of functionals*, Arch. Hist. Exact Sci. **27**, no. 3, (1982), 233–295; II. Mainly about his work on general topology, 1909–1928, *ibid.* **34**, no. 4, (1985), 279–380; III. Fréchet as analyst, 1909–1930, *ibid.* **37**, no. 1, (1987), 25–76. MR816369

Taylor, Angus E. (2008). *Fréchet, René Maurice*, in Dictionary of scientific biography (C. C. Gillispie, ed.), Charles Scribner's Sons, Detroit, vol. 17, p. 309–311.

Tazzioli, Rossana (2011). *I matematici tra scienza, politica e istituzioni, in Francesco Cassata and Claudio Pogliano* (eds.), Scienza e cultura dell'Italia unita, Storia d'Italia, Annali 26, Giulio Einaudi editore, pp. 385–415.

Tazzioli, Rossana (2013). *Les mathématiciens italiens face à l'interventionnisme dans la Première Guerre mondiale*, in Antonin Durand, Laurent Mazliak and Rossana Tazzioli (eds.), Des mathématiciens et des guerres; Histoires de confrontations (XIXe–XXe siècles). Actes du colloque tenu à l'Institut des sciences de la communication du CNRS (ISCC) le 8 février 2012, CNRS Éditions, Paris, pp. 27–39.

Thalberg, Olaf (1943). *Norsk Matematisk Forening gjennom 25 år*, Norsk mat. tidsskr. **25**, 65–75.

Tricomi, F. G. (2008). *Pincherle, Salvatore*, in Dictionary of scientific biography (C. C. Gillispie, ed.), Charles Scribner's Sons, Detroit, vol. 10, p. 610.

Tropp, Henry S. (1992). *Jesse Douglas, the Plateau problem, and the Fields Medal: some personal reflections*, in (**Rassias, 1992**), pp. 43–49, DOI 10.1142/9789812812773_0007. MR1209208

Tucker, A. W. (1935). *The topological congress in Moscow*, Bull. Amer. Math. Soc. **41**, no. 11, 764, DOI 10.1090/S0002-9904-1935-06174-9. MR1563184

Turner, Laura E. (2011). *Cultivating mathematics in an international space: roles of Gösta Mittag-Leffler in the development and internationalization of mathematics in Sweden and beyond, 1880–1920*, PhD thesis, Department of Science Studies, Aarhus University.

Turner, Laura E. and Henrik Kragh Sørensen (2013). *Cultivating the herb garden of Scandinavian mathematics: the congresses of Scandinavian mathematicians, 1909–1925*, Centaurus **55**, no. 4, 385–411, DOI 10.1111/1600-0498.12011. MR3138534

Ulam, S. (1946). *Stefan Banach 1892-1945*, Bull. Amer. Math. Soc. **52**, no. 7, 600–603, DOI 10.1090/S0002-9904-1946-08622-X. MR1564845

Veblen, O. (1929). *Differential invariants and geometry*, in (**ICM, 1928**), vol. I, pp. 181–189.

Veblen, O. (1933a). *Projektive Relativitätstheorie*, Ergebnisse der Mathematik und ihrer Grenzgebiete, Band 2, no. 1, Springer, Berlin.

Veblen, O. (1933b). *Spinors in projective relativity*, Proc. Nat. Acad. Sci. USA **19**, 979–989.

Veblen, O. (1936/7). *Spinors and projective geometry*, in (**ICM, 1936**), vol. I, pp. 111–127. [JFM 63.0693.02; Zbl 0018.32604]

Veblen, O. (1946). *George David Birkhoff. March 21, 1884 – November 12, 1944*, Proc. Amer. Phil. Soc. (Yearbook), 279–85; reprinted in Biographical Memoirs, vol. 80, National Academy of Sciences, Washington, DC.

Veblen, O. and J. H. C. Whitehead (1932). *The foundations of differential geometry*, Cambridge Tract No. 29, Cambridge University Press.

Volkert, Klaus (2018). *Vorher—nachher: zwei Briefe des Mathematikers William Threlfall (1933 und 1945)*, Math. Semesterber. **65**, no. 1, 1–14, DOI 10.1007/s00591-017-0213-4. MR3763957

Volterra, Vito (1920). *Saggi scientifici*, N. Zanichelli, Bologna.

Vucinich, Alexander (2000). *Soviet mathematics and dialectics in the Stalin era*, Historia Math. **27**, no. 1, 54–76, DOI 10.1006/hmat.1999.2261. MR1745515

van der Waerden, B. L. (1930). *Moderne Algebra*, Teil I, Springer-Verlag; Teil II, 1931.

van der Waerden, B. L. (1935). *Nachruf auf Emmy Noether*, Math. Ann. **111**, no. 1, 469–476, DOI 10.1007/BF01472233. MR1513008

van der Waerden, B. L. (1971). *The foundation of algebraic geometry from Severi to André Weil*, Arch. History Exact Sci. **7**, no. 3, 171–180, DOI 10.1007/BF00357215. MR1554142

van der Waerden, B. L. (1986). *Francesco Severi and the foundations of algebraic geometry*, Symposia mathematica, Vol. XXVII (Cortona, 1983), Sympos. Math., XXVII, Academic Press, London, pp. 239–244. MR863151

Wavre, Rolin (1948/1971). *The International Congresses of Mathematicians*, in (**Le Lionnais, 1948/1971**), vol. I, pp. 312–318.

Weckbecker, Arno (1985). *"Gleichschaltung" der Universität? Nationalsozialistische Verfolgung Heidelberger Hochschullehrer aus rassischen und politischen Gründen*, in Karin Buselmeier (ed.), Auch eine Geschichte der Universität Heidelberg, Quadrat, Mannheim, pp. 273–291.

Weil, André (1929). *L'arithmétique sur les courbes algébriques*, Acta Math. **52**, no. 1, 281–315, DOI 10.1007/BF02547409. MR1555278

Weil, André (1950). *The future of mathematics*, Amer. Math. Monthly **57**, 295–306, DOI 10.2307/2306198. MR34296

Werkmeister, William H. (1936). *The Second International Congress for the Unity of Science*, Phil. Rev. **45**, no. 6, 593–600.

Weyl, Hermann and F. Joachim Weyl (1943). *Meromorphic Functions and Analytic Curves*, Annals of Mathematics Studies, no. 12, Princeton University Press, Princeton, N. J. MR0009057

Whitehead, J. H. C. (1952). *Obituary: Elie Joseph Cartan, 1869–1951*, Obit. Notices Roy. Soc. London **8**, 71–95 (1 plate). MR0051758

Whitney, Hassler (1936). *Differentiable manifolds*, Ann. of Math. (2) **37**, no. 3, 645–680, DOI 10.2307/1968482. MR1503303

Whitney, Hassler (1988). *Moscow 1935: topology moving toward America*, in (**Duren, 1988/1989**), vol. 1, pp. 97–117. MR1003166

Wiener, Norbert (1914). *A simplification in the logic of relations*, Proc. Camb. Phil. Soc. **17**, 387–390; reprinted in Jean van Heijenoort (ed.), From Frege to Gödel: a source book in mathematical logic, 1879–1931, Harvard University Press, 1967, pp. 224–227.

Wiener, Norbert (1936/7). *Gap theorems*, in (**ICM, 1936**), vol. I, pp. 284–296. [JFM 63.0974.01; Zbl 0019.03402]

Wiener, Norbert (1948). *Cybernetics, or Control and Communication in the Animal and the Machine*, Actualités Sci. Ind., no. 1053, Hermann et Cie., Paris; The Technology Press, Cambridge, Mass.; John Wiley & Sons, Inc., New York. MR0025096

Wiener, Norbert (1949). *Extrapolation, Interpolation, and Smoothing of Stationary Time Series. With Engineering Applications*, The Technology Press of the Massachusetts Institute of Technology, Cambridge, Mass; John Wiley & Sons, Inc., New York, N. Y.; Chapman & Hall, Ltd., London. MR0031213

Wiener, Norbert (1953). *Ex-prodigy. My childhood and youth*, Simon and Schuster, New York. MR0057817

Wiener, Norbert (1956). *I am a mathematician. The later life of a prodigy*, Doubleday and Co., Garden City, N. Y. MR0077455

Yushkevich, A. P. (2007). *Encounters with mathematicians*, 2nd ed., in S. Zdravkovska and P. L. Duren (eds.), Golden years of Moscow mathematics, Hist. Math., vol. 6, Amer. Math. Soc., Providence, RI, pp. 1–33. MR1246564

Yushkevich, A. P. (2008a). *Gelfond, Alexandr Osipovich*, in Dictionary of scientific biography (C. C. Gillispie, ed.), Charles Scribner's Sons, Detroit, vol. 5, pp. 342–343.

Yushkevich, A. P. (2008b). *Khinchin, Aleksandr Yakovlevich*, in Dictionary of scientific biography (C. C. Gillispie, ed.), Charles Scribner's Sons, Detroit, vol. 7, pp. 351–352.

Zappa, G. (1984). *The Francesco Severi school of mathematics around 1940* (Italian), Riv. Mat. Univ. Parma (4) **10**, 11–14. MR777309

Zariski, O. (1935). *Algebraic surfaces*, Ergebnisse der Mathematik und ihrer Grenzgebiete, Band 3, no. 5, Springer, Berlin.

Name Index

Subject Index